GENES IN
MAMMALIAN
REPRODUCTION

MODERN CELL BIOLOGY

RECENT VOLUMES PUBLISHED IN THE SERIES

GENES IN MAMMALIAN REPRODUCTION

Ralph B. L. Gwatkin, Editor

Department of Reproductive Biology
Case Western Reserve University
Cleveland, Ohio

and

ReproGene
Beachwood, Ohio

 **WILEY-LISS**

A JOHN WILEY & SONS, INC., PUBLICATION
NEW YORK • CHICHESTER • BRISBANE • TORONTO • SINGAPORE

1993

Address all Inquiries to the Publisher
Wiley-Liss, Inc., 605 Third Avenue, New York, NY 10158-0012

Printed in the United States of America.

While the authors, editors, and publisher believe that drug selection and dosage and the specification and usage of equipment and devices, as set forth in this book, are in accord with current recommendations and practice at the time of publication, they accept no legal responsibility for any errors or omissions, and make no warranty, express or implied, with respect to material contained herein. In view of ongoing research, equipment modifications, changes in governmental regulations and the constant flow of information relating to drug therapy, drug reactions, and the use of equipment and devices, the reader is urged to review and evaluate the information provided in the package insert or instructions for each drug, piece of equipment, or device for, among other things, any changes in the instructions or indication of dosage or usage and for added warnings and precautions.

Library of Congress Cataloging-in-Publication Data

Genes in mammalian reproduction / edited by Ralph B.L. Gwatkin.
        p.   cm. -- (Modern cell biology ; v. 12)
    Includes bibliographical references and index.
    ISBN 0-471-56146-0
    1. Reproduction--Regulation. 2. Molecular genetics. I. Gwatkin, Ralph B.L. II. Series.
    [DNLM: 1. Genetics, Biochemical. 2. Mammals--genetics.
3. Reproduction--genetics.]
QH573.M63 vol. 12
[QP251]
574.87 s--dc20
[599' .016]
DNLM/DLC
for Library of Congress
                                                    92-18110
                                                    CIP

The text of this book is printed on acid-free paper.

Contents

Contributors

John P. Adelman, Vollum Institute for Advanced Research, L474, Oregon Health Sciences University, Portland, OR 97201-3098 **[229]**

Chris T. Bond, Vollum Institute for Advanced Research, L474, Oregon Health Sciences University, Portland, OR 97201-3098 **[229]**

Richard L. Cate, Department of Molecular Biology, Biogen Inc., Cambridge, MA 02142 **[185]**

Robert P. Erickson, Departments of Pediatrics and Cellular Biology, University of Arizona Health Sciences, Tucson, AZ 85724 **[1]**

John R. Gosden, MRC Human Genetics Unit, Western General Hospital, Edinburgh, EH4 2XU, Scotland **[73]**

Nancy H. Ing, Department of Cell Biology, Baylor College of Medicine, Houston, TX 77030 **[271]**

Gerald M. Kidder, Department of Zoology, The University of Western Ontario, London, Ontario N6A 5B7, Canada **[45]**

Ross A. Kinloch, Department of Cell and Developmental Biology, Roch Institute of Molecular Biology, Roch Research Center, Nutley, NJ 07110 **[27]**

Jean J. Latimer, Radiobiology and Health, University of California, San Francisco, CA 94143-0750 **[131]**

William L. Miller, Department of Biochemistry, North Carolina State University, Raleigh, NC 27695-7622 **[247]**

Takuya Murata, Department of Anatomy and Cell Biology, University of Southern California School of Medicine, Los Angeles, CA 90033 **[207]**

Roger A. Pedersen, Department of Anatomy, University of California, San Francisco, CA 94143-0750 **[131]**

Christine Pourcel, U.R.E.G. (Inserm U 163), Institut Pasteur, Paris 75015, France **[173]**

Ming-Jer Tsai, Department of Cell Biology, Baylor College of Medicine, Houston, TX 77030 **[271]**

Sophia Y. Tsai, Department of Cell Biology, Baylor College of Medicine, Houston, TX 77030 **[271]**

Paul M. Wassarman, Department of Cell and Developmental Biology, Roch Institute of Molecular Biology, Roch Research Center, Nutley, NJ 07110 **[27]**

John D. West, Department of Obstetrics and Gynaecology, Centre for Reproductive Biology, University of Edinburg, Edinburg, EH3 9EW, Scotland **[73]**

Cheryl A. Wilson, Department of Molecular Biology, Biogen Inc., Cambridge, MA 02142 **[185]**

Shao-Yao Ying, Department of Anatomy and Cell Biology, School of Medicine, University of Southern California, Los Angeles, CA 90033 **[207]**

The numbers in brackets are the opening page numbers of the contributors' articles.

PREFACE

Molecular biology, the study of genetics and biochemistry at the molecular level, is transforming modern biology beyond recognition. We now have some understanding of genes: not only how their expression is regulated at the cellular level, but even how genes control the development of such complex structures as limbs and eyes. Yet the techniques of this new discipline are only now beginning to be applied to reproductive processes, and few reproductive biologists are presently contributing to these developments.

The aim of *Genes in Mammalian Reproduction* is to describe many of these new developments and to make them available to a broad range of reproductive biologists, providing them with the current picture, from which future research projects may be developed. Towards this end the authors were asked to reference key research techniques, including the use of transgenic mouse lines expressing antisense constructs under the control of inducible promoters to determine the role of specific genes in reproductive processes, the polymerase chain reaction to amplify specific gene sequences, and the techniques of specific gene targeting. They were also asked to indicate future research trends and needs.

Among the topics covered are the many unique mechanisms involved in spermatogenesis that may provide opportunities for contraception. It was previously thought that gene expression did not occur in gametes, but it is now apparent that many genes are active in spermatids, including the oncogenes and those genes for protamines, LDH-X, and sex-determination. The role of oncogenes may have practical importance considering that if anti-oncogene drugs are developed their side effect may be male infertility. Oocytes, even when arrested in the first meiotic prophase, are also active in transcription and translation. The genes that are active include *oct-3*, which codes for a protein essential for cell division; the *c-kit* oncogene, which may play a role in oocyte maturation; and ZP-3, a single copy gene that codes for the sperm receptor on the zona pellucida and is located on chromosome 5.

We know that in mammals the switch to embryonic control of development starts at ovulation, as oocyte mRNAs begin to decline, and is essentially complete after 1–3 cell divisions, at which time embryonic transcription begins. Most of our knowledge derives from the laboratory mouse, in which a number of genes have been identified and their expression determined during embryonic compaction and blastocyst formation. However, we know little of other species, including our own. This knowledge is badly needed because the availability of in vitro fertilization therapy and of the polymerase chain reaction to amplify specific DNA sequences now makes preimplantation diagnosis possible. These techniques have already been used to sex preimplantation human embryos, so that expression of x-linked diseases may be prevented by transfering only female conceptuses. Although complex, these procedures are appropriate for such high risk pregnancies.

Following implantation the role of specific genes in mammalian development is less clear. There is evidence for the action of various growth factors—including transforming growth factor and the insulin-like growth factors—in development, but a precise role awaits the application of anti-sense techniques, dominant negative transgenics, and gene targeting. Another fascinating aspect of development described in this book is the imprinting of developmental genes as they pass through the testis, thus demonstrating that the contribution of the male, as well as of the female, is essential for complete mammalian development. How such imprinting occurs is not known, but it probably involves the binding of gamete-specific factors that protect the gene from the methylation that appears to occur between the spermatogonial stage and the first meiotic prophase.

Molecular biology is also beginning to transform our understanding of the hormones that are involved in mammalian reproduction. The gene for anti-müllerian hormone (also known as müllerian-inhibiting substance or MIS) has been identified, as has its location on the short arm of human chromosome 19. The function of MIS was once thought to be the regression of the müllerian ducts in the male fetus, but now it also appears to play a role in testis development, regulation of germ cell development, and the descent of the testis. Because of these many functions MIS should probably be renamed. Considerable knowledge is also accumulating concerning the multiple roles of the gonadal proteins (inhibin, activin and follistatin), gonadotropin-releasing hormone and the reproductive steroids (estradiol and progesterone). As described in *Genes in Mammalian Reproduction*, their genes and those of some of their receptors have been cloned and expressed in several cell lines. Future research should provide an understanding of how these genes are controlled, opening up new avenues for the treatment of infertility and new approaches to contraception.

I thank all of the contributors to this volume for their excellent coverage of this rapidly developing field and for their willingness to update their chapters as we waited for the final ones to arrive. I also thank Brian Crawford, William Curtis, and other members of the editorial staff at Wiley-Liss for their help, as well as Joe Harford for including this book in the *Modern Cell Biology* series.

Ralph B.L. Gwatkin
Beachwood, Ohio
August 1992

ABOUT THE EDITOR

RALPH B.L. GWATKIN is Adjunct Professor in the Department of Reproductive Biology, School of Medicine, Case Western Reserve University, and President of ReproGene, Beachwood, Ohio. Professor Gwatkin received his B.A. and M.A. from the University of Toronto and his Ph.D. in microbiology from Rutgers University. After postdoctoral studies at the University of Illinois, he became a staff member of the Wistar Institute and Assistant Professor at the University of Pennsylvania. He then joined the Merck Institute for Therapeutic Research where he spent most of his career as a research director. In 1982 he joined McMaster University Medical School as Professor and from 1986–1989 was Director of Reproduction and Development Biology at the Cleveland Clinic Foundation. Professor Gwatkin has made a number of notable research contributions (among them the first cloning of mammalian cells from liquid cultures, the first studies of the growth of viruses in mammalian eggs and early embryos, and elucidation of the nutritional requirements for post-blastocyst development) and was the first to demonstrate the existence of specific sperm receptors in the solubilized zona pellucida. He has been the recipient of a research career development award from the National Institutes of Health and of awards from the Society for the Study of Reproduction and the American Fertility Society. Dr. Gwatkin is the author of *Fertilization Mechanisms in Man and Mammals* and editor of *Manipulation of Mammalian Development*. He is also Editor-in-Chief of the international research journal *Molecular Reproduction and Development*.

Genes in Mammalian Reproduction: 1–26
© 1993 Wiley-Liss, Inc.

Molecular Genetics of Mammalian Spermatogenesis

Robert P. Erickson

I. INTRODUCTION

Spermatogenesis, the process that produces the male gamete, was originally intensively studied by anatomists, electron microscopists, and endocrinologists. Their classic studies have been summarized in a number of papers [Fawcett, 1979; Leblond and Clermont, 1952; Oakberg, 1956]. Geneticists, biochemists, and molecular biologists have subsequently been attracted to the process, and their work, too, has been multiply reviewed [Beatty, 1970; Bellvé, 1979; Erickson et al. 1981; Erickson, 1990; Hecht, 1986; Willison and Ashworth, 1987]. I believe that the special attraction of the system of spermatogenesis for the latter group of investigators, who could be characterized as molecular geneticists, is the realization that it provides a complete developmental sequence from resting germ cells to a highly differentiated mature cell type that is accessible in one, reasonably abundant adult tissue. The development of cell separations for purifying the cells in various stages of this developmental process [Meistrich, 1972] and the availability of mutants, especially in mice, that block the differentiative process at various stages enhance the profitability of studying the system.

Despite the frequent reviews in the past, the rate of progress is such that this further one seems warranted at this time. More-over, the subject has expanded to such a degree that this review cannot be considered to be exhaustive—a book will soon be required if one is to cover the subject fully. This review attempts to emphasize the extent to which spermatogenesis involves unique molecular pathways of development that might be targets for chemical, i.e., drug, inhibition with the goal of male fertility control.

II. DNA SYNTHESIS DURING SPERMATOGENESIS

A. Replicative and Methylation

In mammals, as in other diploid eukaryotes, major replicative DNA synthesis occurs in pre-leptotene spermatocytes [Kofman-Alfaro and Chandley, 1970]. This is associated with high levels of DNA polymerase activities, which decline to low levels during spermiogenesis but do not decline further during sperm maturation [Daentl et al., 1977; Hecht et al., 1979]. DNA methylation, however, is not completely replicative at this time. While many single copy genes are methylated throughout spermatogenesis [Rahe et al., 1983; Groudine and Conklin, 1985], repetitive genes such as satellite sequences remain undermethylated throughout spermatogenesis [Feinstein et al., 1985; Ponzetto-Zimmerman and Wolgemuth, 1984; Sanford et al., 1984]. On the other hand, the pattern of methylation for genes expressed during spermatogenesis changes during the process: Transition protein 1 becomes progressively less methylated, protamine 1 and 2 genes become progressively more methylated [Trasler et al., 1990], and *Pgk*-2 is demethylated and remethylated during spermatogenesis [Ariel et al., 1991].

B. Possibly Related to Genetic Exchange

A low level of DNA synthesis occurs during pachytene [Meistrich et al., 1975], which may be related to alterations in gene expression or to repair and/or meiotic pairing [Smyth and Stern, 1973]. As reviewed by Roeder [1990], it is now believed that DNA strand exchange reactions catalyzed by recombination enzymes are responsible for homologous pairing in meiosis, while the synaptonemal complex is believed to have a less active role. Several DNA repair synthesis-related enzymes reach their peak of specific activity at pachytene during mouse spermatogenesis [Orlando et al., 1984] while DNA helicase activity is also at its peak at this time [David et al., 1982].

Hotta et al. [1985] have found meiosis-specific transcripts of the fraction of DNA that has delayed replication and that they believe have a role in chromosome pairing. The properties of these transcripts are surprising as they are apparently diverse, unique sequences, yet they maintain significant homology between plants and mammals. There is clearly a need for a more detailed examination of recombinases and other enzymes related to genetic recombination in mammalian spermatogenesis. While DNA synthesis has frequently been the target of anticancer drugs, it has not usualy been considered a target for fertility control. However, this pachytene DNA synthesis may involve testis-specific enzymes that could be safe targets for fertility contol.

III. RNA SYNTHESIS DURING SPERMATOGENESIS

A. General

Studies on RNA synthesis started with histochemical descriptions and have progressed to the current detailed investigations of specific base sequences that are expressed. The application of autoradiography, following the administration of isotopic precursors, to studies of testicular RNA synthesis at first concluded that little RNA was transcribed after meiosis, since spermatids exposed to short pulses of ^{3}H-uridine showed only a small amount of uridine incorporation by comparison to pachytene spermatocytes [Monesi, 1967]. However, quantitative autoradiographic studies demonstrated that the rate of RNA synthesis per cell decreased only fourfold during meiosis, i.e., the RNA synthesis to DNA ratio did not change during the meiotic divisions [Loir, 1972]. Biochemical analyses of the rate of incorporation of labeled precursors into specific macromolecules using testicular cells separated by sedimentation techniques also showed that high rates of RNA synthesis continued in postmeiotic cell stages [Meistrich, 1972; Lee and Dixon, 1972]. Such results were confirmed by a direct in situ method for studying RNA polymerase activity on microscopic samples [Moore, 1971]. This latter study, and estimates of pool size indicating that intracellular uridine pools are increasing rather than decreasing after meiosis [Geremia et al., 1977], indicate that the results are not artifacts of altering uridine pool size.

B. Ribosomal RNA Synthesis

The question of the continuity of ribosomal RNA synthesis through meiosis has been extensively addressed. Examination of testicular RNA synthesis by a variety of electromicroscopic techniques [Kierszenbaum and Tres, 1974a,b, 1975] detected ribosomal RNA transcription prior to but not following meiosis. Studies of the time of synthesis of ribosomal RNA found in maturing mouse spermatozoa suggested that its synthesis is premeiotic [Betlach and Erickson, 1976], while characterization of newly synthesized RNA from separated spermatogenic cells has led to discrepant findings. Geremia et al. [1978] characterized the newly synthesized RNA from spermatids by sucrose density gradient centrifugation and concluded that it included ribosomal RNA, while Erickson et al. [1980a] characterized such RNA by electrophoresis and found ribosomal RNA synthesis to decrease markedly in postmeiotic cells. An alternative approach to the study of ribosomal RNA synthesis during spermatogenesis uses silver staining techniques that reflect the activity of nucleolus organizing regions. (NORs) in the previous cell cycle. Schmid et al. [1977] concluded that ribosomal RNA was synthesized postmeiotically by this technique, but their results may have been influenced by the changing composition of chromatin proteins that markedly affect silver binding [Bloom and Goodpasture, 1976].

C. Postmeiotic Gene Expression

1. Overview. Genetic dogma indicated that gene expression did not occur in animal gametes. Postmeiotic gene expression appeared to run counter to the dominant interpretations of evolutionary theory, which concluded that genetic selection should only be zygotic: Gene expression in gametes could result in phenotypic differences affecting function and could potentially be subject to selection. This dogma was supported by the early experiments of Muller and Settles [1927]. They found that sperm nullisomic for about 1/40 of the *Drosophila* genome (created with chromosomal translocations) still fertilized eggs and that normal embryonic development continued as the missing material was contributed to the zygote by eggs disomic for this genetic material. Similar experiments

yielded the same result in mice [Lyon et al., 1972]. However, in both cases the investigators were unaware of the syncytial nature of spermatogenesis, which could allow the products of any genes that might be expressed postmeiotically to be shared among haploid nuclei [Erickson, 1973]. Current molecular genetic techniques have allowed ample confirmation of the presence of postmeiotic gene expression. Two-dimensional gel electrophoresis analyses of the products of in vitro translation of RNA purified from separated spermatocytes and spermatids showed twice as many spermatid-specific as spermatocyte-specific gene products, with only a relatively small number of proteins programed by RNAs from both cell types [Fujimoto and Erickson, 1982].

Assays of the mRNA for specific proteins performed by in vitro translation from purified RNA demonstrated that the mRNA for protamine and phosphoglycerate kinase 2 (PGK-2) increased after meiosis [Erickson et al., 1980b]. These results did not prove that there was postmeiotic transcription, since the mRNAs might have been transcribed earlier and only processed postmeiotically. Therefore several groups made cDNA libraries and found clones for specific RNAs that increased or first appeared after meiosis [Kleene et al., 1983; Dudley et al., 1984; Fujimoto et al., 1984]. A survey of testicular cDNAs showed that about half increased in abundance after meiosis [Thomas et al., 1989]. About half of these (one-fourth of the total) first appeared after meiosis [Thomas et al., 1989]. Thus ample confirmation of postmeiotic gene expression has been obtained, and the sharing of a postmeiotically expressed gene product between spermatids of a heterozygous transgenic mouse has now been shown [Braun et al., 1989a]. In what follows, I will present the data on postmeiotic expression of a number of classes of genes that have been studied in detail and that have at least one member that is expressed postmeiotically. This will be followed by a short summary of several genes that have been studied in less detail.

a. Protamines. The evidence for postmeiotic expression of protamines and details of their regulation have been extensively reviewed by Hecht [1990]. They provide another example of the sharing of transcripts between spermatids [Caldwell and Handel, 1991]. Several aspects of their regulation will be discussed in later sections. In addition, transgenic mice have been used to show that 4.8 kb 5' of protamine 1 [Peschon et al., 1987] and 859 5' base pairs (bp) of protamine 2 [Stewart et al., 1988] can elicit spermatid expression. In contrast, and unexpectedly, 185 5' bp of the mouse metallothionein I gene promoted meiotic expression of ornithine transcarbamylase, usually only expressed in liver, as is the metallothionein gene [Kelley et al., 1988]. This result needs to be pursued to determine what combination of DNA sequences and transacting factors were responsible.

b. Lactate dehydrogenase X. Lactate dehydrogenase X (LDH-X) activity is first found in the testes of 19-day-old mice—the time when spermatocytes are first observed [Goldberg and Hawtrey, 1967]. Radiopulse immunoprecipitation experiments performed on fractionated cells concluded that LDH-X synthesis continues postmeiotically, with about half of the total synthesized in spermatids [Meistrich et al., 1977]. Studies with the cloned LDH-X gene have confirmed and extended these findings. We originally studied a cDNA from spermatids that was shown to have postmeiotic transcription by the very sensitive method of nuclear run-off [Fujimoto et al., 1984]. When this cDNA was sequenced it was discovered that it encoded LDH-X [Tanaka and Fujimoto, 1986]. Studies with this and other LDH-X clones have shown a high abundance of LDH-X message in round spermatids that decreases in later spermatids [Fujimoto et al., 1988; Jen et al.,

1990]. There are also interesting changes in the polyadenylation of the LDH-X message during spermatogenesis, which is discussed further below. While many attempts to use LDH-X for a sperm vaccine have been unsuccessful, metabolic inhibitors targeted for the unique substrate specificity of LDH-X could be more extensively studied.

c. PGK-2. The developmental appearance of PGK-2 initially indicated that the protein was not synthesized until after meiosis [VandeBerg et al., 1976; Kramer and Erickson, 1981]. Immunofluorescent studies confirmed the localization of PGK-2 antigen to postmeiotic cells [Kramer, 1981; Blüthmann et al., 1982]. When the X-linked PGK-1 gene was cloned, it was found to cross-hybridize to PGK-2 and thus was used to study transcription in the testis—demonstrating transcription at pachytene with further transcription occurring after meiosis [Erickson et al., 1985]. Little or none of the premeiotically transcribed message is found on polysomes, while the specific mRNA is abundantly found on polysomes after meiosis [Gold et al., 1983].

The more sensitive the assay used, the earlier the time that expression was detected. Less sensitive functional assays for PGK-2 mRNA first detected this message after meiosis [Erickson et al., 1980b], whereas reverse transcription-polymerase chain reaction (RTPCR) detected transcription significantly earlier than that for protamines, with message abundance peaking in early spermatids [Robinson and Simon, 1991]. Analyses in transgenic mice demonstrated that 323 bp of 5' sequence is sufficient for this pattern of expression [Robinson et al., 1989]. Genomic cloning of PGK-2 has shown that it is a reverse transcriptase (i.e., retroposon) mediated, processed gene insert [McCarrey and Thomas, 1987]. This must have been a sporadic event that was then maintained by selection, because PGK-2 can compensate for the extinction of expression of X-linked PGK-1 with X inactivation during spermatogenesis (which is discussed further below).

d. Heat shock proteins. Not surprisingly, since it has long been known that spermatogenesis is quite heat sensitive, requiring 32°C in the scrotum and being ablated at the intraabdominal site at 37°C, a number of studies of heat shock proteins in testes have been made. Krawczyk and collaborators [1987] pointed out that levels of heat shock protein 70 (hsp 70) were about 100-fold higher in testes of both mouse and rats than in other tissues. Analyses by two-dimensional gel electrophoresis and with antibodies have demonstrated that members of the hsp 70 family are synthesized in response to heat stress and also constitutively in the mouse testes, while there are unique members of the family that are synthesized late in spermatogenesis [Allen et al., 1988a,b; Maekawa et al., 1989].

With the development of the subclones of hsp 70 family members, which detected different transcripts in testes, and with cloning and sequencing of members of the hsp 70 family, it became clear that an hsp 70 family member detected by pMHS213 and cDNA clone for the *hec70t* is preferentially expressed in postmeiotic cells [Zakeri and Wolgemuth, 1987; Matsumoto and Fujimoto, 1990], while the cloned gene HSP 70.2 is abundantly expressed in pachytene spermatocytes and decreases during the latter stages of spermatogenesis [Zakeri et al., 1988]. Both genes are constitutively expressed, and the predicted amino acid sequence is closely related to that of an hsp 70 family member involved in uncoating the clathrin network surrounding coated pits. Postmeiotic expression of one member of the family has been confirmed by in situ mRNA studies in rats [Krawczyk et al., 1988]. There are also two members of the hsp 90 family that are expressed in testes, but the germ cell specificity of HSP 86 is not related to postmeiotic expression [Lee, 1990; Gruppi et al., 1991].

e. Oncogenes. The expression of oncogenes during spermatogenesis has been a popular topic for molecular studies and has been recently reviewed [Probst et al., 1988a]. The c (cellular) oncogenes, which are expressed abundantly in premeiotic cells, are different from those that are expressed postmeiotically. The c-*ras* gene family consists of three related members, and their expression has been extensively studied in mouse testis [Wolfes et al., 1989]. They, and the c-*raf* protooncogene, are expressed in meiotically active spermatogonial stem cells, during meiotic prophase, and to some extent in postmeiotic spermatids [Wolfes et al., 1989]. They involve different functions, since the c-*raf* oncogene possesses serine and threonine protein kinase activity and the c-*ras* family encodes the plasma membrane GTP-binding proteins. High levels of c-*myc* and c-*jun* mRNAs are found in spermatocytes [Wolfes et al., 1989]. Both encode nuclear proteins, and the c-*jun* gene product complexes with the c-*fos* product, which is also expressed at high levels in spermatogonia [Wolfes et al., 1989] to form a DNA-binding complex whose target sequences are the activator protein-1 (AP-1)–binding site.

The expression of these transacting factors, which have the potential to bind to the AP-1 site, may be involved in the transcriptional control of genes that begin to be expressed at or near the time of meiosis. Interestingly, studies on the expression of *jun*B and c-*jun* mRNAs in male germ cells were complicated by the rapid rise in the levels of their messages that occurs when the testicular cells are dissociated [Alcivar et al., 1990]. This effect is maximal in the testis of 8-day-old mice. This probably corresponds to the fact that the mRNAs for *jun* and *jun*B are rapidly induced, following the addition of serum growth factors.

While the c-*raf* protein kinase is predominantly expressed premeiotically, the c-*mos* protein kinase is predominantly expressed in haploid spermatids [Probst et al., 1988b]. Although the c-*mos* transcript length in spermatids is unusual, a protein product is translated from the message [Herzog et al., 1988]. c-*abl*, another tyrosine kinase protoncogene, is abundantly expressed postmeiotically—its short message results from the use of an alternative polyadenylation site [see below; Oppi et al., 1987; Meijer et al., 1987] and is associated with unique phosphoproteins in postmeiotic cells [Ponzetto et al., 1989]. c-*pim*-1 is another c-oncogene that encodes a kinase and is expressed postmeiotically [Sorrentino et al., 1988]. It maps to the *t*-region (see below).

The *ret* transforming gene is a hybrid molecule that was formed by recombination between a tyrosine kinase and a zinc finger-like protein. It, too, is expressed in postmeiotic round spermatids [Takahashi et al, 1988]. The reason for the expression of these multiple oncogenes postmeiotically is not clear. Perhaps the identification of the c-*mos* gene product as the cytostatic factor, CSF, involved in cleavage arrest [Sagata et al., 1989] is relevant to its transcription postmeiotically when spermatids lose their potential to divide. Another postmeiotically expressed c-oncogene, c-*int*-1 [Schakleford and Varmus, 1987], codes for a secreted protein, perhaps a growth factor, that is expressed in the embryonic nervous system (and its homolog in *Drosophila*, *wingless* has a role in early development), but it has no obvious role in spermiogenesis. In the future, if anti-oncogene drugs are developed for cancer prevention, male sterility could be an outcome (if the drugs cross the blood–testis barrier) since the testes is exceptional for the degree to which c-oncogenes are expressed in the adult.

2. Sex determination–related genes. The first gene cloned as a candidate for the testes-determining factor, *ZFY*, was found after an exhaustive search for Y sequences found in XX males and missing in an XY female, which were highly likely to include sequences for sex determination [Page et

al., 1987]. Its evolutionary conservation and expression in testes of almost all mammals studied gave credence to the idea that this was the proper gene. However, the discovery of "XX males" (they all had some degree of sexual ambiguity) that did not have the *ZFY* sequence [Verga and Erickson, 1989] but had other Y chromosomal sequences [Palmer, et al, 1989] and the discovery that the XY female with a deletion of *ZFY* was lacking more Y sequences than originally thought [Page et al., 1990] made it very unlikely that *ZFY* was, indeed, the testes-determining factor. The inappropriate expression of the mouse homolog of *ZFY*, *Zfy*-1, in the germ cells of differentiating gonads, instead of Sertoli cells (since germ cells are not essential for testicular differentiation [Koopman et al., 1989])also cast doubt on the sex-determining role of *ZFY*. Indeed, an alternative sex-determining gene, *SRY*, was found in the human Y material present in the ZFY^- "XX males" [Sinclair et al., 1990], and its role as necessary, but perhaps not sufficient, for testes determination has been confirmed by the detection of mutations in it in XY females [Jäger et al., 1990; Berta et al., 1990], by its expression in Sertoli cells in the primordial testes [Koopman et al., 1990], and by the creation of male XX mice that carry a transgene for it [Koopman et al., 1991]. Nonetheless the close linkage of *ZFY* to *SRY* in both mice and humans, despite the quite different location of these genes in regard to the pseudoautosomal region in the two species, suggests that *ZFY* has an important role in male development, even if it is not the primary gene involved in sex determination.

Despite the close linkage to *SRY*, *ZFY* and its homologs have quite different patterns of gene expression in humans and mice. In humans there are two members of this family, *ZFY* and *ZFX*, the latter a precise homolog at 99% amino sequence identify [Schnëider-Gädicke et al., 1989]. In mice, there are two Y-linked copies, *Zfy*-1 and *Zfy*-2, an X-linked copy, *Zfx*, and an autosomal copy, *Zfa* [Mardon et al., 1989; Nagamine et al., 1989]. *Zfa* is a recruited, retroposon product, i.e., processed but functional, copy of *Zfx* [Ashworth et al., 1990]. *ZFX* and *ZFY* seem to be expressed very widely in human tissues. Specifically, *ZFY* was expressed in all human male (Y chromosome positive) cell lines studies by Northern analyses, including T-cell leukemia, neuroblastoma, fibroblast, and lymphoid blastoid lines, whereas *ZFX* was expressed in these and adult testes [Schnëider-Gädicke et al., 1989]. In another study, *ZFY* was expressed as a 3 kb transcript in testis and as a 5.7 kb transcript in somatic cells—primarily tumor lines [Lau and Chan, 1989]. *ZFY* was found to be expressed in fetal brain and adult liver, while in another study *ZFX* was expressed in all male, female, adult, and fetal tissues examined [Palmer et al, 1990].

Although it is clear that in mice *Zfx* is expressed in all tissues [Mardon et al., 1990], similar to the expression pattern of *ZFX* in man, *Zfy* has much more limited expression. *Zfy* was found by Northern analysis and RTPCR to be expressed in adult testis but not in brain, heart, kidney, liver, and lungs [Nagamime et al., 1990]. In the work of Mardon and Page [1989] it was shown that *Zfy*-1 and *Zfy*-2 are expressed in adult testis because a cDNA for each was found in adult testis, but these authors, using Northern analysis, did not detect *Zfy*-1 or *Zfy*-2 expression in adult brain, heart, kidney, liver, lung, or spleen.

The precise timing and cell type of expression for *Zfy* in testis is less clear cut. *Zfy* was not detected by Northern analysis in pachytene spermatocytes but was detected by RTPCR—this result was confirmed by finding a positive signal in *Tfm* testis, in which spermatogenesis is arrested at meiosis I [Nagamime et al., 1990]. Kallikin et al., [1989] found a marked increase in *Zfy* transcripts postmeiotically, which appeared, using the low sensitivity technique of allele-specific oligodeoxynucleotide hybridization, to be predominantly *Zfy*-1. Others have found that *Zfy*-2 transcripts only become evident in testis at the pachytene stage and are present at a level about three-fold higher than *Zfy*-1 [Gubbay et al., 1990]. It seems probable that *Zfy* transcripts have an important role in spermatogenesis. No studies of *Sry* expression during spermatogenesis have yet been published. It is known that *Sry* is expressed in Sertoli cells in the developing testis [Koopman et al., 1990]. The cell-type specificity of expression in adult testis is yet to be described.

3. Proopiolmelanocortin and related genes. Proopiomelanocortin (POMC) is a precursor protein that contains the sequences for adrenocorticotropin, β-endorphin, and the melanocyte-stimulating hormones, and it is expressed in a variety of tissues other than pituitary. The findings that the POMC transcript was expressed at as high a level in the testes as in the hypothalamus, but that the peptides coded by the mRNA were at a lower level (by 2–3 orders of magnitude) than those found in the hypothalamus, were surprising [Chen et al., 1984]. At the same time it was found that the POMC transcripts were about 150 bases shorter than those in the pituitary or hypothalamus, which raised the question of whether the lack of expression was related to the change in transcript size.

While in situ RNA localization showed POMC messenger RNA to be located in Leydig cells of the rat, Northern blots performed on mice of different ages (and with mutations eliminating germ cells) showed that POMC is also expressed in germ cells, at least in the latter species [Pintar et al., 1984]. The likelihood that POMC might be involved in paracrine functions in the testes was supported by the results from detailed in situ mRNA localization, which showed that only Leydig cells that were surrounded by tubules in stages IX–XII of the cycle of spermatogenesis expressed the transcript, i.e., stages in which the characteristic grouping of cells includes spermatocytes in early stages of meiosis (leptotene and zygotene), forming the central layer while type A spermatogonia are present in the periphery [Gizang-Ginzberg and Wolgemuth, 1985]. These authors also occasionally saw transcripts in spermatogonia and spermatocytes, establishing that germ cell expression also occurs in mice. Furthermore, they extended the evidence for a paracrine role of POMC in testis by showing that the POMC transcripts were dependent on the presence of germ cells, while a mouse mutant in which spermatogenesis is arrested in early spermiogenesis did not have an effect on the POMC transcripts [Gizang-Ginzberg and Wolgemuth, 1987]. It was found that the shorter testicular POMC transcript resulted from the use of an alternative transcription start site closer to the initiation codon [Jeannotte et al., 1987].

Preproenkephalin, a protein-like POMC that can give rise through proteolytic digestion to a number of different peptides, some of which have opioid activity, is also expressed in the testis. In rat testis, as with POMC, the levels of message are similar to those in brain, while the amount of metenkephalin was less than 4% of that in brain [Kilpatrick, et al., 1985]. Cell separations showed that the greatest concentrations of the preproenkephalin mRNA are in pachytene spermatocytes and round spermatids [Kilpatrick and Millette, 1986]. In fact, in rats the highest abundance of preproenkephalin mRNA was localized to spermatids, suggesting postmeiotic expression [Yoshikawa and Aizawa, 1988]. The major transcript in both rat [Kilpatrick, et al., 1985] and mouse [Kilpatrick, et al., 1986] was larger than that found in the pituitary; Kilpatrick et al. [1986] also found the pituitary-sized transcript in somatic cells of the testis. The larger testicular preproenkephalin transcript is the result of alternative RNA splicing in which an alternative acceptor site in intron A is used so that the germ cell–specific preproenkephalin transcript has a 491 bp untranslated leader [Garett, et al., 1989].

The role of the proenkephalin peptides may be in fertilization rather than to render a paracrine function in the testis—the peptides are stored in the sperm acrosome and are released from it during the acrosome reaction [Kew, et al., 1990]. A similar localization has been found for cholecystokinin, a neurotransmitter/neuromodulator, and this too may have a role in fertilization [Persson, et al., 1989]. In contrast, choline acetyltransferase (which is the key enzyme in the synthesis of the neurotransmitter acetylcholine) is transcribed postmeiotically, but is localized in the postacrosomal region of the head and in the annulus of the midpiece [Ibáñez et al., 1991]. Given the large number of pharmacological agents targeted at neurotransmitters, their potential effects on male fertility should be considered.

4. Other genes. Expression of a number of other genes has been studied in spermatogenesis, sometimes inspired by "Let's have a look-see" and sometimes by hypotheses related to the potential developmental importance of the gene. Two homeobox-containing genes have been found to be expressed during mammalian spermatogenesis. The MH-3 homeobox-containing gene, which seems most homologous to the *Deformed* locus of *Drosophila*, is abundantly expressed at pachytene with possible postmeiotic expression [Rubin et al., 1986]. *Hox*-1.4, the homeobox-containing gene most homologous to the *Antennapedia* locus of *Drosophila*, was weakly expressed before meiosis and abundantly expressed after meiosis, strongly suggesting postmeiotic expression [Wolgemuth, et al., 1987].

The screening of a mouse testicular cDNA library with an α-tubulin probe detected two testes-specific isoforms that were not very different in sequence than the general somatic isoforms [Viliasaute et al., 1986]. More recently, Hecht's group found a quite divergent α-tubulin that is also expressed in testes. Its amino acid sequence differs by 30% from the general somatic and previously described testicular forms [Hecht et al., 1988]. Hecht et al., [1988] were able to make an antibody to a synthetic peptide of this unique α-tubulin sequence that confirmed the testes specificity and demonstrated specific localization to the meiotic spindle and to the sperm manchette [Hecht et al., 1988].

The general approach of screening testicular cDNA libraries with probes of interest, or by using degenerate PCR primers, has led to the identification of other apparently unique, testicular sequences, for instance, those for the POUbox containing transacting factor [Goldsborough et al., 1990] and a zinc finger gene [Cunliffe et al., 1990]. Although such screening is frequently done at low stringency, an alternative approach is to use a degenerate oligodeoxynucleotide probe that might better pick up a homologous sequence. This approach was used to detect a testis specific tyrosine kinase [Fischman et al., 1990] which is predominantly expressed in spermatocytes [Keshet et al., 1990]. Differential cDNA screening is another approach to obtain postmeiotically expressed genes. Differential cDNA cloning using seminiferous tubules versus total testicular RNA led to the isolation of a postmeiotically expressed gene that has some features of a leucine zipper [Van der Hoorn et al., 1990].

There are many recent papers about the cloning of genes for sperm-specific antigens; one in particular is of interest because of its relevance to the *T* region (the *T* region is discussed in detail below). Clones have been isolated from the chromosomal site of the *T* region, i.e., proximal chromosome 17. One clone that was characterized and turned out to map outside the *T* region coded for a sequence very similar to the sperm-coating glycoprotein gene [Kasahara et al., 1989]. Since this is a gene

that is expressed in the epididymis in the rat but was found to be expressed in the testes of mice, it suggests an interesting evolutionary divergence. The reason for the postmeiotic expression of preproacrosin [Adham et al., 1989] is easily understood, since the gene product is a sperm acrosome enzyme stored for later use. This may also be the reason for haploid expression of a rabbit sperm membrane protein [Wang et al., 1990] whereas the expression of the *mak* protein kinase [Matsushime et al., 1990] may reflect a function possibly similar to that of postmeiotically expressed protein kinase c-oncogenes (see above). The reasons for postmeiotic transcription of calmodulin [Slaughter et al., 1989], a transferrin-like mRNA [Stallard et al., 1991], and male enhanced antigen [Lau et al., 1989] are unclear.

D. Evidence for Altered Transcriptional Patterns in Germ Cells

As Table I shows, genes that are expressed in both testicular germ cells and other tissues often produce different transcripts in the tissues. Multiple transcripts from single genes have frequently been found to be due to alternative splicing [Breitbart et al., 1987] but multiple promoters are frequently used for developmental control [Schibler and Sierra, 1987]. The examples in Table I include both mechanisms.

Some apparent examples in the testes have turned out to be due to transcription from different genes. For example, in the case of hsp 70 there was cross-hybridization of the probe to the transcript of a different gene that was being uniquely expressed in testes [Zakeri and Wolgemuth, 1987] (see above) In the case of histone 2B [Moss et al., 1989], it is possible that the 500 bp testis transcript is from a testis-specific gene, but not the one that gives rise to the 800 bp polyadenylated transcript. It is more likely that these are different transcripts from the same gene and the difference may indicate that the U7 small nuclear ribonucleo-

TABLE I. Genes With Altered Transcripts in Testicular Germ Cells

Gene	Somatic Tissue expression	Normal transcript (kb)	Testiclar transcript	References
c-*abl*	Many	5.5, 6.5	4 kb shortened by use of unusual poly adenylation site; protein not different	Oppi et al. [1987], Meijer et al. [1987], Ponzetto et al. [1989]
Angiotensin-converting enzyme	"Pulmonary" but widely distributed	5.1, 4.2	2.7 kb due to initiation in 12th intron	Howard et al. [1990], Kumar et al. [1991]
Cytochrome c_S	All	0.7, 1.1, and 1.3	1.7 kb that is not polysomal	Hake et al. [1990]
Farnesyl pyrophosphate synthetase	Liver > most tissues	1.2	1.4 kb due to use of alternative promoter	Teruya et al. [1990]
β1,4-galacto syltransferase	Many	4.1	2.9, 3.1 kb by use of alternative polyadenylation site	Shaper et al. [1990]
Histone 2B	Only testes	0.5; not adenylated	0.8 kb polyadenylated plus 12 extra C-terminal amino acids	Moss et al. [1989]
Hox-1.4	Embryonic spinal cord	1.7, 2.4	1.35 kb in spermatocytes; 1.45 in spermatids	Wolgemuth et al. [1987]
c-*mos*	Many	1.3, 1.4, and 2.3	1.7 kb	Propst and Vande Woude [1985]
Phosphoribosyl pyrophosphate synthetase I	Brain and adrenal	2.3	1.4kb	Taira et al. [1989]
c-*pim*-1	Immune tissues	2.8	2.4 kb	Meijer et al. [1987], Sorrentino et al. [1988]
Pre-proenkephalin	Brain	1.5	1.9kb contains large portion of first intron	Garrett et al. [1989], Kipatrick et al. [1990]
N-*ras*	Many	1.3, 2.4, and 5.0	1.3 kb	Wolgemuth et al. [1987]
ret finger protein	Many	2.4, 3.4	2.8 kb	Takahishi et al. [1988]

proteinmediated cleavage reactions are not utilized in spermatids.

However, in many other cases there is clear evidence for alternative splicing and/or processing from the same gene. Per-haps the altered transcripts for the c-oncogenes are related to the variation in translational efficiency associated with alternative promoters and leaders seen for many c-oncogenes [Kozak, 1988]. A detailed analysis of the proenkephalin germ cell promoter (see above) has identified a GC-rich stretch lacking a TATA sequence but containing a concensus SP1-binding site downstream [Kilpatrick, et al, 1990]. Some of these sequences share homology with the germ cell promoter of rat cytochrome c_T, the spermatogenic cytochrome c [Kilpatrick et al., 1990]. Altered transcription of rat farnesyl pyrophosphate synthetase in testes is due to the utilization of an upstream promoter that leads to a high level of poorly translated message [Teruya et al., 1990]. Specific 5'-binding factors are being explored for protamine-2 by gel retardation [Johnson et al., 1991] and in vitro transcriptional systems [Bunick et al., 1990]. It is possible that the postmeiotic expression of c-oncogenes that are transcriptional activators (see above) are responsible for this utilization of alternative promoters. The novel thyroid receptor detected in testis [Benbrook and Pfahl, 1987] has not been shown to be in germ cells, but if it is it might be involved in the use of different promoters.

Angiotensin-converting enzyme provides a particularly instructive example. The most common isozyme of this dipeptidyl exopeptidase (the activity is carboxylterminal) has been frequently studied because of its role in the control of blood pressure. The characterization of the testicular protein had shown differences in size from the common isozyme, but immunological relatedness. Cloning studies have shown that the carboxyterminal sequences of the somatic and testicular isozymes are identical [Kumar et al., 1989; Ehlers et al., 1989]. more recently, the difference has been shown to be due to the initiation of transcripton of the testicular isozyme in the 12th intron of the somatic angiotensin-converting enzyme gene [Howard et al, 1990; Sen et al., 1990; Kumar et al., 1991]

Altered splicing may reflect different forms of RNA helicase-like proteins that are involved in release of mRNA from spliceosomes [Company et al., 1991]. The maternal (oocyte-synthesized) messenger RNA An3 found in the animal pole of *Xenopus* seems to encode an ATP-dependent RNA helicase [Gururajan et al., 1991] with 74% identity to the mouse testes-specific PL10 [Leroy et al., 1989]. The possibility that testis-unique transcription initiation factors exist could provide targets for chemical (i.e., drug) inhibition of spermatogenesis.

E. Evidence for Altered Posttranscriptional Regulation in Germ Cells

Gold and Hecht [1981] studied compartmentalization of Poly(A)$^+$ RNA in total mouse testes and found that an unusually high proportion was nonpolysomal. This observation may be related to findings that suggest that polyadenylation of mRNA seems to occur by two different mechanisms during spermatogenesis. The first is the "classic" shortening of poly(A) tracts that is generally seen with mRNAs during translation. These changes are thought to be mediated by the poly(A)-binding protein, which determines mRNA stability in vitro. It is required for 60S ribosomal subunit-dependent translation initiation [Bernstein et al., 1989: Sachs and Davis 1989]. Examples of these (presented in Table II) would include protamine 1, transition proteins 1 and 2, HEM1050, and the mitochondrial capsule selenoprotein.

Several genes translated during spermatogenesis have contrary changes in polyadenylation that represents the second mechanism. These include LDH-X, which shows an increase in polyadenylation at about the time of meiosis without major changes in its translation at that time and cytochrome c_T, the testes-specific cytochrome c, which has a highly polyadenylated form while on polysomes and shows decreased adenylation later. Cytochrome c_T has a cytoplasmic polyadenyla-

TABLE II. Polyadenylation Changes During Spermatogenesis

Genes	Time of Transcription	Polyadenylation changes	Ribosome locational changes	References
LDH-X	Pachytene, continuing to spermatid	+100 bp at meiosis	Both forms on and off polysomes	Fujimoto et al. [1988]
Cytochrone c_T	High in pachytene, decreased in round spermatids	− 100 bp at meiosis	Polyadenylated form on polysomes	Hake et al. [1990]
Transition proteins 1 & 2; protamines 1 & 2; HEM 1050	Round spermatids	Decreased in late spermatids	Associated with ribosomes when shortening occurs	Kleene [1989]
Mitochondrial capsule seleno-protein	Late meiotic cells to spermatid	Decreased in late spermatids	Associated with ribosomes when shortening occurs	Kleene et al. [1990]

tion element [Hake et al., 1990] and it may be that LDH-X does as well. These genes may be regulated by a mechanism seen in maturing *Xenopus* oocytes in which polysome recruitment occurs by polyadenylation in response to a cytoplasmic polyadenylation element [Paris and Richter, 1990]. However, in *Xenopus* oocytes and early embryos, this "element" seems to be the usual polyadenylation sequence now signaling continued adenylation and thereby preventing deadenylation [Varnum and Wormington, 1990; Fox and Wickens, 1990] while other RNA sequences may signal post-midblastula transistion degradation of maternal mRNAs [Bouvet et al., 1991]. It is interesting that the *c-mos* gene, which is expressed both in maturing oocytes and during spermatogenesis (see above), has such a cytoplasmic polyadenylation element [Paris and Richter, 1990]. Sequences involved in posttranscriptional regulation are also being explored in transgenic mice. While it has been demonstrated in transgenic mice

that the 5′ regions of some postmeiotically expressed genes contain the sequence information required for correct timing of expression (see above), only recently have such studies started to explore the role of 3′-untranslated sequences in translational regulation. Very interestingly, fusion with 156 nucleotides of 3′-untranslated sequence from the mouse protamine gene delayed the translation of a human growth hormone recorder gene [Braun et al., 1989] in transgenic mice from early in spermatogenesis to the elongating spermatid stage when the protamine 1 gene is normally translated. In addition, whereas the control transgenic product was located in the acrosome, the product of the fusion gene was still intracellular, but not in the acrosome. A 18 kD protein specifically binding to the 3′-untranslated region of protamine 2 mRNA has been found [Kwon and Hecht, 1991].

The 3′ regulatory mechanisms-seen in mammalian testes are in contrast to those in *Drosophila* spermatogenesis in which translational control and cytoplasmic poly-adenylation are controlled by a 12 nucleotide sequence element in the 5' leader [Schäfer et al., 1991]. However, sequences 5' to the coding region may also be involved in vertebrate gametogenesis—5' sequences involved in intramolecular secondary structure can greatly affect changes in translational efficiency between oocytes and newly fertilized zygotes [Fu et al., 1991]. As in the case of transcriptional initiation, the possibility that testes-specific posttranscriptional regulatory mechanisms might exist could provide targets for a male fertility control medication.

IV. X-INACTIVATION DURING SPERMATOGENESIS

The inactivation of one X-chromosome in the somatic cells of mammalian females is well known and a much studied phenomenon. Single X-chromosome inactivation during mammalian male gamete formation [Lifschytz, 1972] is less well known, and the implications of this X-inactivation for the biochemical genetics of spermatogenesis have not generally been explored.

This phenomenon explains three unusual aspects of male meiosis: 1) Male sterility of X-autosome translocations [Russell and Montgomery, 1969]—this is sometimes associated with XY chromosome dissociation, which also is a cause of sterility [Beechey, 1973]. It may be related to abnormal RNA synthesis in the sex vesicle, which can be seen with autosomal–autosomal translocations [Speed, 1986]. 2) The difference between XX, *Sxr* (sex-reversed) mice, which are aspermatogenic, and XO, *Sxr* mice, which unergo spermatogenesis but immotile sperm result [Cattanach et al., 1971]. The lack of inactivation of a second X is also the likely explanation of infertility in Klinefelter syndrome, XXY. 3) The recruitment of autosomal copies of X-linked genes by retroposon activities, which are only expressed during spermatogenesis.

These include PGK-2 (discussed above), *Zfa* (discussed above) and a pyruvate dehydrogenase subunit [Dahl et al., 1990].

A study of spermatozoal glucose-6-phosphate dehydrogenase (G6PD) in mice found it to be identical by several criteria to the erythrocytic, X-linked form rather than the autosomal, hexose-6-phosphate dehydrogenase [Erickson 1975]. A study of the specific activities of G6PD in mouse testes during the first wave of spermatogenesis and in spermatogenic cells separated by the staput technique suggested that it might well be synthesized premeiotically and persist into sperm [Erickson, 1976]. Thus, on the basis of this one enzyme, it can be postulated that X-inactivation during spermatogenesis involves the same loci that are inactivated in female somatic cells.

It was originally believed that X-chromosome inactivation was complete. This belief, and the lack of clarity of results, delayed acceptance of the finding that X-linked steroid sulfatase remains active on the inactive chromosome in man [Shapiro et al., 1979]. Steroid sulfatase is on Xp near the terminal pseudoautosomal region. A more general acceptance of lack of inactivation of the Xp terminus developed with the finding of expressed genes, including *mic*2 [Goodfellow et al., 1984] and the GMCSF receptor [Gough, et al., 1990] in the pseudoautosomal region. Steroid sulfatase appears to remain active during spermatogenesis as well [Raman and Das, 1991].

More recently, the discovery that another region of non-X-inactivation was separated from Xp by a region of X-inactivation demonstrated that human X-inactivation is patchy [Brown and Willard 1990]. *Zfx* provides another "patch" of non-X-inactivation in man but not in mice. In fact, a gene has now been found near the X-inactivation center that is only expressed on the inactive X [Brown et al., 1991] . It remains to be seen if this "patchy", X-inactivation is also present during spermatogenesis, but the finding that mice show *Zfx* inactivation and have an almost identical autosomal homologue, *Zfa*, recruited by retroposon-like activity that is only expressed during spermatogenesis suggests that the pattern of X-inactivation in males during spermatogenesis will be similar to that of females in somatic cells.

V. *t*-ALLELE TRANSMISSION RATIO DISTORTION

Of the genetic variants in a variety of organisms that result in nonmendelian genetic segregation ratios, perhaps the most interesting are mutations at the *t*-region of the mouse. In addition to unorthodox features of genetic transmission, *t*-alleles have remarkable effects on early embryonic development and differentiation [Glueksohn-Waelsch and Erickson, 1970; Bennett, 1975; Erickson et al., 1980c]. The tail, from which the "T" derives, has served as the indicator phenotype for mutations with more profound effects. The dominant *T* (brachyury) mutation causes a short tail in the heterozygous state. A large number of recessive mutations, t^n, interact with *T*, to produce taillessness in compound heterozygotes (T/t^n). Such tailless mice are maintained via a balanced lethal system with the two homozygous types eliminated prenatally (both T/T and t^n/t^n are embryonic lethals). Males (but not females) heterozygous for a *t* allele usually transmit it to many more of their progeny than the 50% dictated by Mendel's laws. Many *t* alleles recently recovered from wild populations have segregation ratios of greater than 0.95, while laboratory populations are usually characterized by lower ratios. The ratios vary considerably between individual males carrying a particular *t* allele as well as between different *t* alleles [Braden, 1972]; there are low distorters as well. However, males who are compound heterozygotes for two different *t* alleles that complement each other (for viability) are sterile.

It can now be stated that the essential attributes of the *t* complex are the prezygotic transmission ratio distortion and crossover suppression, not the developmental recessive mutations leading to embryonic arrest. Many of the early analyses of *t*-complex "developmental lethals", were interpreted on the assumption that the mutations were very closely linked to each other and thus likely to be biochemically related. However, improved understanding of the crossover suppression property of the *t* complex reveals that this is not the case. A 12–15 cM length of proximal chromosome 17 that carries a complete *t* haplotype will almost never cross over with a wild-type chromosome. The observation that recombination occurs in compound heterozygotes containing two different *t* haplotypes has allowed recombinational analysis [Artzt et al. 1982] to supplement deletional analysis [Lyon and Bechtol, 1977; Erickson et al., 1978; Babiarz, 1982] for genetic dissection of the *t* region. These analyses, when combined with molecular cloning of the region, have demonstrated the presence of two large, and several smaller, adjacent inverted regions of sufficient size to inhibit crossing-over [Shin et al. 1983; Herrmann et al. 1986]. Rare recombination events between the wild-type chromosome and the paired inversions occur predominantly between a smaller inverted duplication in this region on the wild-type chromosome and homolgous DNA on the *t* chromosome [Herrmann et al., 1987].

Transmission ratio distortion can be accounted for by two different mechanisms: Either unequal numbers of two kinds of spermatozoa or equal numbers of functionally different spermatozoa are produced. Selective cell death during spermatogenesis could be hypothesized to be involved in the generation of unequal numbers of functionally different spermatozoa. In such an hypothesis, unequal numbers of spermatozoa could result because some chromosomes preferentially segregate to cytoplasm that is destined to die (a hypothesis for *segregation distorter* in *Drosophila* [Peacock and Erickson, 1965]. However, there is a deficiency of only about 13% of the cells expected on the basis of mouse spermatogenic kinetics, and these include pre- and postmeiotic cells [Oakberg, 1956a, b]. Thus nonmendelian ratios of the magnitude seen with *t* alleles cannot be explained by this mechanism.

Alternatively, there is evidence that suggests that the transmission ratio distortion controlled by *t* alleles involves functional differences between spermatozoa bearing different *t* alleles. Delayed mating, in which fertilization occurs as soon as the spermatozoa reach the fallopian tube, compared with normal mating, which requires that sperm remain in the female reproductive tract for several hours before fertilization, nullifies the transmission ratio distortion in the case of several *t* alleles [Braden, 1958; Yanagisawa et al., 1961; Erickson, 1973]. The fact that the *t* allele effect can be altered by changing the time of insemination relative to ovulation suggests that there are two classes of spermatozoa with unequal physiological characteristics, which could arise as the result of haploid gene expression. Molecular probes have confirmed the presence of equal numbers of the two kinds of spermatozoa even when recovered from the female reproductive tract [Silver and Olds-Clark, 1984]. The transmission ratios for a number of *t* alleles found with in vitro fertilization were similar to those found with delayed mating [McGrath and Hillman, 1980a, 1981; Garside and Hillman, 1989a,]. This result suggests that it is the physiological environment rather than the time element that is crucial. Artificial insemination and chimera experiments can be interpreted as showing that the segregating non-*t*-bearing sperm are dysfunctional [Olds-Clarke and Peitz, 1985; Seitz and Bennett, 1985].

A unifying hypothesis to explain the multiple effects of *t* alleles might predict that proteins coded for in the *T* region are related to membrane transport, hormone receptors, adenylate cyclase, or other membrane features affecting intracellular metabolism. Such membrane proteins might not be shared between spermatids by the intracellular bridge (Fig. 1). For in-

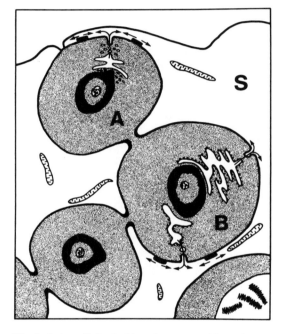

Fig. 1. *Intercellular bridges in spermatids and proposed inhibition of cell membrane–protein motility. Here early spermatids, with acrosomal caps commencing to form adjacent to nuclei, are surrounded by Sertoli cells (S) (a single line indicates the two opposed plasma membranes). Whether a cell membrane protein is processed by the Golgi complex and enters the plasma membrane with the fusion of a Golgi vacuole, as in cell A, or whether direct secretion from the endoplasmic reticulum might occur, as in cell B, it is postulated that the mRNA for a membrane-bound protein would not be free in the cytoplasm but attached to membrane-bound ribosomes. Once in the membrane, lateral motility of membrane proteins is depicted, while an inhibition of motility across intercellular bridges by specialized structures or steric hindrance is indicated by a thickening of the membrane structures at this point. (Reproduced from Erickson, 1978, with permission of the publisher.)*

stance, Meruelo and Edidin [1975] have detected *H-2*–determined differences in hepatic cAMP levels due to altered glucagon binding [Lafuse and Edidin, 1980], which we hypothesized might be relevant to sperm function. The possibility that *H-2* and *T* are functionally related had been suggested because of their linkage disequilibrium [Snell, 1968; Hammerberg and Klein, 1975]. Neither *H-2* nor the *T* region seems to have an effect on spermatozoal cAMP levels [Erickson et al., 1979] measured under physiological conditions, although differences have been reported [Nadjicka and Hillman, 1980] when measured in the presence of a phosphodiesterase inhibitor. While biochemical differences related to sperm activation have not been found, reproducible differences in sperm movement [Katz et al., 1979; Olds-Clarke, 1983] and a premature acrosome reaction [Brown et al., 1989] were associated with *t* alleles.

The biological complexity of the fascinating *t* complex has attracted a number of investigators interested in studying the molecular genetics of the region. Early studies used two-dimensional gel electrophoresis to focus on protein products expressed in the testes that were encoded by genes in the *t* complex. These analyses detected nine polypeptides that were expressed in testes and were encoded by this region [Silver et al., 1983]. Two of these were unique to testis and expressed strongly in postmeiotic cells, while three others were not testes unique but were expressed in postmeiotic cells [Silver et al., 1987]. One of these has attracted particular attention: TCP-1. Sanchez et al. [1985] demonstrated nonequivalent expression of wild-type and *t*-complex encoded forms of TCP-1 in germ cell fractions isolated subsequent to ^{35}S-methionine labeling, and this was confirmed by Silver et al. [1987]. TCP-1 expression was not limited to testes and was found in a variety of cell culture lines, tissues, and developmental stages with the exception of

two-cell embryos [Sanchez and Erickson, 1985; Silver et al., 1983]. The identification of TCP-1 as a Golgi membrane–associated protein whose synthesis is enhanced with the enlargement of the Golgi complex during spermatogenesis, contributing to the synthesis of the acrosome, explained the wide tissue distribution of this gene product and the reason for enhanced synthesis postmeiotically [Willison et al., 1989].

Microdissection of the proximal half of chromosome 17 isolated probes [Rohme et al. 1984] that were used to isolate cosmids from the *t* complex region. Further search of the cosmids for CpG-rich islands led to the discovery of one postmeiotically expressed gene from the *Tcd*-3 (distorter/sterility region of the *t* complex [Rappold et al., 1987]. Other cosmids derived by this approach from the *t* complex region were studied for a possible role in sex determination with negative results [Durbin et al., 1989]. Differential screening of DNA libraries to look for testes-specific clones mapping to chromosome 17 detected an unusual stretch of DNA with overlapping open reading frames and a stretch of alternating purine and pyrimidine residues [Sarvetnick et al., 1989] and a gene coding for a novel protein with a predicted coiled-coil structure mapping near the t^{h20} deletion [Mazarakis et al., 1991]. Other developmentally interesting genes, such as *Oct*-4 (a transacting factor), were mapped to the *t* complex [Scholer et al., 1990].

The above examples of cloning genes from the *t* region are examples of reverse genetics. However, they were not highly targeted on a single gene of interest. The single gene of interest that has been highly targeted for reverse genetic approaches is *Tcr*, the *t*-complex response element. Lyon [1984] put forth a model to explain transmission ratio distortion in which three transacting distorters would act in a harmful way on the wild-type form of *Tcr*, while the *Tcr* in the *t* complex (*Tcr*^{t}) would be resistant to this effect. Lyon had summarized a large number of experiments using various translocation chromosomes that provided "pieces" of the *t* complex to study their effects on transmission ratio distortion. As Lyon herself admitted [Lyon and Zenthon, 1987], her model did not agree with important data of Hammerberg [1982]. In further work by Lyon and Zenthon [1987] a number of *cis* effects on the transmission ratio distortion were found. Others found major effects of the genetic background [Gummere et al., 1986] and more regions of the *t* complex with effects on transmission ratio distortion [Silver and Remis, 1987]. Molecular cloning of the region between recombinant chromosome breakpoints that define the hypothetical *Tcr* locus was performed [Rosen et al., 1990] and T66 genes that encode male germ cell–specific transcripts were localized in this segment [Bullard and Schimenti, 1990]. One of these genes undergoes alternative splicing during spermatogenesis such that an allele-specific and haploid-specific transcript occurs [Cebra-Thomas et al., 1991]. A second group [Willison et al., 1990] have also identified a postmeiotic transcript from this region. However, evidence that these genes function in transmission ratio distortion has been lacking. A number of transgenic mice have been created with the gene, but apparently the transgene has had little effect on transmission ratio distortion. Given the complexities of the genetic effects in this region, it may well be that the cloned gene(s) do not represent an essential element of transmission ratio distortion.

Reverse genetics has also been applied to the problem of the sterility that results in mice compound heterozygosis for *t* alleles. A gene has been cloned, *tctex-1*, that produces a germ cell–specific transcript that is eight fold overexpressed in *t* homozygotes [Lader et al., 1989]. An interesting case of transmission ratio distortion in a transgenic mouse (nontransmission of the transgene by fertile males, normal transmission by females [Palmiter, et al., 1984]) is explained by the toxicity of postmeiotic expression of herpes simplex virus thymidine kinase [Wilkie, et al., 1991]. In summary, no definitive reasons for the complexity of the *t* region are yet available. These exciting recent cloning results make it likely that we will soon understand some of these complex phenomena.

VI. CONCLUSION

It should now be obvious that there are many unique molecular mechanisms involved in spermatogenesis. Some of the reasons for altered enzyme expression are easily understood, e.g., X-inactivation of the only X requires autosomal isozymes to replace essential enzymatic activities. In the case of LDH-X, altered substrate specificity related to the unique physiological environment of the spermatozoa is a reasonable explanation. other changes do not seem to have satisfactory explanations at this time, e.g., the activation of multiple oncogenes postmeiotically.

Nonetheless, the many molecular changes which have been described in this review provide potential avenues for male fertility regulation. While it is difficult to deliver drugs to the seminiferous tubule because of the blood–testis barrier, a male "pill" not involving alteration of androgen metabolism (and consequent changes in libido) is a highly sought after goal. It is hoped that this description of molecular changes during spermatogenesis will encourage further research into the unique biochemical pathways involved in sperm maturation such that, one day, "designer drugs" for male fertility control will be a reality.

REFERENCES

Adham IM, Klemm U, Maier, WM, Hoyer-Fender S, Tsaousidou S, Engel W (1989): Molecular cloning of preproacrosin and analysis of its expression pattern in spermatogenesis. Eur J Biochem 182:563–568.

Alcivar AA, Hake LE, Hardy MP, Hecht NB (1990): Increased levels of jun β and c-*jun* m RNAs in male germ cells following testicular cell dissociation. J Biol Chem 265:20160–20165.

Allen RL, O'Brien DA, Eddy Em (1988a): A novel hsp 70-like protein (P70) is present in mouse spermatogenic cells. Mol Cell Biol 8:828–832.

Allen RL, O'Brien DA, Jones CC, Rockett DL, Eddy EM (1988b): Expression of heat shock proteins by isolated mouse spermatogenic cells. Mol Cell Biol 8:3260–3266.

Ariel M, McCarrey J, Cedar H (1991): Methylation patterns of testis-specific genes. Proc Natl Acad Sci USA 88:2317–2321.

Artzt K, McCormick P, Bennett D (1982): Gene mapping with the T/t complex of the mouse. I. *t*-Lethal genes are nonallelic. Cell 28:463–470.

Ashworth A, Skene B, Swift S, Lovell-Badge R (1990): *Zfa* is an expressed retroposon derived from an alternative transcript of the *Zfx* gene. EMBO J 9:1529–1534.

Babiarz D, Garrisi GJ, Bennett D (1982): Genetic analysis of t^{w73} haplotype of the mouse using deletion mutations: evidence for a parasite lethal mutation. Genet Res 39:111–120.

Beatty RA (1970): The genetics of the mammalian gamete. Biol Rev 45:73–120.

Beechey CV (1973): X–Y chromosome dissociation and sterility in the mouse. Cytogenet. Cell Genet, 12:60–67.

Bellvé AR (1979): The molecular biology of mammalian spermatogenesis. Oxford Rev Reprod Biol 1:159–261.

Benbrook D, Pfahl M (1987): A novel thyroid hormone receptor encoded by a cDNA clone from a human testis library. Science 238:788–791.

Bennett D (l975): The *T*-locus of the mouse. Cell 6: 441–454.

Bernstein P, Peltz L, Ross J (1989): The poly(A)–binding protein complex is a major determinant of mRNA stability in vitro. Mol Cell Biol 9:659–670.

Berta P, Hawkins JR, Sinclair AH, Taylor A, Griffiths BL, Goodfellow PN, and Fellous M (1990): Genetic evidence equating SRY and the testis-determining factor. Nature 348:448–450.

Betlach CJ, Erickson RP (1976): 28S and 18S ribonucleic acid from mammalian spermatozoa. J Exp Zool 198:49–56.

Bloom SE, Goodpasture C (1976): An improved technique for selective silver staining and nucleolar organizer regions in human chromosomes. Hum Genet 34:199–206.

Blüthmann H, Cicurel L, Kuntz GWK, Haedenkamp G, Illmensee K (1982): Immunohistochemical lo-

calization of mouse testis-specific phosphoglycerate kinase (PGK-2) by monoclonal antibodies. EMBO J 1:479–484.

Bouvet P, Paris J, Phillipe M, Osborne HB (1991): Degradation of a developmentally regulated mRNA in *Xenopus* embryos is controlled by the 3'-region and requires the translation of another maternal mRNA. Mol Cell Biol 11:3115–3124.

Braden AWH (1958): Influence of time of mating on the segregation ration of alleles at the *T* locus in the house mouse. Nature 181:786–789.

Braden AWH (1972): *T*-locus in mice. Segregation distortion and sterility in the male. In Beatty RA, Gluecksohn-Waelsch S (eds): Proceedings of the International Symposium on The Genetics of the Spermatozoon. Edinburgh, pp 289–305.

Braun RE, Behringer RR, Peschon JJ, Brinster RL, Palmiter RD (1989a): Genotypically haploid spermatids are phenotypically diploid. Nature 337:373–376.

Braun RE, Peschon JJ, Behringer RR, Brinster RL, Palmiter RD (1989b): Protamine 3'-untranslated sequences regulate temporal translational control and subcellular localization of growth hormone in spermatids of transgenic mice. Genes Dev 3:793–802.

Breitbart RE, Andreadis A, Nadal-Ginard B (1987): Alternative splicing: A ubiquitous mechanism for the generation of multiple protein isoforms from single genes. Annu, Rev Biochem 56:467–495.

Brown CJ, Willard HF (1990): Localization of a gene that escapes inactivation to the X chromosome proximal short arm: implications for X inactivation. Am J Hum Genet 46:273–279.

Brown CJ, Ballabio A, Rupert JL, Lafreniere RG, Grompe M, Tonlorenzi R, Willard, HF (1991): A gene from the region of the human X inactivation centre is expressed exclusively from the inactive X chromosome. Nature 349:38–44.

Brown T, Cebra-Thomas JA, Bleil JD, Wassarman PM and Silver LM (1989): A premature acrosome reaction is programmed by mouse *t* haplotypes and could play a role in transmission ratio distortion. Development 106:769–773.

Bullard DC and Schimenti JC (1990): Molecular cloning and genetic mapping of the *t complex responder* candidate gene family. Genetics 124:957–966.

Bunick D, Johnson PA, Johnson TR, Hecht NB (1990): Transcription of the testis-specific mouse protamine 2 gene in a homologous *in vitro* transcription system. Proc Natl Acad Sci USA 87:891–895.

Caldwell KA, Handel MA (1991): Protamine tran script sharing among postmeiotic spermatids. Proc Natl Acad Sci USA 88:2407–2411.

Cattanach BM, Pollard CE and Hawkes SG (1971):

Sex reversed mice: XX and XO males. Cytogenetics 10:318–337.

Cebra-Thomas JA, Decker CL, Snyder LC, Pilder SH Silver LM (1991): Allele-and haploid-specific product generated by alternative splicing from a mouse *t complex responder* locus candidate. Nature 349:239–241.

Chen C-LC, Mather JP, Morris PL, and Bardin CW (1984): Expression of pro-opiomelanocortin-like gene in the testis and epididymis. Proc Natl Acad Sci USA 81:5672–5675.

Company M, Arenas J, Abelson J (1991): Requirement of the RNA helicase-like protein PRP22 for release of messenger RNA from spliceosomes. Nature 349:487–493.

Cunliffe V, Koopman P, McLaren A, Trowsdale J (1990): A mouse zinc finger gene is transiently expressed during spermatogenesis. EMBO J 9:197–205.

Daentl D, Erickson RP, Betlach CJ (1977): DNA synthetic capabilities of differentiating sperm cells. Differentiation 8:159–166.

Dahl, H-H M, Brown RM, Hutchison WM, Maragos C, Brown GK (1990): A testis-specific form of the human pyruvate dehydrogenase EI subunit is coded for by an intronless gene on chromosome 4. Genomics 8:225–232.

David JC, Vinson D, Loir M (1982): Developmental changes of DNA ligase during ram spermatogenesis. Exp Cell Res 141:357–364.

Dudley K, Potter J, Lyon MF, Willison KR (1984): Analysis of male sterile mutations in the mouse using haploid stage expressed cDNA probes. Nucleic Acids Res 12:4281–4293.

Durbin E J, Erickson R P, Craig A (1989): Characterization of GATA/GACA-related sequences on proximal chromosome 17 of the mouse. Chromosoma 97:301–306.

Ehlers MRW, Fox EA, Strydom DJ, Riordan JF (1989): Molecular cloning of human testicular angiotensin-converting enzyme: The testis isozyme is identical to the C-terminal half of endothelial angiotensin-converting enzyme. Proc Natl Acad Sci USA 86:7741–7745.

Erickson RP (1973): Haploid gene expression versus meiotic drive: the relevance of intercellular bridges during spermatogenesis. Nature [New Biol] 243:210–212.

Erickson RP (1975): Mouse spermatozoal glucose-6-phosphate dehydrogenase is the X-linked form. Biochem, Biophys Res Commun 63:1000–1004.

Erickson RP (1976): Glucose-6-phosphate dehydrogenase activity changes during spermatogenesis: possible relevance to X-chromosome inactivation. Dev Biol 53:134–137.

Erickson RP (1978): t-Alleles and the possibility of

post-meiotic gene expression during mammalian spermatogenesis. Fed Proc 37:2517–2521.

Erickson RP (1990): Post-meiotic gene expression. Trends in Genetics 6:264–269.

Erickson RP, Lewis SE, Slusser, KS (1978): Deletion mapping of the t-complex of chromosome 17 of the mouse. Nature 274:163–168.

Erickson RP, Butley MS, Martin SR and Betlach CJ (1979): Variation among inbred strains of mice in adenosine 3':5' cyclic monophosphate levels of spermatozoa. Genet Res 33, 129–136.

Erickson RP, Erickson, J M, Betlack CJ Meistrich ML (1980a): Further evidence for haploid gene expression during spermatogenesis: Heterogeneous, poly (A)–containing RNA is synthesized post-meiotically. J Exp zool 214:13–20.

Erickon RP, Kramer JM, Rittenhouse J, Salkeld A (1980b): Quantitation of messenger RNAs during mouse spermatogenesis: protamine-like histone and phosphoglycerate kinase-2 mRNAs increase after meiosis. Proc Natl Acad Sci USA 77:6086–6090.

Erickson RP, Hammerberg C, Sanchez E (1980c): t-mutants in the mouse and alterations in early development. In Proceedings of Current Research Trends in Prenatal Craniofacial Development(Pratt RM, Christiansen RL (eds): New York: Elsevier North-Holland, pp 103–118.

Erickson RP, Lewis SE, Butley M (1981): Is haploid gene expression possible for sperm antigens? J Reprod Immunol 3:195–217.

Erickson RP, Michelson AM, Rosenberg MP, Sanchez E, Orkin SH (1985): Post-meiotic transcription of phophoglyceratekinase 2 in mouse testes. Biosci Rep 5:1087–1091.

Fawcett DW (1979): The cell biology of gametogenesis in the male. Perspect Biol Med 31: 556–573.

Feinstein SI, Racaniello VR, Ehrlich M, Gehrke CW, Miller DA, Miller OJ (1985): Pattern of undermethylation of the major satellite DNA of mouse sperm. Nucleic Acids 13:3969–3978.

Fischman K, Edman JC, Shackleford GM, Turner JA, Rutter WJ, Nir U (1990): A murine *fer* testis-specific transcript (*fer*T) encodes a truncated fer protein. Mol Cell Biol 10:146–153.

Fox LA, Wickens M (1990): Poly (A) removal during oocyte maturation: A default reaction selectively prevented by specific sequences in-the 3' UTR of certain maternal mRNAs. Genes Dev, 4:2287–2298.

Fu L, Ye R, Browder LW, Johnston RN (1991): Translational potentiation of messenger RNA with secondary structure in *Xenopus*. Science 251:807–810.

Fujimoto H, Erickson RP (1982): Functional assays for mRNA detect many new messages after male meiosis in mice. Biochem Biophys Res. Commun 108:1369–1375.

Fujimoto H, Erickson RP, Quinto M, Rosenberg MP (1984): Post-meiotic transcription in mouse testes detected with spermatid cDNA clones. Biosci Rep 4:1037–1044.

Fujimoto H, Erickson RP, Toné S (1988): Changes in polyadenylation of lactate dehydrogenase-X mRNA during spermatogenesis in mice. Mol Reprod Dev 1:27–34.

Garrett JE, Collard M E, Douglass JO (1989): Translational control of germ cell-expressed mRNA imposed by alternative splicing: opioid peptide gene expression in rat testis. Mol Cell Biol 9:4381–4389.

Garside W, Hillman N (1989a): The in vivo and in vitro transmission frequencies of the t^{w5}-haplotype in mice Genet Res 53:21–24.

Garside W, Hillman N (1989b): The transmission ratio distortion of the t^{h2}-haplotype in vivo and in vitro. Genet Res 53:25–28.

Geremia R, Biotani C, Conti M, Monesi V (1977): RNA synthesis in spermatocytes and spermatids and preservation of meiotic RNA during spermiogenesis in the mouse. Cell Differ 5:343–355.

Geremia R, D'Agostino A, Monesi V (1978): Biochemical evidence of haploid gene activity in spermatogenesis of the mouse. Exp Cell Res 111:23–30.

Gizang-Ginsberg E, Wolgemuth DJ (1985): Localization of mRNAs in mouse testes by in situ hybridization: distribution of alpha-tubulin and developmental stage specificty of pro-opiomelanocortin transcripts. Dev Biol 111:293–305.

Gizang-Ginsberg E, Wolgemuth DJ (1987): Expression of the proopiomelanocortin gene is developmentally regulated and affected by germ cells in the male mouse reproductive system. Proc Natl Acad Sci US A 84:1600–1604.

Glucksohn-Waelsch S, Erickson RP (1970): The T-locus of the mouse: Implications for mechanisms of development. Curr Top Dev Biol 5:281–316.

Gold B, Hecht NB (1981): Differential compartmentalization of messenger ribonucleic acid in murine testis. Biochemistry 20:4871–4877.

Gold B, Fujimoto H, Kramer JM, Erickson RP, Hecht NB (1983): Haploid accumulation and translational control of phosphoglycerate kinase 2 messenger RNA during mouse spermatogenesis. Devel Biol 98:392–399.

Goldberg E, Hawtrey C (1967): The ontogeny of sperm specific lactate dehydrogenase in mice. J Exp Zool 164:309–316.

Goldsborough A, Ashworth A, and Willison K (1990): Cloning and sequencing of POU-boxes expressed in a mouse testis. Nucleic Acids Res 18:1634.

Goodfellow P, Pym B, Mohandas T, Shapiro LJ (1984): The cell surface antigen, MIC2X, escapes X-inactivation. AM J Hum Genet 36:777–782.

Gough NM, Gearing DP, Nicola NA, Baker E, Pritchard M, Callen DF, Sutherland GR (1990): Localization of the human GM-CSF receptor gene to the X–Y pseudoautosomal region. Nature 345:734–736.

Groudine M, Conklin KF (1985): Chromatin structure and de novo methylation of sperm DNA: implications for activation of the paternal genome. Science 228:1061–1068.

Gruppi CM, Zakeri ZF, Wolgemuth DJ (1991): Stage and lineage-regulated expression of two HSP 90 transcripts during mouse germ cell differentiation and embryogenesis. Mol Reprod Dev 28:209–217.

Gubbay J, Kiipman P, Collignon J, Burgoyne P, Lovell-Badge R (1990): Normal structure and expression of Zfy genes in XY female mice mutant in Tdy. Development 109:647–653.

Gummere GR, McCormick PJ, Bennett D (1986): The influence of genetic background and the homologous chromosome 17 on t-haplotype transmission ration distortion in mice. Genetics 114:235–245.

Gururajan R, Perry-O'Keefe H, Melton DA, weeks DL (1991): The Xenopus localized messenger RNA An3 may encode an ATP-dependent RNA helicase. Nature 349:717–719.

Hake LE, Alcivar AA, Hecht NB (1990): Changes in mRNA length accompany translational regulation of the somatic and testis-specific cytochrome c genes during, spermatogenesis in the mouse. Development 110:249–257.

Hammerberg C (1982): The effects of the t-complex upon male reproduction are due to complex interactions between its several regions. Genet Res 39:219–226.

Hammerberg C, Klein J (1975): Linkage disequilibrium between H-2 and t complexes in chromosome 17 of the mouse. Nature 258:296–299.

Hecht NB (1986): Regulation of gene expression during mammalian spermatogenesis in Rossant J Pedersen R (eds): "Experimental Approaches to Mammalian Embryonic Development. New York: Cambridge University Press, pp 151–193.

Hecht NB (1990): Regulation of "haploid expressed genes" in male germ cells. J Reprod Fert: 88:679–693.

Hecht NB, Farrell D, Williams JL (1979): DNA polymerases in mouse spermatogenic cells separated by sedimentation velocity. Biochm Biophys Acta 561:358–368.

Hecht NB, Distel RJ, Yelick PC, Tanhauser SM, Driscoll, CE, Goldberg E, Tung KSK (1988): Localization of a highly divergent mammalian testicular alpha tubulin that is not detectable in brain. Mol Cell Biol 8:996–1000.

Herrmann B, Bucan M, Mains PD, Frischauf AM, Silver LM, Lehrach H (1986): Genetic analysis of the proximal portion of the mouse *t* complex: evidence for a second inversion within *t* haplotypes. Cell 44:469–476.

Herrmann BG, Barlow DP, Lehrach H (1987): A large inverted duplication allows homologous recombination between chromosomes heterozygous for the proximal *t*-complex inversion. Cell 48:813–825.

Herzog NK, Singh B, Elder J, Lipkin I, Trauger RJ, Millette CF, Goldman DS, Wolfes H, Cooper GM, Arlinghaus RB (1988): Identification of the protein product of the c-*Mos* proto-oncogene in mouse testes. Oncogene 3:225–229.

Hotta Y, Tabata S, Stubbs L, Stern H (1985): Meiosis-specific transcripts of a DNA component replicated during chromsome pairing: homology across the phylogentic spectrum. Cell 40: 785–793.

Howard TE, Shai S-Y, Langford KG, Martin BM, Bernstein KE (1990): Transcription of testicular angiotensinconverting enzyme (ACE) is initiated within the 12th intron of the somatic ACE gene. Mol Cell Biol 10:4294–4302.

Ibáñez, CF, Pelto-Huikko M, Soder O, Ritzen E M, Hersh B, Hokfelt T, Persson H (1991): Expression of choline acetyltransferase mRNA in spermatogenic cells results in an accumulation of the enzyme in the postacrosomal region of mature spermatozoa. Proc Natl Acad Sci USA 88:3676–3680

Jäger RJ, Amvret M, Hall K, Scherer G (1990): A human XY female with a frame shift mutation in the candidate testis-determining gene *SRY*. Nature 348:452–453.

Jeannotte L, Burbach JPH, Drouin J (1987): Unusual proopiomelanocortin ribonucleic acids in extrapituitary tissues: intronless transcripts in testes and long poly(A) tails in hypothalamus. Mol Endocrinol 1:749–757.

Jen J, Deschepper CF, Shackleford GM, Lee CYG, Lau Y-FC (1990): Stage-specific expression of the lactate dehydrogenase-X gene in adult and developing mouse testes. Mol Reprod Develop, 25:14–21.

Johnson PA, Bunick D, Hecht NB (1991): Protein binding regions in the mouse and rat protamine-2 genes. Biol Reprod 44:127–134.

Kallikin LM, Fujimoto H, Witt MP, Verga V, Erickson RP (1989): A genomic clone of *Zfy*-l from a YDOM mouse strain detects post-meiotic gene expression of Zfy in testes. Biochem Biophys Res Commun 165:1286–1291.

Kasahara M, Gutknecht J, Brew K, Spurr N, Goodfellow PN (1989): Cloning and mapping of a testis-specific gene with sequence similarity to a sperm-coating glycoprotein gene. Genomics 5:527–534.

Katz DF, Erickson RP, Nathanson M (1979): Beat

frequency is bimodally distributed in spermatozoa from T/t^{l2} mice. J Exp Zool 210:529–535.

Kelley KA, Chamberlain JW, Nolan JA, Horwich AL, Kalousek F, Eisenstadt J, Herrup K, Rosenberg LE (1988): Meiotic expression of human ornithine transcarbamylase in the testes of transgenic mice. Mol Cell Biol 8:1821–1825.

Keshet E, Itin A, Fischman K, Nir U (1990): The testis-specific transcript (*ferT*) of the tyrosine kinase *FER* is expressed during spermatogenesis in a stagespecific manner. Mol Cell Biol 10:5021–5025.

Kew D, Muffly KE, Kilpatrick DL (1990): Proenkephalin products are stored in the sperm acrosome and may function in fertilization. Proc Natl Acad Sci USA 86:6166–6170.

Kierszenbaum AL, Tres LL (1974a): Transcription sites in spread meiotic prophase chromosomes from mouse spermatocytes. J Cell Biol 63:923–925.

Kierszenbaum AL, Tres LL (1974b): Nucleolar and perichromosomal RNA synthesis during meiotic prophase in the mouse testis. J Cell Biol 60:39–53.

Kierszenbaum AL, Tres LL (1975): Structural and transcriptional features of the mouse spermatid genome. J Cell Biol 65:258–270.

Kilpatrick DL, Howells RD, Noe M, Bailey LC, Udenfriend S (1985): Expression of preproenkephalin-like mRNA and its peptide products in mammalian testis and ovary. Proc Nat l Acad Sci USA 82:7467–7469.

Kilpatrick DL, Millette CF (1986): Expression of proenkephalin messenger RNA by mouse spermatogenic cells. Proc Nat l Acad Sci USA 83:5015–5018.

Kilpatrick DL, Zinn SA, Fitzgerald M, Higuici H, Sabol SL, Meyerhardt J (1990): Transcription of the rat and mouse proenkephalin genes is initiated at distinct sites in spermatogenic and somatic cells. Mol Cell Biol 10:3717–3726.

Kleene KC (1989): Poly (A) shortening accompanies the activation of translation of five mRNA's during spermiogenesis in the mouse. Development 106:367–373.

Kleene KC, Kistel RJ, Hecht NB (1983): cDNA clones encoding cytoplasmic poly (A) + RNAs which first appear at detectable levels in haploid phases of spermatogenesis in the mouse. Dev Biol 98:455–464.

Kleene KC, Smith J, Bozorgzaoleh A, Harris M, Hahn L, Karimpour I, Gerstel J (1990): Sequence and developmental expression of the mRNA encoding the seleno-protein of the sperm mitochondrial capsule in the mouse. Dev Biol 137: 305–402.

Kofman-Alfaro S, Chandley AC (1970): Meiosis in the male mouse: An autoradiographic investigation. Chromosoma 31:404–420.

Koopman P, Gubbay J, Collignon J, Lovell-Badge R

(1989): *Zfy* gene expression patterns are not compatible with a primary role in mouse sex determination. Nature 342:940–942.

Koopman P, Münsterberg A, Capel B, Vivian N, Lovell-Badge R (1990): Expression of a candidate sex-determining gene during mouse testis differentiation. Nature 348:450–452.

Koopman P, Gubbay J, Vivian N, Goodfellow P, Lovell-Badge R, (1991): Male development of chromosomally female mice transgenic for Sry. Nature 351:117–121.

Kozak M (1988): A profusion of controls. J Cell Biol 107:1–7.

Kramer JM (1981): Immunoflourescent localization of PGK-1 and PGK-2 isozymes within specific cells of the mouse testis. Dev Biol 87:30–36.

Kramer JM, Erickson RP (1981): Developmental programs of PGK-1 and PGK-2 isozymes in spermatogenic cells of the mouse: Specific activities and rates of synthesis. Dev Biol 87:30–36.

Krawczyk Z, Wisniewski J, Biesiada (1987): A hsp 70-related gene is constitutively highly expressed in testis of rat and mouse. Mol Biol Rep 12:27–34.

Krawczyk Z, Mali P, Parvinen M (1988): Expression of a testis-specific hsp 70 gene-related RNA in defined stages of rat seminiferous epithelium. J Cell Biol 107:1317–1323.

Kumar RS, Kusari J, Roy N, Soffer RL, Sen GC (1989): Structure of testicular angiotensin-converting enzyme. J Biol Chem 264:16754–16758.

Kumar RS, Thekkumkara TJ, Sen GC (1991): The mRNAs encoding the two angiotensin-converting isozymes are transcribed from the same gene by a tissue-specific choice of alternative transcription initiation sites. J Biol Chem 266:3854–3862.

Kwon YK, Hecht NB (1991): Cytoplasmic protein binding to highly conserved sequences in the 3' untranslated region of mouse protamine 2 in mRNA, a translationally regulated transcript of male germ cells. Proc Natl Acad Sci USA 88:3584–3588.

Lader E, Ha H-S, O'Neill M, Artzt K, Bennett D (1989): *tctex*-1: A candidate gene family for a mouse *t* complex sterility locus. Cell 58:969–979.

Lafuse W, Edidin M (1980): Influence of the mouse major histocompactibility complex, H-2, on liver adenylate cyclase activity and on glucagon binding to liver cell membranes. Biochemistry 19:49–54.

Lau Y-FC, Chan K, Sparkes R (1989): Male-enhanced antigen gene is phylogenetically conserved and expressed at late stages of spermatogenesis. Proc Natl Acad Sci USA 86:8462–8466.

Lau Y-FC, Chan K (1989): The putative testis-determining factor and related genes are expressed as discrete-sized transcripts in adult gonadal and somatic tissues. Am J Hum Genet 45:942–952.

Leblond CP, Clermont Y (1952): Spermiogenesis of rat, mouse, hamster and guinea pig as revealed by the "periodic acid-fuschin sulfurous acid" technique. Am J Anat 90:167–216.

Lee IP, Dixon RL (1972): Antioneoplastic drug effects on spermatogenesis studies by velocity sedimentations cell separation. Toxicol Appl Pharmacol 23:20–41.

Lee S-J (1990): Expression of HSP 86 in male germ cells. Mol Cell Biol 10:3239–3242.

Leroy P, Alzari P, Sasson D, Wolgmuth D, Fellous M, (1989): The protein encoded by a murine male germ cell–specific transcript is a putative ATP-dependent RNA helicase. Cell 57:549–559.

Lifschytz E (1972): X-chromosome inactivation: An essential feature of normal spermiogenesis in male heterogametic organisms. In Beatty RA, Gluechsohn-Waaelsch S (eds): Proceedings of the International Symposium on The Genetics of the Spermatozoon. Edinburgh, pp. 223–232.

Loir M (1972): Metabolism de l'acide ribonucleique et des proteines dans les spermatocytes et les spermatides du belier (Ovis aries). Ann Biol Anim Biochim Biophys 12:203–219.

Lyon MF (1984): Transmission ratio distortion in mouse *t*-haplotypes is due to multiple distorter genes acting on a responder locus. Cell 37:621–628.

Lyon MF, Glennister PH, Hawkes SG (1972): Do the *H-2* and *T*-loci of the mouse have a function in the haploid phase of sperm? Nature 240:152–153.

Lyon MF, Bechtol KB (1977): Derivation of mutant *t*-haplotypes of the mouse by presumed duplication or deletion. Genet Res 30:63–76.

Lyon MF, Zenthon J (1987): Differences in or near the responder region of complete and partial *t*-haplotypes. Genet Res 50:29–34.

Maekawa M, O'Brien DA, Allen RL, Eddy EM (1989): Heat-shock cognate protein (hsc 71) and related proteins in mouse spermatogenic cells. Biol Reprod 40:843–852.

Mardon G, Mosher R, DiSteche CM, Nishioka Y, McLaren A, Page DC (1989): Duplication, deletion, and polymorphism in the sex-determining region of the mouse Y chromosome. Science 243:78–80.

Mardon G, Page DC (1990): The sex-determining region of the mouse Y chromosome encodes a protein with a highly acidic domain and 13 zinc fingers. Cell 56:765–770.

Mardon G, Luoh S-W, Simpson EM, Gill G, Brown LG, Page DC (1990): Mouse Zfx protein is similar to Zfy-2: Each contains an acid activating domain and 13 zinc fingers. Mol Cell Biol 10:681–688.

Matsumoto M, Fujimoto H (1990): Cloning of a hsp 70-related gene expressed in mouse spermatids. Biochem Biophys Res Commun 166:43–49.

Matsushime H, Jinno A, Takagi N, Shibuya M (1990): A novel mammalian protein kinaose gene (*mak*) is highly expressed in testicular germ cells at and after meiosis. Mol Cell Biol 10:2261–2268.

Mazarakis ND, Nelki D, Lyon MF, Ruddy S, Evans EP, Freemont P, Dudley K (1991): Isolation and characterization of a testis-expressed developmentally regulated gene from the distal inversion of the mouse *t*-complex. Development 111:561–571.

McCarrey JR, Thomas K (1987): Human testis-specific PGK gene lacks introns and possesses characteristics of a processed gene. Nature 326:501–505.

McGrath J, Hillman N (1980a): The in vitro transmission frequency of the t^{12} mutation in the mouse. J Embryol Exp Morphol 60:141–151.

McGrath J, Hillman N (1980b): Sterility in mutant (t^{Lx}/t^{Ly}) male mice III: *In vitro* fertilization. J Embryol Exp Morphol 59:49–58.

McGrath J, Hillman N (1981): The in vitro transmission frequency of the t^6 allele. Nature 283:479–481.

Meijer D, Hermans A, von Lindern M, van Agthoven T, de Klein A, Mackenbach P, Grootegoed A, Talarico D, Della Valle G, Grosveld G, (1987): Molecular characterization of the mouse testis specific c-*abl* mRNA in mouse. EMBO J 6:4041–4048.

Meistrich ML (1972): Separation of mouse spermatogenic cells by velocity sedimentation. J Cell Physiol 80:299–312.

Meistrich ML, Reid BO, Barcellona WJ (1975): Meiotic DNA synthesis during mouse spermatogenesis. J Cell Biol 64:211–222.

Meistrich ML, Trostle, PK, Fraport M, Erickson RP (1977): Biosynthesis and localization of lactate dehydrogenase X in pachytene spermatocytes, and spermatids of mouse testes. Dev Biol 60:428–441.

Meruelo D, Edidin M (1975): Association of mouse liver adenosine 3′: 5′-cyclic monophosphate (cyclic AMP) levels with histocompatibility-2 genotype. Proc Natl Acad Sci USA 72:2644–2648.

Monesi V (1967): Ribonucleic acid and protein synthesis during differentiation of male germ cells in the mouse. Arch Anat Microsc Morphol Exp 56:61–74.

Moore GPM (1971): DNA-dependent RNA synthesis in fixed cells during spermatogenesis in mouse. Exp Cell Res 68:462–465.

Moss SB, Challoner PB, Groudine B (1989): Expression of a novel histone 2B during mouse spermiogenesis. Dev Biol 133:83–92.

Muller HJ, Settles F (1927): The non-functioning of the genes in spermatozoa. Z Induk Abstam Vererb 43:285–312.

Nadjicka M, Hillman N (1980): Differences in cAMP levels in t^n and non-t^n-bearing mouse spermatozoa. Biol Reprod 22:1102–1105.

Nagamine CM, Chan K, Kozak CA, Lau, Y-F (1989): Chromosome mapping and expression of a putative testis-determining gene in mouse. Science 243:80–83.

Nagamine CM, Chan K, Hake LF, Lau Y-FC (1990): The two candidate testis-determining Y genes (*Zfy*-1 and *Zfy*-2) are differentially expressed in fetal and adult mouse tissues. Genes Dev 4:63–74.

Oakberg EF (1956a): A description of spermiogenesis in the mouse and its use in analysis of the cycle of the seminiferous epithelium and germ cell renewal. Am J Anat 99:391–413.

Oakberg EF (1956b): Duration of spermatogenesis in the mouse and timing of stages of the cycle of the seminiferous epithelium. Am J Anat 99:507–516.

Olds-Clarke P (1983): Nonprogressive sperm motility is characteristic of most complete *t* haplotypes in the mouse. Genet Res 42:151–157.

Olds-Clarke P, Peitz B (1985): Fertility of sperm from *t*/+ mice: evidence that +− bearing sperm are dysfunctional. Genet Res 47:49–52.

Oppi C, Shore SK, Reddy EP (1987): Nucleotide sequence of testis-derived c-*abl* cDNAs: implications for testis-specific transcription and *abl* oncogen activation. Proc Natl Acad Sci USA 84:8200–8204.

Orlando P, Grippo P, Geremia R (1984): DNA repair synthesis-related enzymes during spermatogenesis in the mouse. Exp Cell Res 154:499–505.

Page DC, Mosher R, Simpson EM, Fisher EMC, Mardon G, Pollack J, McGillivray B, de la Chapelle A, Brown LG (1987): The sex-determining region of the human Y chromosome encodes a finger protein. Cell 51:1091–1104.

Page DC, Fisher EMC, McGillivray B, Brown LG, (1990): Additional deletion in sex-determining region of human Y chromosome resolves paradox of X,t (Y; 22) female. Nature 346:279–281.

Palmer MS, Sinclair AH, Berta P, Ellis NA, Goodfellow PN, Abbas NE, Fellous M (1989): Genetic evidence that ZFY is not the testis-determining factor. Nature 342:937–939.

Palmer MS, Berta P, Sinclair AH, Pym B, Goodfellow PN (1990): Comparison of human ZFY and ZFZ transcripts. Proc Natl Acad Sci USA 87:1681–1685.

Palmiter RD, Wilkie TM, Chen HY, Brinster RL (1984): Transmission distortion and mosaicism in an unusual transgenic mouse pedigree. Cell 36:869–877.

Paris J, Richter JD (1990): Maturation-specific polyadenylation and translational control: diversity of cytoplasmic polyadenylation elements, influence of poly (A) tail size, and formation of stable polyadenylation complexes. Mol Cell Biol 10:5634–5645.

Peacock WJ, Erickson J (1965): Segregation-distortion and regularly nonfunctional products of spermatogenesis in Drosophila melanogaster. Genetics 51: 313–328.

Persson H, Rehfeld JF, Ericsson A, Schalling M, Pelto-Huikko M, Hokfelt T (1989): Transient expression of the cholecystokinin gene in male germ cells and accumulation of the peptide in the acrosomal granule: possible role of cholecystokinin in fertilization. Proc Natl Acad Sci USA 86:6166–6170.

Peschon JJ, Behringer RR, Brinster RL, Palmiter RD (1987): Spermatid-specific expression of protamine 1 in transgenic mice. Proc Natl Acad Sci USA 84:5316–5319.

Pintar JE, Schachter BS, Herman AB, Durgerian S, Krieger DT (1984): Characterization and localization of proopiomelanocortin messenger RNA in the adult rat testis. Science 225:632–634.

Ponzetto C, Wadewitz AG, Pendergast AM, Witte ON, Wolgemuth DJ (1989): P150$^{c\text{-}abl}$ is detected in mouse male germ cells by an in vitro kinase assay and is associated with stage-specific phosphoproteins in haploid cells. Oncogene 4:685–690.

Ponzetto-Zimmerman C, Wolgemuth DJ (1984): Methylation of satellite sequences in mouse spermatogonia and somatic DNAs. Nucleic Acids Res 12:2807–2822.

Probst F, Vande Woude GF (1985): Expression of c-mos proto oncogene transcripts in mouse tissues. Nature 315:516–518.

Probst F, Rosenberg MP, Vande Woude GF (1988a): Proto-oncogene expression in germ cell development. Trends Genet 4:183–188.

Probst F, Rosenberg MP, Oskarsson MK, Russell LB, Nguyen-Huu MC, Nadeau J, Jenkins NA, Copeland NG, Vande Woude GF (1988b): Genetic analysis and developmental regulation of testis-specific RNA expression of Mos, Abl, actin and Hox-1.4. Oncogene 2:227–233.

Rahe B, Erickson RP, Quinto M (1983): Methylation of unique sequence DNA during spermatogenesis in mice. Nucleic Acids Res 11:7947–7959.

Raman R, Das P (1991): Mammalian sex chromosomes III. Activity of pseudoautosomal steroid sulfatase enzyme during spermatogenesis in Mus musculus. Somatic Cell Mol Genet (in press).

Rappold GA, Stubbs L, Labeit S, Crkrenjakor RB, Lehrach H (1987): Identification of a testis-specific gene from the mouse t-complex next to a CpG-rich island EMBO J 6:1975–1980.

Robinson MO, McCarrey JR, Simon MI (1989): Transcriptional regulatory regions of testis-specific PGK2 defined in transgenic mice. Proc Natl Acad Sci USA 86: 8437–8441.

Robinson MO, Simon MI (1991): Determining transcript number using the polymerase chain reaction: Pgk-2, mPZ, and PGK-2 transgene mRNA levels during spermatogenesis. Nucleic Acids Res 19:1557–1562.

Roeder GS (1990): Chromosome synapsis and genetic recombination. Trends Genet 6:385–389.

Rohme D, Fox H, Hermann B, Frischauf A-M, Edstrom JE, Mains P, Silver LM, Lehrach H (1984): Molecular clones of the mouse t complex derived from microdissected metaphase chromosomes. Cell 36:783–788.

Rosen LL, Bullard DC, Silver LM, Schimenti JC (1990): Molecular cloning of the t complex responder genetic locus. Genomics 8:134–140.

Rubin MR, Toth LE, Patel MD, D'Eustachio P, Nguyen-Huu MC (1986): A mouse homeo box gene is expressed in spermatocytes and embryos. Science 233:663–667.

Russell LB, Montgomery CS (1969): Comparative studies on X-autosome translocations in the mouse. I. Origin, viability, fertility, and weight of five T (X; 1) 'S'. Genetics 63:103–120.

Sachs AB, Davis RW (1989): The poly (A) binding protein is required for poly (A) shortening and 60S ribosomal subunit-dependent translation initiation. Cell 58:857–867.

Sagata N, Watanabe N, Vande Woude GF, Ikawa Y (1989): The c-mos proto-oncogene product is a cytostatic factor responsible for meiotic arrest in vertebrate eggs. Nature 342:512–518.

Sanchez ER, Erickson RP (1985): Expression of the Tcp-1 locus of the mouse during early embryogenesis. J Embryol Exp Morphol 89:113–122.

Sanchez ER, Hammerberg C, Erickson RP (1985): Quantitation of two-dimensional gel proteins reveals unequal amounts of Tcp-1 gene products during mouse spermatogenesis but no correlations with transmission ratio distortion. J Embryol Exp Morphol. 89:123–131.

Sanford J, Forrester L Chapman V, Chandley A, Hastie N (1984): Methylation patterns of repetitive DNA sequences in germ cells of Mus musculus. Nucleic Acids Res 12:2823–2836.

Sarventnick N, Tsai J-Y, Fox H, Pilder SH, Silver LM (1989): A mouse chromosome 17 gene encodes a testes-specific transcript with unusual properties. Immunogenetics 30:34–41.

Schäfer M, Kuhn R, Bosse F, Schäfer U (1990): A conserved element in the leader mediates postmeiotic translation as well as cytoplasmic polyadenylation of a Drosophila spermatocyte mRNA. EMBO J 9:4519–4525.

Schibler U, Sierra F (1987): Alternative promoters in developmental gene expression. Annu Rev Genet 21:237–257.

Schmid M, Hofgartner FJ, Zenzes MT, Engel W (1977): Evidence for postmeiotic expression of ribosomal RNA genes during male gametogenesis. Hum Genet 38:279–284.

Schnëider-Gädicke A, Beer-Romero P, Brown LG, Nussbuam R, Page DC (1989): ZFX has gene structure similar to ZFY, the putatitve sex determinant, and escapes X inactivation. Cell 57:1247–1258.

Scholer HR, Dressler GR, Balling R, Rohdewohld H, Gruss P (1990): *oct-4*: A germline-specific transcription factor mapping to the mouse *t*-complex. EMBO J1 9:2185–2195.

Seitz AW, Bennett D (1985): Transmission distortion of t-haplotypes is due to interactions between meiotic partners. Nature 313:143–144.

Sen GC, Thekkumkara TJ, Kumar RS (1990): Angiotensin-converting enzyme: structural relationship of the testicular and the pulmonary forms. J Cardiovasc Pharmacol 16:S14–S18.

Shackleford GM, Varmus HE (1987): Expression of the proto-oncogene *int*-1 is restricted to postmeiotic male germ cells and the neural tube of mid-gestational embryos. Cell 50:89–95.

Shaper NL, Wright WW, Shaper JH (1990): Murine β1, 4-galactosyltransferase: Both the amounts and structure of the mRNA are regulated during spermatogenesis. Proc Natl Acad Sci USA 87:791–795.

Shapiro LJ, Mohandas T, Weiss T, Romeo G (1979): Non-inactivation of an X-chromosome locus in man. Science 204:1224–1226.

Shin HS, Flaherty L, Artzt K, Bennett D, Ravetch J (1983): Inversion of the *H-2* complex of *t*-haplotypes in mice. Nature 306:380–383.

Silver LM, Uman J, Danska T, Garrels JI (1983): A diversified set of testicular cell proteins specified by genes within the mouse *t* complex. Cell 35:35–45.

Silver LM, Olds-Clarke P (1984): Transmission ratio distortion of mouse *t* haplotypes is not a consequence of wild-type sperm degeneration. Dev Biol 105:205–212.

Silver LM, Remis D (1987): Five of the nine genetically defined regions of mouse t-haplotypes are involved in transmission ratio distortion. Genet Res 49:51–56.

Silver LM, Kleene KC, Distel RJ, Hecht NB (1987): Synthesis of mouse *t* complex proteins during haploid stages of spermatogenesis. Dev Biol 119:605–608.

Sinclair AH, Berta P, Palmer MS, Hawkins JR, Griffiths BL, Smith MJ, Foster JW, Frischauf A-M, Lovell-Badge R, Goodfellow PN (1990): A gene from the human sex-determining region encodes a protein with homology to a conserved DNA-binding motif. Nature 346:245–250.

Slaughter GR, Meistrich ML, Means AR (1989): Expression of RNAs for calmodulin, actins, and tubulins in rat testis cells. Biol Reprod. 40:395–405.

Smyth DR, Stern H (1973): Repeated DNA synthesized during pachytene in *Lilium henryi*. Nature [New Biol] 245:94–94.

Snell GD (1968): The H-2 locus of the mouse: Observations and speculations concerning its comparative genetics and its polymorphisms. Folia Biol 14:335–358.

Sorrentino V, McKinney MD, Giorgi M, Geremia R, Fleissner E (1988): Expression of cellular proto-oncogenes in the mouse male germ line: A distinctive 2.4 kilobase *pim*-1 transcript is expressed in haploid postmeiotic cells. Proc Natl Acad Sci USA 85:2191–2195.

Speed RM (1986): Abnormal RNA synthesis in sex vesicles of tertiary trisermic male mice. Chromosoma 93:267–270.

Stallard BJ, Collard MW, Griswold MD (1991): A transferrinlike (hemiferrin) mRNA is expressed in the germ cells of rat testis. Mol Cell Biol 11:1448–1453.

Stewart TA, Hecht NB, Hollingshead PG, Johnson PA, Leong JC, Pitts SL (1988): Haploid-specific transcription of protamine-*myc* and protamine-T-antigen fusion genes in transgenic mice. Mol Cell Biol. 8:1748–1755.

Taira M, Iizasa T, Yamada K, Shimada H, Tatibana M (1989): Tissue-differential expression of two distinct genes for phosphoribosyl pyrophosphate synthetase and existence of the testis-specific transcript. Biochim Biophys Acta 1007:203–208.

Takahashi M, Inaguma Y, Hiai H, Hirose F (1988): Developmentally regulated expression of a human "finger"-containing gene encoded by the 5' half of the *ret* transforming genes. Mol Cell Biol 8:1853–1856.

Tanaka S, Fujimoto H (1986): A postmeiotically expressed clone encodes lactate dehydrogenase isozyme X. Biochem Biophys Res Commun 136:760–766.

Teruya JH, Kutsunai SY, Spear DH, Edwards PA, Clarke CF (1990): Testis-specific transcription initiation sites of rat farnesyl pyrophosphate synthetase mRNA. Mol Cell Biol 10:2315–2326.

Thomas KH, Wilke TM, Tomashefsky P, Bellve AR, Simon MI (1989): Differential gene expression during mouse spermatogenesis. Biol Reprod 41:729–739.

Trasler JM, Hake LE, Johnson PA, Alcivar AA, Millette CF, Hecht NB (1990): DNA methylation and demethylation events during meiotic prophase in the mouse testis. Mol Cell Biol 10:1828–1834.

VandeBerg JL, Cooper DW, Close PJ (1976): Testis specific phosphoglycerate kinase B in mouse. J Exp Zool 198:231–240.

Van der Hoorn FA, Tarnasky HA, Nordeen SK (1990): A new rat gene RT7 is specifically expressed during spermatogenesis. Dev Biol 142:147–154.

Verga V, Erickson RP (1989): An extended long-range restriction map of the human sex-determining region on Y$_p$, including ZFY, finds marked homology on X$_p$ and no detectable Y sequences in an XX male. Am J Hum Genet 44:756–765.

Varnum SM, Wormington WM (1990): Deadenylation of maternal mRNAs during *Xenopus* oocyte maturation does not require *cis*-sequences: A default mechanism for translational control. Genes Dev 4:2278–2286.

Viliasaute A, Wang D, Dobner P, Dolph P, Lewis SA, Cowan NJ (1986): Six mouse alpha-tubulin mRNAs encode five distinct isotypes: Testis-specific expression of two sister genes. Mol Cell Biol 6:2409–2419.

Wang L, Miao S, Yan Y, Li Y, Zong C, Koide SS (1990): Expression of a sperm protein gene during spermatogenesis in mammalian testis: An in situ hybridization study. Mol Reprod Dev 26:1–5.

Wilke TM, Braun RE, Ehrman WJ, Palmiter RD, Hammer RE (1991): Germ-line intrachromosomal recombination restores fertility in transgenic MyK-103 male mice. Genes Dev 5:38–48.

Willison K, Ashworth A (1987): Mammalian spermatogenic gene expression. Trends Genet 3:351–355.

Willison K, Lewis V, Zuckerman KS, Cordell J, Dean C, Miller K, Lyon MF, Marsh M (1989): The *t* complex polypeptide (TCP-1) is associated with the cytoplasmic aspect of Golgi membranes. Cell 57:621–632.

Willison KR, Hynes G, Davies P, Goldsborough A, Lewis VA (1990): Expression of three *t*-complex genes, *Tcp*-1, D17Leh117c3, and D17Leh66, in purified spermatogenic cell populations. Genet Res 56:193–201.

Wolfes H, Kogawa K, Millette CF, Cooper GM (1989): Specific expression of nuclear proto-oncogenes before entry into meiotic prophase of spermatogenesis. Science 245:740–743.

Wolgemuth DJ, Viviano CM, Gizang-Ginsberg E, Frohman MA, Joyner AL, Martin GR (1987): Differential expression of the mouse homeobox-containing gene *Hox-1.4* during male germ cell differentiation and embryonic development. Proc Natl Acad Sci USA 84:5813–5817.

Yanagisawa K, Dunn LC, Bennett D (1961): On the mechanism of abnormal transmission ratios at *T* locus in the house mouse. Genetics 46:1635–1644.

Yoshikawa K, Aizawa T (1988): Enkephalin precursor gene expression in postmeiotic germ cells. Biochem Biophys Res Commun 151:664–671.

Zakeri ZF, Wolgemuth DJ (1987): Developmental-stage–specific expression of the hsp 70 gene family during differentiation of the mammalian male germ line. Mol Cell Biol 7:1791.

Zakeri ZF, Wolgemuth DJ, Hunt CR (1988): Identification and sequence analysis of a new member of the mouse hsp 70 gene family and characterization of its unique cellular and developmental pattern of expression in the male germ line. Mol Cell Biol 8:2925–2932.

ABOUT THE AUTHOR

ROBERT P. ERICKSON is the Holsclaw Family Professor of Human Genetics and Inherited Diseases in the Departments of Pediatrics and Molecular and Cellular Biology at the University of Arizona, where he teaches pediatric genetics and molecular biology. After receiving his B.A. from Reed College in 1960, he received his M.D. at Stanford University School of Medicine, where he also actively pursued research in the Department of Genetics. Dr. Erickson's internship at Cornell University School of Medicine and residency at Albert Einstein College of Medicine were followed by two years at the National Institutes of Health as a researcher in the laboratory of Dr. Christian B. Anfinsen. This was followed by a postdoctoral fellowship with N.A. Mitchison at the National Institute of Medical Research, London, where he commenced his studies on the role of cell surface antigens in development. This research avenue was continued during a sabbatical with François Jacob at the Pasteur Institute, Paris, 1975-76; a shift in techniques to those of molecular genetics occurred with a sabbatical at the Imperial Cancer Research Fund Laboratories, London, 1983-84. Dr. Erickson's current research involves the use of molecular genetical differentiation in mice and humans and the use of antisense techniques to study gene expression in the preimplantation embryo and during spermatogenesis. He is author or coauthor of over 150 research papers and is on the editorial boards of *Molecular Reproduction and Development* and *Antisense Research and Development*. He has been the recipient of a research career development award from the NIH, a Guggenheim Fellowship, and an Eleanor Roosevelt Cancer Research Fellowship.

Genes in Mammalian Reproduction: 27–43
© 1993 Wiley-Liss, Inc.

Specific Gene Expression During Oogenesis in Mice

Ross A. Kinloch and Paul M. Wassarman

I. INTRODUCTION

The subject of overall gene expression during oogenesis in mice has been addressed several times in recent years [e.g., Wassarman, 1983, 1988b; Bachvarova, 1985; Schultz, 1986; Davidson, 1986] and will be only summarized here. Rather, this article focuses on expression of a few specific genes, *c-mos, oct-3, c-kit, mZP3, tissue-plasminogen activator* (t-PA), and *lactate dehydrogenase* (LDH), during oogenesis in mice. For the most part, the pattern of expression of these genes has been examined only relatively recently. Although few in number, these genes illustrate different routes taken by developing mouse eggs in order to regulate accumulation of specific gene products during oogenesis for use during oogenesis, fertilization, and/or early embryogenesis.

II. OOGENESIS IN THE MOUSE

The unfertilized mouse egg is the end-product of *oogenesis,* a process that begins during fetal development and ends months to years later in the sexually mature adult [Biggers and Schuetz, 1972; Zuckerman and Weir, 1977; Jones, 1978; Austin and Short, 1982; Knobil and Neill, 1988; Wassarman, 1991]. Oogenesis begins with the appearance of *primordial germ cells* (days 7–9 fetus), which become the *oogonia* that populate fetal ovaries (days 11–12 fetus) and which, in turn, become nongrowing *oocytes* (days 12–14 fetus) that populate ovaries of neonatal mice. The transition from oogonia to oocytes involves a change from mitotic to meiotic cells. Progression through the first meiotic prophase (leptotene, zygotene, pachytene, diplotene), with pairing of homologous chromosomes, crossing-over, and recombination, takes 4–5 days. Shortly after birth (day 21 postcoitus [p.c.]), nearly all oocytes are arrested in late diplotene (dictyate-stage), where they remain until stimulated to resume meiotic progression at the time of ovulation. This pool of small, nongrowing oocytes is the sole source of unfertilized eggs in the sexually mature mouse (about 6 weeks of age). It should be noted that as much as 50% of the oocyte population present in the ovary at birth is lost during the first week following birth.

In sexually mature mice, each ovary contains about 8,000 nongrowing oocytes. Each nongrowing oocyte (12–15 μm in diameter) is contained within a cellular follicle that grows concomitantly with the oocyte, from a single layer of a few epithelial-like cells to three layers of cuboidal *granulosa* cells (~900 cells; ~125 μm in diameter) by the time the oocyte has completed its growth (80–85 μm in diameter). During this growth phase (~2 weeks), oocytes are continually arrested at the *dictyate stage* of the first meiotic prophase. The dictyate stage is characterized by very diffuse chromosomes, and oocyte growth is characterized by high rates of transcription and translation. The *theca* is first distinguishable, outside of and separated by a basement membrane, from the granulosa cells when the granulosa region is two cell layers thick (~400 cells; ~100 μm in diameter). During several days, while the oocyte remains a constant size, the follicular cells undergo rapid division, increasing to more than 50,000 cells and resulting in a *Graafian follicle* more than 600 μm in diameter. The follicle exhibits an incipient *antrum* when it is several layers thick (~6,000 cells; ~250 μm in diameter) and, as the antrum expands, the oocyte takes up an acentric position surrounded by two or more layers of granulosa cells (*cumulus cells*). The innermost layer becomes columnar in shape and constitutes the *corona radiata*. These innermost follicle cells communicate both with the oocyte and other follicle cells through an extensive network of *gap junctions*. Apparently, many metabolic precursors and other small molecules (less than ~1,000 M_r) required by the growing and fully grown oocyte, pass through these gap junctions from follicle cells into the oocyte.

Fully grown oocytes in Graafian follicles resume meiosis and complete the first meiotic reductive division (*meiotic maturation*) just prior to ovulation. Resumption of meiosis can be mediated by a hormonal stimulus in vivo (surge in the level of luteinizing hormone [LH]) or simply by release of oocytes from their ovarian follicles into a suitable culture medium in vitro. Meiotic maturation takes 12–14 hours and involves germinal vesicle breakdown and nuclear progression from the dictyate stage of the first meiotic prophase to metaphase II (second meiotic division). Unfertilized eggs display 20 chromosomes, each composed of two chromatids, aligned on the metaphase II spindle and a small *polar body* containing 20 homologous chromosomes. The ovulated eggs complete meiosis, with separation of chromatids and emission of a second polar body containing one-half the chromosomal complement, upon fertilization or artificial activation.

Under normal laboratory housing conditions, a sexually mature mouse ovulates once every 4 days. In a natural ovulation, a mouse releases 8–12 eggs, whereas a superovulated mouse (injected with pregnant mare's serum [PMS] followed by human chorionic gonadotropin [hCG] releases 20–60 eggs (these numbers are very dependent on the mouse strain). Eggs are released from the ovarian follicle, enter the opening (ostium) of the oviduct (fallopian tube), and move to the lower ampulla region of the oviduct, where fertilization probably takes place. It has been estimated that mouse eggs remain capable of being fertilized and giving rise to normal offspring for about 8–12 hours following ovulation.

III. OVERALL GENE EXPRESSION DURING OOGENESIS

A. Transcription

A fully grown mouse oocyte contains 0.4–0.6 ng of RNA, or about 200 times the amount found in a typical somatic cell [Olds et al., 1973; Bachvarova, 1974; Schultz and Wassarman, 1980; Sternlicht and Schultz, 1981]. Approximately 10%–15%, 20%–25%, and 60%–65% of the RNA is poly(A)$^+$, transfer, and ribosomal RNA, respectively [Wassarman, 1983; Bachvarova, 1985; Schultz, 1986]. The steady state level of RNA increases dramatically during oocyte growth, exhibiting biphasic kinetics with respect to oocyte volume [Sternlicht and Schultz, 1981]. RNA synthesized during oocyte growth is quite stable. For example, RNA synthesized in vivo during early stages of oocyte growth is present in ovulated eggs some 10–20 days later [Bachvarova, 1974; Jahn et al., 1976]. Similarly, poly(A)$^+$ RNA synthesized during oocyte growth in vitro exhibits a half-life of more than 10 days [Bachvarova, 1981; Brower et al., 1981]. It can be estimated that, overall, there is a 300-fold increase in RNA content of oocytes during their growth phase (2–3 weeks). During early and middle stages of oocyte growth, changes in nucleolar ultrastructure and RNA polymerase activity are consistent with high rates of ribosomal RNA synthesis [Chouinard, 1971; Moore and Lintern-Moore, 1978; Wassarman and Josefowicz, 1978]. The rate of RNA accumulation during latter stages of the oocyte growth phase is significantly lower than during early and middle stages of oocyte growth [Sternlicht and Schultz, 1981]. Although RNA continues to be synthesized in fully grown oocytes [Wassarman and Letourneau, 1976], it is at a diminished rate compared with midgrowth oocytes [Sternlicht and Schultz, 1981] and declines to barely detectable levels after the onset of meiotic maturation (germinal vesicle breakdown and chromosome condensation) [Bloom and Mukherjee, 1972; Rodman and Bachvarova, 1976; Wassarman and Letourneau, 1976]. As much as 50% of poly(A)$^+$ RNA accumulated during oocyte growth is either degraded or deadenylated during meiotic maturation [Bachvarova and DeLeon, 1980; Brower et al., 1981; DeLeon et al., 1983; Bachvarova et al., 1985]. Furthermore, it is possible that bona fide mRNA constitutes only about 3% of total RNA, or about 15% of poly(A)$^+$ RNA in fully grown oocytes [Roller et al., 1989] The unfertilized egg and one-cell embryo contain about 1.7 × 10^7 and 2.4 × 10^7 poly(A)$^+$ RNA molecules, respectively [Piko and Clegg, 1982; Clegg and Piko, 1983; Davidson, 1986].

B. Translation

In addition to being extremely active transcriptionally, growing mouse oocytes are extremely active translationally [Wassarman, 1983; Schultz, 1986; Wassarman, 1988]. Overall, the absolute rate of protein synthesis increases markedly during oocyte growth, from about 1.1 pg/hr per nongrowing oocyte to about 41.8 pg/hr per fully grown oocyte [Schultz et al., 1979a, b], and then decreases somewhat during meiotic maturation to about 33 pg/hr per unfertilized egg [Schultz et al., 1978]. It has been estimated that as much as 15% of total poly(A)$^+$ RNA is associated with polysomes in fully grown oocytes [DeLeon et al., 1983]. Fully grown oocytes contain about 30 ng of protein (25 ng of protein, exclusive of the zona pellucida) or about 50–60 times more protein than a typical somatic cell [Lowenstein and Cohen, 1964; Brinster, 1967; Schultz and Wassarman, 1977]. The complement of proteins synthesized is extremely diverse, such that several hundred "spots" can be visualized on fluorograms of high-resolution, two-dimensional gels containing ^{35}S-methionine–labeled oocyte extracts [Wassarman, 1983]. Fully grown oocytes are particularly rich in certain structural proteins, such as tubulin [Schultz et al., 1979a,b] and actin [Kaplan et al., 1982], as well as in certain enzymes, such as lactate dehydrogenase [Mangia and Epstein, 1975; Mangia et al., 1976; Cascio and Wassarman, 1982] (discussed below), glucose-6-phosphate dehydrogenase [Mangia and Epstein, 1975], and creatine kinase [Iyengar et al., 1983].

C. Summary

Mouse oocytes, arrested in the dictyate stage of the first meiotic prophase, are extremely active with respect to both transcription and translation throughout their 2–3 week growth phase. Fertilized mouse eggs inherit relatively large amounts of a wide variety of transcripts and proteins that are synthesized and stored during oogenesis. In this context, oocyte growth in the mouse (and probably in mammals, in general) resembles oocyte growth in a wide variety of nonmammalian animals [Browder, 1985; Davidson, 1986].

IV. SPECIFIC GENE EXPRESSION DURING OOGENESIS

A. *c-mos*

The *c-mos* protooncogene is the cellular homolog of the transforming gene *v-mos* from Moloney murine sarcoma virus [Oskarsson et al., 1980]. It is a member of the serine/threonine protein kinase family.

Unlike many protooncogenes, expression of *c-mos* in mice is restricted to a few tissues, most notably ovary and testis [Propst and Vande Woude, 1985]. Results of in situ hybridization analyses suggest that *c-mos* expression occurs specifically in growing and fully grown oocytes within the ovary [Goldman et al., 1987; Mutter and Wolgemuth, 1987]. Quantification of these results indicates that *c-mos* mRNA accumulates to a steady-state level of about 10^5 copies per fully grown oocyte during oocyte growth. *c-mos* is also expressed specifically within the germ cell compartment of the testis.

Northern blot and in situ hybridization analyses have revealed the pattern of *c-mos* expression during meiotic maturation of oocytes, as well as during early embryogenesis [Mutter et al., 1988; Goldman et al., 1988]. During meiotic maturation (i.e., conversion of fully grown oocytes into unfertilized eggs), the level of *c-mos* transcripts decreases by about 20%. This decrease does not occur until eggs have entered metaphase II and the first polar body has been extruded [Mutter et al., 1988]. In addition, the size of *c-mos* transcripts increases from 1.40 kb to 1.65 kb due to posttranscriptional polyadenylation of preexisting cytoplasmic transcripts [Goldman et al., 1988]. The level of *c-mos* transcripts continues to fall following fertilization, such that *c-mos* mRNA is undetectable at the two-cell stage of development and remains undetectable through the blastocyst stage of development. Overall, the pattern of *c-mos* expression described is consistent with the behavior of

the bulk of maternal poly(A)$^+$ RNA in mammalian eggs. Furthermore, cytoplasmic polyadenylation of stored maternal mRNA has been reported for a variety of animals, including mice [Clegg and Piko, 1983; Bachvarova et al., 1985; Davidson, 1986; Huarte et al., 1987; Rosenthal and Ruderman, 1987], and is apparently generally related to temporal translation of processed mRNA.

The *c-mos* oncoprotein p39^{mos} is first synthesized by meiotically arrested, fully grown oocytes having an intact germinal vesicle, is present during meiotic maturation and in unfertilized eggs [Paules et al., 1989; Zhao et al., 1991], but cannot be detected in growing oocytes or in fertilized eggs [Paules et al., 1989].

Certain lines of evidence suggest that p39^{mos} plays a role in meiotic maturation. Fully grown oocytes microinjected with *c-mos* antisense RNA [Paules et al., 1989; O'Keefe et al., 1989] or exposed to antibodies directed against p39^{mos} [Zhao et al., 1991] undergo germinal vesicle breakdown and either do not extrude a first polar body [Paules et al., 1989; Zhao et al., 1990] or fail to enter the second meiotic division [O'Keefe et al., 1989]. Results of analogous experiments carried out in *Xenopus laevis* [Sagata et al., 1989] suggest that *Xenopus* p39^{mos} is cytostatic factor (CSF) itself or is an essential component of CSF. CSF is an activity in amphibian cells that prevents inactivation of maturation-promoting factor (MPF), and there is evidence to suggest that p39^{mos} may accomplish this by phosphorylation of cyclin [Roy et al., 1990]. MPF is a key regulatory component of the G2/M transition in both meiotic and mitotic cells in all eukaryotic organisms [Murray and Kirschner, 1991]. Therefore, it seems likely that p39^{mos} plays a similar role in meiotic maturation of mouse oocytes by interacting directly with MPF, with other components of the MPF pathway, and/or with other proteins. For instance, *c-mos* may be involved in regulating spindle formation and/or function during meiotic maturation of mouse oocytes [Zhao et al., 1991]. Degradation of p39^{mos} is apparently not involved in the release of ovulated eggs from metaphase II arrest [Weber et al., 1991].

Thus *c-mos* mRNA apparently is a stored maternal message that is polyadenylated and translated in a temporally specific manner during oogenesis, thereby permitting p39^{mos} to play a vital role during meiotic maturation.

B. *oct-3*

oct-3 is a maternally expressed octamer-binding protein that is encoded by the murine *oct-3* gene [Rosner et al., 1990; Scholer et al., 1990a; Okamoto et al., 1990]. *oct-3* is a relatively new member of the POU domain family of regulatory genes. Such genes share a region, the POU domain, that consists of a POU-specific domain and a POU homeobox domain, linked by a short variable region [Herr et al., 1988]. Like all other POU domain proteins [Ruvkun and Finney, 1991] *oct-3* protein is capable of transactivating promoters containing an octamer motif through the DNA-binding properties of its POU domain [Lenardo et al., 1989; Scholer et al., 1989a,b; Rosner et al., 1990; Okamoto et al., 1990; Scholer et al., 1990a,b].

oct-3 was identified as the first example of a transcription factor that is specific for the earliest stages of mammalian development. Clues to the pattern of *oct-3* expression were provided by studies of *oct-3*-binding activity in extracts prepared from both male and female primordial germ cells (PGCs), ovulated eggs, embryonic stem (ES) cells, and embryonic carcinoma (EC) cell lines [Scholer et al., 1989a,b; Lenardo et al., 1989]. The *oct-3* gene is expressed as both a maternal and an embryonic mRNA, and *oct-3* expression has been studied by a combination of Northern blot and in situ hybridization analyses, as well as by RNase

protection analyses [Rosner et al., 1990; Scholer et al., 1990a,b].

oct-3 mRNA is detected as a 1.55 kb transcript in growing mouse oocytes and ovulated eggs, but is not found in nongrowing oocytes [Rosner et al., 1990; Scholer et al., 1990b]. *oct-3* mRNA is also present in fertilized eggs [Rosner et al., 1990]. Zygotic expression of *oct-3* is first detected at the morula stage of development, and throughout early embryogenesis *oct-3* expression is detectable in cells that are pluripotent or totipotent [Rosner et al., 1990; Scholer et al., 1990b.]. At day 8.5 of development expression of *oct-3* is restricted to PGCs [Rosner et al., 1990; Scholer et al., 1990b]. In adult animals, *oct-3* expression is limited to oocytes within the ovary [Rosner et al., 1990; Scholer et al., 1990b] and to testis, but not to sperm [Rosner et al., 1990]. *oct-3* mRNA is found in RNA prepared from undifferentiated but not differentiated EC cells, as well as in RNA prepared from ES cells [Rosner et al., 1990].

Recent evidence suggests a nontranscriptional role for maternally inherited *oct-3* protein. Rather, the protein is likely to be involved in DNA replication in fertilized eggs, leading to successful division of one-cell embryos [Rosner et al., 1991]. In this context, fertilized eggs microinjected with *oct-3* antisense oligonucleotides arrest at the one-cell stage of development (Fig. 1) and can be rescued by subsequent injection of *oct-3* mRNA. Furthermore, microinjection of DNA fragments containing the octamer motif into fertilized eggs also results in a block at the one-cell stage of development. Analysis of ^{3}H-thymidine incorporation into DNA, following microinjection of either of the inhibitors of one-cell cleavage, revealed a significant inhibition of thymidine incorporation, as compared with controls [Rosner et al., 1991]. A role for zygotic *oct-3* protein has yet to be described, but it may act in maintaining the differentiation capability of cells that ex-

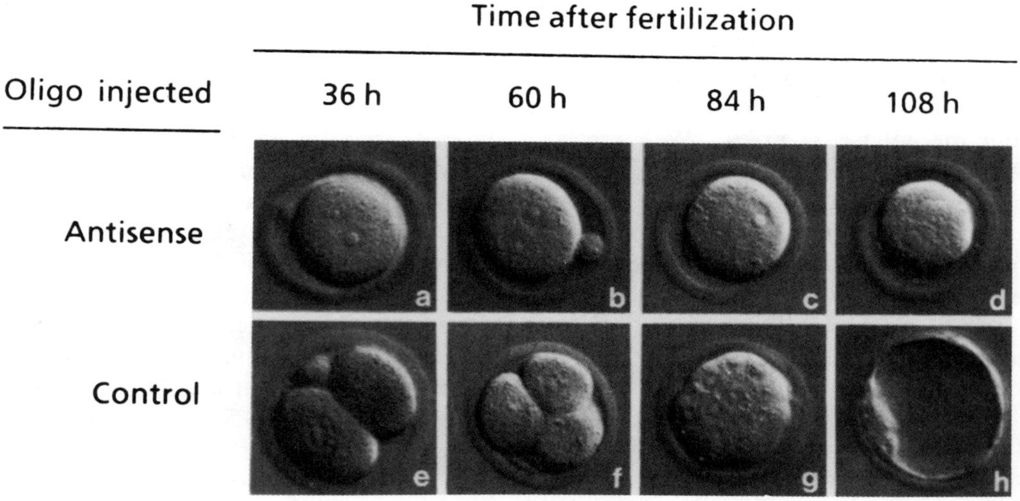

Time after fertilization

Oligo injected	36 h	60 h	84 h	108 h

Fig. 1. *Antisense* oct-3 *oligonucleotide blocks the first embryonic cell division. Photomicrographs of embryos injected with either antisense* oct-3 *oligonucleotides (**a–d**) or control oligonucleotides (**e–h**) are shown. The time after fertilization when the photomicrographs were taken is shown at the top. a–d show one-cell embryos, e is a two-cell embryo, f is a four-cell embryo, g is a compacted morula, and h is a late blastocyst. (Reproduced from Rosner et al., 1991,© Cell Press)*

press the protein [Rosner et al., 1990; Scholer et al., 1990b].

In summary, *oct-3* expression during early mouse development is specific to cells that are either totipotent or pluripotent and thus possess the ability to differentiate. *oct-3* expression is specific to cells constituting the germline lineage.

C. *c-kit*

The *c-kit* protooncogene is the cellular homolog of *v-kit*, an oncogene present in HZ4 feline sarcoma virus [Besmer et al., 1986]. *c-kit* is a member of the tyrosine kinase receptor family [Besmer et al., 1986; Yarden et al., 1987; Qiu et al., 1988; Majumder et al., 1988], and its ligand, stem cell factor (SCF) or kit ligand (KL), has been identified [Zsebo et al., 1990; Martin et al., 1990].

Interest in *c-kit* and SCF intensified when they were shown to be encoded at the murine genetic loci *white-spotting* (*W*) and *steel* (*Sl*) respectively [Chalbot et al., 1988;

Geissler et al., 1988; Nocka et al., 1989; Zsebo et al., 1990]. Mutations at both of these loci have been of interest for many years, since they both produce a very similar range of pleiotrophic effects, affecting development of hematopoietic, melanocyte, and germ cell lineages [Russell, 1979; Silvers, 1979].

Embryonic expression of *c-kit* in mice is very complex [Orr-Urtreger et al., 1990]. In this context, it should be noted that, in general, there is good correlation between sites of *c-kit* expression and the three major cell types affected by mutations at the *W* locus. However, *c-kit* is also expressed in tissues not known to be affected in *W* mutants (e.g., CNS, craniofacial structures, and intestinal tract). In normal adult mice, *c-kit* mRNA (5.5 kb) is found in RNA prepared from placenta, brain, bone marrow, lung, ovary, and testis [Nocka et al., 1989].

Sites of *c-kit* expression in mouse ovary have been identified by Northern blot and in situ hybridization analyses [Orr-Urtreger et al., 1990; Manova et al., 1990]. Northern blot hybridization analyses of RNA prepared from growing oocytes, ovulated eggs, two-cell embryos, and blastocysts revealed that c-kit transcripts are present in nongrowing oocytes, and their steady-state level increases during oocyte growth. It is estimated that fully grown oocytes contain about 15 fg of c-kit mRNA. The level of c-kit mRNA declines severalfold following fertilization, and c-kit transcripts are undetectable by the blastocyst stage of development [Manova et al., 1990]. In situ hybridization analyses of ovarian sections, prepared from pre- and postnatal mice of various ages, confirmed this pattern of c-kit expression and provided additional information. For example, they revealed that c-kit expression is first detected in ovarian sections of prenatal mice, specifically within oocytes that have reached diplotene of the first meiotic prophase where, they remain throughout growth [Manova et al., 1990]. c-kit expression is also detected in interstitial tissue of ovaries from 14 to 17 day animals, but never in follicle cells at any stage of their development [Orr-Urtreger et al., 1990; Manova et al., 1990]. Spermatogenic cells are a site of c-kit expression in the testis [Manova et al., 1990].

The presence or absence of c-kit protein during early mouse development correlates very well with c-kit expression during this period. Results of indirect immunofluorescence analyses, using antibodies directed against c-kit protein [Nocka et al., 1990], indicate that the protein is present on the surface of oocytes, ovulated eggs, and one- and two-cell embryos, but not on blastocysts (Fig. 2) [Manova et al., 1990].

These findings suggest a role for c-kit in postnatal development of female germ cells, in addition to its well established role in proliferation and migration of primordial germ cells. It has been suggested that c-kit plays a role in initiation and/or maintenance of oocyte growth [Manova et al., 1990] and in meiotic maturation of oocytes [Manova et al., 1990; Orr-Urtreger et al., 1990]. Phenotypes of certain W and Sl mutations provide a basis for these suggestions. For example, in W/W^v and W^v/W^v mice the rate of development of oocytes and spermatogenic cells is slower than in wild-type mice [Coulombre and Russell, 1954], and in juvenile Sl/Sl^t infertile females ovaries lack follicles with growing oocytes, despite the presence of abundant primordial follicles [Kuroda et al., 1988].

D. mZP3

mZP3 is one of three glycoproteins that make up the mouse egg extracellular coat, or zona pellucida. mZP3 is an 83,000 M_r glycoprotein that consists of a 44,000 M_r polypeptide (402 amino acids), three or four complex-type asparagine (N)–linked oligosaccharides, and an unknown number of serine/threonine (O)–linked oligosaccharides [Wassarman et al., 1985; Wassarman, 1988a]. During the initial stages of fertilization, mZP3 serves as both a primary sperm receptor, involved in species-specific binding of sperm to eggs, and as inducer of the acrosome reaction, a form of exocytosis in sperm [Wassarman, 1987a,b, 1990; Kopf and Gerton, 1991].

mZP3 is a single-copy gene, located on mouse chromosome 5 [Lunsford et al., 1990], that encodes a 1.5 kb, polyadenylated mRNA [Kinloch et al., 1988, 1990a; Ringuette et al., 1988; Roller et al., 1989; Kinloch and Wassarman, 1989a,b]. Results of Northern blot and in situ hybridization analyses, as well as RNase protection analyses, suggest that mZP3 is expressed exclusively in growing oocytes (Fig. 3) [Philpott et al., 1987; Roller et al., 1989; Kinloch and Wassarman, 1989a,b]. In adult female mice, mZP3 mRNA is found only in RNA prepared from ovaries and, within ovaries, is found only in growing oocytes. RNase protection assays indicate that nongrowing oocytes (12–15 μm) contain undetectable levels of mZP3 mRNA (<1

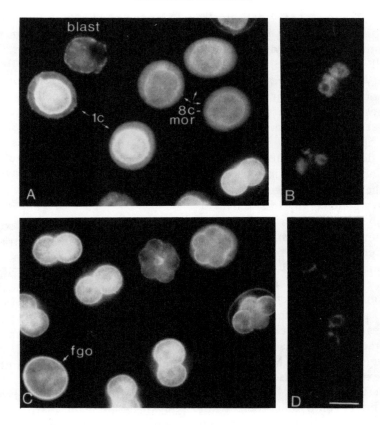

Fig. 2. *Expression of* c-kit *protein on the surface of fully grown oocytes and embryos. Fluorescence images of embryos obtained 1.5, 2.5, or 3.5 days after fertilization.* **A:** *Embryos incubated with* c-kit *immune serum absorbed with W/W mast cells. 1c, One-cell fertilized embryo; 8c-mor, compact eight-cell embryos or morulae; blast, early blastocyst.* **B:** *Two-cell embryos incubated with immune serum absorbed with +/+ mast cells.* **C:** *A fully grown oocyte (fgo) and two-cell to eight-cell embryos incubated with* c-kit *immune serum absorbed with W/W mast cells.* **D:** *Two-cell embryos incubated with second antibody only. Bar = 50 μm for all panels. (Reproduced from Manova et al., 1990, with permission of the Company of Biologists, Ltd.)*

$\times$ 10^3 copies/oocyte) (Fig. 4). However, the steady-state level of *mZP3* mRNA increases markedly during oocyte growth (2–3 weeks), reaching 2.5–3 $\times$ 10^5 copies per 70–80 μm oocyte [Roller et al., 1989; Kinloch and Wassarman, 1989b]. A dramatic fall in *mZP3* mRNA levels occurs during ovulation (meiotic maturation), when as much as 98% of *mZP3* mRNA is destroyed (5 $\times$ 10^3 copies/unfertilized egg). RNase protection assays indicate that unfertilized eggs and cleavage stage embryos contain undetectable levels of *mZP3* mRNA (<1 $\times$ 10^3 copies/zygote). These findings are consistent with results of assays of *mZP3* synthesis and secretion during oogenesis and early development [Bleil and Wassarman, 1980;

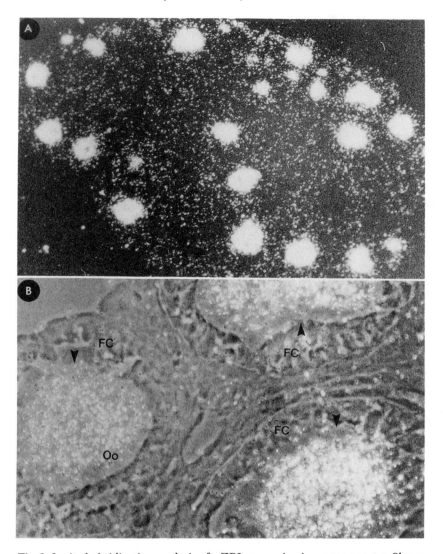

Fig. 3. *In situ hybridization analysis of* mZP3 *expression in mouse oocytes. Shown are photomicrographs (**A**, bright-field; **B**, darkfield) of the same ovarian section hybridized with a radiolabeled* mZP3 *probe and subjected to autoradiography. FC, follicle cells; Oo, oocyte.*

Salzmann et al., 1983; Wassarman et al., 1985; Wassarman, 1988a].

Expression of *mZP3* is regulated by *cis*-acting sequence(s) in the 5'-flanking region of the gene, together with oocyte-specific protein(s) [Lira et al., 1990; Schickler et al., 1992]. It has been demonstrated that only 470 bp of *mZP3* gene 5'-flanking region is sufficient to target expression of a reporter gene, encoding firefly luciferase, to growing oocytes in transgenic mice. An oocyte-specific protein ($\sim$60,000 M_r), OSP-1, binds to the sequence 5'-TGATAA-3' located within the first 100 bp of the *mZP3* gene

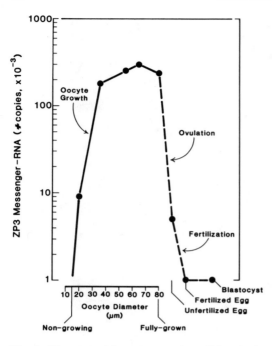

Fig. 4. *Diagrammatic representation of the steady-state levels of* mZP3 *mRNA in mouse oocytes, eggs, and preimplantation embryos, as determined by quantitative RNase protection assays [see Roller et al., 1989, for details]. The number of copies of* mZP3 *mRNA ($\times 10^{-3}$) is plotted as a function of the stage of mouse development.*

promoter. Changes in levels of this protein during oogenesis and early cleavage are consistent with the pattern of *mZP3* gene expression during these stages of mouse development. Thus OSP-1 may interact with other ubiquitous or cell-specific proteins to activate *mZP3* gene expression during oocyte growth.

E. t-PA

t-PA (72,000 M_r) is a member of the family of serine proteases that convert the serum protein plasminogen into plasmin [Strickland, 1980]. While *t-PA* activity cannot be detected in fully grown mouse oocytes, unfertilized and fertilized eggs exhibit relatively high levels of *t-PA* activity

in vitro [Huarte et al., 1985]. The biological function of such high levels of *t-PA* during these stages of development is not clear. *t-PA* first appears during meiotic maturation of oocytes and is dependent on germinal vesicle breakdown, but not on polar body emission (Fig. 5). *t-PA* activity can be detected 5 hours after germinal vesicle breakdown. Furthermore, the appearance of *t-PA* during meiotic maturation is not dependent on concomitant transcription (i.e., occurs in the presence of either actinomycin D or α-amanitin), suggesting the presence of untranslated *t-PA* mRNA in fully grown oocytes.

Several lines of evidence suggest that appearance of *t-PA* activity in unfertilized mouse eggs is regulated posttranscriptionally during oogenesis. Although *t-PA* activity cannot be detected in growing or fully grown oocytes, Northern blot hybridization analyses reveal that *t-PA* mRNA is present

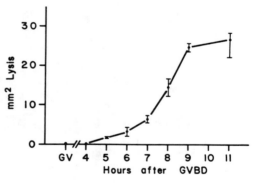

Fig. 5. *Time-course of* t-PA *production by in vitro maturing mouse oocytes. Primary oocytes, having an intact germinal vesicle, were allowed to resume meiosis in culture. Individual oocytes were collected at the times indicated after germinal vesicle breakdown, and their* t-PA *content was assayed by zymography in the presence of plasminogen. Two to four oocytes were tested for each time point. The zymogram was allowed to develop for 43 hours. The area of substrate lysis catalyzed by each oocyte extract was measured; the mean area and range are shown. (Reproduced from Huarte et al., 1985, © Cell Press.)*

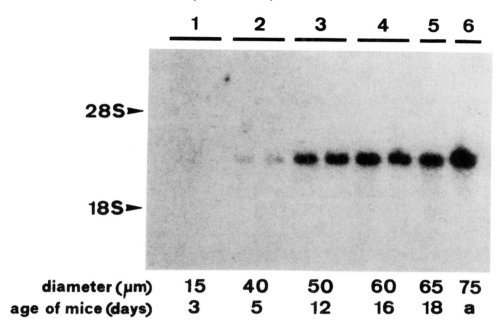

| diameter (µm) | 15 | 40 | 50 | 60 | 65 | 75 |
| age of mice (days) | 3 | 5 | 12 | 16 | 18 | a |

Fig 6. *Accumulation of* t-PA *mRNA during oocyte growth. RNA from primordial, growing, and fully grown primary oocytes was analyzed by Northern blot hybridization using a cRNA probe transcribed from pSP65-MT₁. Each lane contains the RNA from 40 oocytes of the indicated size, derived from mice of the indicated age. a, Adult. (Reproduced from Huarte et al., 1987, with permission of the publisher.)*

in growing oocytes (Fig. 6) and accumulates to a steady-state level of about 10^5 copies in the cytoplasm of fully grown oocytes [Huarte et al., 1987]. Translation of stored *t-PA* mRNA is triggered by meiotic maturation of oocytes. Furthermore, *t-PA* mRNA is destroyed during the latter stages of meiotic maturation, such that undetectable levels of the message are found in fertilized eggs.

Concomitant with the appearance of *t-PA* activity in oocytes that have resumed meiosis, *t-PA* mRNA undergoes a substantial increase in size (~400–600 nt) due to increased polyadenylation at the 3'-end of the molecule [Huarte et al., 1987]. This structural change is initiated within 3 hours of germinal vesicle breakdown. Microinjection of antisense RNAs complementary to 103 nt of the extreme 3'-untranslated region of *t-PA* mRNA into fully grown oo-

cytes leads to hybrid formation and amputation of the 3'-terminal sequences of the message [Strickland et al., 1988]. Such modification of *t-PA* mRNA in oocytes is sufficient to prevent polyadenylation, translational activation, and destabilization of the message during meiotic maturation of oocytes. These and other observations [Vassalli et al., 1989] strongly suggest that cytoplasmic 3'-polyadenylation, regulated by sequences in the 3'-noncoding region of *t-PA* mRNA, is necessary and sufficient for translational activation of the message during meiotic maturation of mouse oocytes. It should be noted that hypoxanthine phosphoribosyltransferase (*HPRT*) mRNA in fully grown mouse oocytes undergoes polyadenylation (~150–200 nt) at its 3'-end during meiotic maturation, and this precedes an increase in *HPRT* activity [Paynton et al., 1988].

F. LDH

Fully grown mouse oocytes possess unusually high levels of LDH activity, with virtually all of the activity attributable to the "heart-type" isozyme LDH-H_4 [Auerbach and Brinster, 1968; Mangia and Epstein, 1975; Mangia et al., 1976; Cascio and Wassarman, 1982]. LDH synthesis, which represents as much as 1.8% of total protein synthesis during oocyte growth, decreases about sevenfold during meiotic maturation of oocytes and about 20-fold in fertilized eggs (compared with fully grown oocytes) [Cascio and Wassarman, 1982]. Fully grown oocytes possess about 150 fg of LDH mRNA, or about 2×10^5 copies per oocyte [Roller et al., 1989]. During meiotic maturation of oocytes, the steady-state level of LDH mRNA falls only by about 20%, whereas LDH synthesis falls about sevenfold. A similar discrepancy has been noted for changes in β-actin synthesis and mRNA levels during meiotic maturation of mouse oocytes and has been attributed to deadenylation (~200 nt) of β-actin mRNA [Bachvarova et al., 1985; Paynton et al., 1988]. Similarly, α-tubulin mRNA undergoes extensive deadenylation and degradation during meiotic maturation [Paynton et al., 1988]. In this context, the size of LDH mRNA decreases in size from about 1.4 kb in fully grown oocytes to about 1.3 kb in unfertilized eggs (R. Roller and P. Wassarman, unpublished results).

G. Summary

1. *c-mos:* Transcribed, but not translated, throughout oocyte growth, polyadenylated and translated in fully grown oocytes, and transcripts largely degraded during meiotic maturation. An example of temporally specific translation during oogenesis of a product required for meiotic maturation.

2. *oct-3:* Transcribed throughout oocyte growth with transcripts present in ovulated eggs and zygotes. An example of a maternally inherited product that is required by the zygote.

3. *c-kit:* Transcribed and translated throughout oocyte growth, with transcripts persisting during meiotic maturation.

4. *mZP3:* Transcribed and translated extensively throughout oocyte growth, transcripts largely degraded and translation terminated during meiotic maturation. The only well-documented example of oocyte-specific gene expression in mice. An example of a product of oogenesis required for fertilization.

5. *t-PA:* Transcribed, but not translated, throughout oocyte growth, polyadenylated and translated during meiotic maturation, transcripts degraded and translation terminated in unfertilized egg. An example of posttranscriptional regulation and temporally specific translation during oogenesis. (*HPRT* may be subject to similar regulation, i.e., activation by polyadenylation of preexisting mRNA, during oogenesis in mice.)

6. *LDH:* Transcribed and translated extensively throughout oocyte growth, transcripts deadenylated and degraded, and translation falls during meiotic maturation. (α-Tubulin and β-actin may be subject to similar regulation, i.e., deadenylation of preexisting mRNA, during oogenesis in mice.)

V. FINAL COMMENTS

mRNA and proteins are synthesized and stored during oogenesis in mice for use during oogenesis, fertilization, and/or preimplantation development. In a number of well-documented cases appearance of specific proteins during oogenesis in mice is temporally regulated. As in other animal systems [Richter, 1991], polyadenylation and deadenylation of preexisting maternal mRNAs serves as the molecular basis, at least in several instances, for translational control during oogenesis in mice. In at least one instance, expression of a gene (*mZP3*) is restricted exclusively to oocytes during their growth phase. Examination of the behavior of a more diverse repertoire of genes will undoubtedly provide additional insights into regulation of gene expression during oogenesis in mice and other mammals.

ACKNOWLEDGMENTS

We are very grateful to Mrs. Alice O'Connor for expert assistance in preparation of this manuscript and to Dr. Ralph Gwatkin for his patience.

REFERENCES

Auerbach S, Brinster RL (1968): Lactate dehydrogenase isozymes in mouse blastocyst cultures. Exp Cell Res 53:313–315.

Austin C, Short R (eds) (1982): Reproduction in Mammals. Cambridge: Cambridge University Press.

Bachvarova R (1974): Incorporation of tritiated adenosine into mouse ovum RNA. Dev Biol 40:52–58.

Bachvarova R (1981): Synthesis, turnover, and stability of heterogeneous RNA in growing mouse oocytes. Dev Biol 86: 384–392.

Bachvarova R (1985): Gene expression during oogenesis and oocyte development in mammals. In Browder L (ed): Oogenesis. New York: Plenum Press, pp 453–524.

Bachvarova R, DeLeon V (1980): Polyadenylated RNA of mouse ova and loss of maternal RNA in early development. Dev Biol 74:1–8.

Bachvarova R, DeLeon V, Johnson A, Kaplan G, Paynton DV (1985): Changes in total RNA, polyadenylated RNA and actin mRNA during meiotic maturation of mouse oocytes. Dev Biol 108:325–331.

Besmer P, Murphy JE, George PC, Qiu FH, Bergold PJ, Lederman L, Snyder HWJ, Brodeur D, Zuckerman EE, Hardy WD (1986): A new acute transforming feline retrovirus and relationship of its oncogene v-kit with the protein kinase gene family. Nature 320:415–421.

Biggers J, Schuetz A (eds) (1972): Oogenesis. Baltimore: University Park Press.

Bleil JD, Wassarman PM (1980): Synthesis of zona pellucida proteins by denuded and follicle-enclosed mouse oocytes during culture in vitro. Proc Natl Acad Sci USA 77:1029–1033.

Bloom A, Mukherjee B (1972): RNA synthesis in maturing mouse oocytes. Exp Cell Res 74:577–582.

Brinster R (1967): Protein content of the mouse embryo during the first five days of development. J Reprod Fertil 10:227–240.

Browder LW (ed) (1985): Developmental Biology: A Comprehensive Synthesis, Vol 1, Oogenesis. New York: Plenum Press.

Brower P, Gizang E, Boreen S, Schultz R (1981): Biochemical studies of mammalian oogenesis: synthesis and stability of various classes of RNA during growth of the mouse oocyte in vitro. Dev Biol 86:373–383.

Cascio SM, Wassarman PM (1982): Program of early development in the mammal: post-transcriptional control of a class of proteins synthesized by mouse oocytes and early embryos. Dev Biol 89:397–408.

Chabot B, Stephenson DA, Chapman VM, Besmer P, Bernstein A (1988): The proto-oncogene c-kit encoding a transmembrane tyrosine kinase receptor maps to the mouse W locus. Nature 335:88–89.

Chamberlin ME, Dean J (1990): Genomic organization of a sex specific gene: the primary sperm receptor of the mouse zona pellucida. Dev Biol 131:207–214.

Chouinard L (1971): A light- and electron-microscope study of the nucleolus during growth of the oocyte in the prepubertal mouse. J Cell Sci 9:637–663.

Clegg KB, Pikó L (1983): Poly(A) length, cytoplasmic adenylation and synthesis of poly(A)$^+$ RNA in early mouse embryos. Dev Biol 95:331–341.

Copeland NG, Gilbert DJ, Cho BC, Donovan PJ, Jenkins NA, Cosman D, Anderson D, Lyman SD, Williams DE (1990): Mast cell growth factor maps near the steel locus on mouse chromosome 10 and is deleted in a number of steel alleles. Cell 63:175–183.

Coulombre JL, Russell ES (1954): Analysis of the pleiotropism at the W-locus in the mouse: the effects of W and W^v substitution upon postnatal development of germ cells. J Exp Zool 126:277–295.

Davidson EH (1986): Gene Activity in Early Development. Orlando, FL: Academic Press.

DeLeon V, Johnson A, Bachvarova R (1983): Half-lives and relative amounts of stored and polysomal ribosomes and poly(A$^+$)-RNA in mouse oocytes. Dev Biol 98:400–408.

Geissler EN, Ryan MA, Housman DE (1988): The dominant-white spotting (W) locus of the mouse encodes the c-kit proto-oncogene. Cell 55:185–192.

Goldman DS, Kiessling AA, Cooper GM (1988): Post-transcriptional processing suggests that c-mos functions as a maternal message in mouse eggs. Oncogene 2:159–162.

Goldman DS, Kiessling AA, Millette CF, Cooper GM (1987): Expression of *c-mos* RNA in germ cells of male and female mice. Proc Natl Acad Sci USA 84:4509–4513.

Herr W, Sturm RA, Clerc RG, Corcoran LM, Baltimore D, Sharp PA, Ingraham HA, Rosenfeld MG, Finney M, Ruvkun G, Horvitz HR (1988): The POU domain: A large conserved region in the mammalian *pit-1, oct-1, oct-2*, and *Caenorhabditis elegans unc-86* gene products. Genes Dev 2:1513–1516.

Huang E, Nocka K, Beier DR, Chu TY, Buck J, Lahm HW, Wellner D, Leder P, Besmer P (1990): The hematopoietic growth factor KL is encoded by the Sl locus and is the ligand of the *c-kit* receptor, the gene product of the *W* locus. Cell 63:225–233.

Huarte J, Belin D, Vassalli J-D (1985): Plasminogen activator in mouse and rat oocytes: Induction during meiotic maturation. Cell 43:551–558.

Huarte J, Belin D, Vassalli A, Strickland S, Vassalli J-D (1987): Meiotic maturation of mouse oocytes triggers the translation and polyadenylation of dormant tissue-type plasminogen activator mRNA. Genes Dev 1:1201–1211.

Iyengar M, Iyengar C, Chen H, Brinster R, Bornslaeger E, Schultz R (1983): Expression of creatine kinase isoenzyme during oogenesis and embryogenesis in the mouse. Dev Biol 96:263–268.

Jahn C, Baran M, Bachvarova R (1976): Stability of RNA synthesized by the mouse oocyte during its major growth phase. J Exp Zool 197:161–172.

Jones R (ed) (1978): The Vertebrate Ovary. New York: Plenum Press.

Kaplan G, Abreu S, Bachvarova R (1982): rRNA accumulation and protein synthetic patterns in growing mouse oocytes. J Exp Zool 220:361–380.

Kinloch RA, Roller RJ, Fimiani CM, Wassarman DA, Wassarman PM (1988): Primary structure of the mouse sperm receptor's polypeptide chain determined by genomic cloning. Proc Natl Acad Sci USA 85:6409–6413.

Kinloch RA, Roller RJ, Wassarman PM (1990a): Organization and expression of the mouse sperm receptor gene. In Davidson EH, Ruderman JV, Posakony JW (eds): Developmental Biology. New York: Wiley-Liss, pp 9–20.

Kinloch RA, Ruiz-Seiler B, Wassarman PM (1990b): Genomic organization and polypeptide primary structure of zona pellucida glycoprotein hZP3, the hamster sperm receptor. Dev Biol 142:414–420.

Kinloch RA, Wassarman PM (1989a): Nucleotide sequence of the gene encoding zona pellucida glycoprotein ZP3—the mouse sperm receptor. Nucleic Acids Res 17:2861–2863.

Kinloch RA, Wassarman PM (1989b): Profile of a mammalian sperm receptor gene. New Biologist 1:232–238.

Knobil E, Neill JD (eds) (1988): The Physiology of Reproduction. New York: Raven Press.

Kopf GS, Gerton GL (1991): The mammalian sperm acrosome and the acrosome reaction. In Wassarman PM (ed): Elements of Mammalian Fertilization, Vol. 1. Boca Raton, FL: CRC Press, pp 153–203.

Kuroda H, Terada N, Nakayama H, Matsumoto K, Kitamura Y (1988): Infertility due to growth arrest of ovarian follicles in *Sl/Sl^t* mice. Dev Biol 126:71–79.

Lenardo MJ, Staudt L, Robbins P, Kuang A, Mulligan RC, Baltimore D (1989): Repression of the IgH enhancer in teratocarcinoma cells associated with a novel octamer factor. Science 243:544–546.

Lira SA, Kinloch RA, Mortillo S, Wassarman PM (1990): An upstream region of the mouse *ZP3* gene directs expression of firefly luciferase specifically to growing oocytes in transgenic mice. Proc Natl Acad Sci USA 87:7215–7219.

Lowenstein J, Cohen A (1964): Dry mass, lipid content, and protein content of the intact and zona-free mouse ovum. J Embryol Exp Morphol 12:113–121.

Majumder S, Brown K, Qiu FH, Besmer P (1988): *c-kit* protein, a transmembrane kinase: Identification in tissues and characterization. Mol Cell Biol 8:4896–4903.

Mangia F, Epstein CJ (1975): Biochemical studies of growing mouse oocytes: Preparation of oocytes and analysis of glucose-6-phosphate dehydrogenase and lactate dehydrogenase activities. Dev Biol 45:211–220.

Mangia F, Erickson RP, Epstein CJ (1976): Synthesis of LDH-1 during mammalian oogenesis and early development. Dev Biol 54:146–150.

Manova K, Nocka K, Besmer P, Bachvarova RF (1990): Gonadal expression of *c-kit* encoded at the *W* locus of the mouse. Development 110:1057–1069.

Martin FH, Suggs SV, Langley KE, Lu HS, Ting J, Okino KH, Morris CF, McNiece IK, Jacobsen FW, Mendiaz EA, Birkett NC, Smith KA, Johnson MJ, Parker VP, Flores JC, Patel AC, Fisher EF, Erjavec HO, Herrera CJ, Wypych J, Sachdev RK, Pope JA, Leslie I, Wen D, Lin D-H, Cupples RL, Zsebo KM (1990): Primary structure and functional expression of rat and human stem cell factors DNAs. Cell 63:203–211.

Moore G, Lintern-Moore S (1978): Transcription of the mouse oocyte genome. Biol Reprod 18:865–870.

Murray AW, Kirschner MW (1991): What controls the cell cycle? Sci Am 264 (3):56–63.

Mutter GL, Grills GS, Wolgemuth DJ (1988): Evidence for the involvement of the proto-oncogene *c-mos* in mammalian meiotic maturation and possibly very early embryogenesis. EMBO J 7:683–689.

Mutter GL, Wolgemuth DJ (1987): Distinct developmental patterns of *c-mos* protooncogene expression in female and male mouse germ cells. Proc Natl Acad Sci USA 84:5301–5305.

Nocka K, Buck J, Levi E, Besmer P (1990): Candidate ligand for the *c-kit* transmembrane kinase receptor: KL, a fibroblast derived growth factor stimulates mast cells and erythroid progenitors. EMBO J 10:3287–3294.

Nocka K, Majumder S, Chabot B, Ray P, Cervone M, Bernstein A, Besmer P (1989): Expression of *c-kit* gene products in known cellular targets of *W* mutations in normal and *W* mutant mice-evidence for an impaired *c-kit* kinase in mutant mice. Genes Dev 3:816–826.

Nocka K, Tan JC, Chiu E, Chu TY, Ray P, Traktman P, Besmer P (1990): Molecular bases of dominant negative and loss of function mutations at the murine *c-kit*/white spotting locus: *W37, Wv, W41* and *W*. EMBO J 9:1805–1813.

Okamoto K, Okazawa H, Okuda A, Sakai M, Muramatsu M, Hamada H (1990): A novel octamer binding transcription factor is differentially expressed in mouse embryonic cells. Cell 60:461–472.

O'Keefe SJ, Wolfes H, Kiessling AA, Cooper GM (1989): Microinjection of antisense *c-mos* oligonucleotides prevents meiosis II in the maturing mouse egg. Proc Natl Acad Sci USA 86: 7038–7042.

Olds P, Stern S, Biggers J (1973): Chemical estimates of the RNA and DNA contents of the early mouse embryo. J Exp Zool 186:39–46.

Orr-Urtreger A, Avivi A, Zimmer Y, Givol D, Yarden Y, Lonai P (1990): Developmental expression of *c-kit,* a proto-oncogene encoded by the *W* locus. Development 109:911–923.

Oskarsson M, McClements WL, Blair DG, Maizel JV, Vande Woude GF (1980): Properties of a normal mouse cell DNA sequence (sarc) homologous to the src sequence of Moloney sarcoma virus. Science 207:1222–1224.

Paules RS, Buccione R, Moschel RC, Vande Woude GF, Eppig JJ (1989): Mouse *mos* protooncogene product is present and functions during oogenesis. Proc Natl Acad Sci USA 86:5395–5399.

Paynton BV, Rempel R, Bachvarova R (1988): Changes in state of adenylation and time course of degradation of maternal mRNAs during oocyte maturation and early embryonic development in the mouse. Dev Biol 129:304–314.

Philpott CC, Ringuette MJ, Dean J (1987): Oocyte-specific expression and developmental regulation of *ZP3,* the sperm receptor of the mouse zona pellucida. Dev Biol 121:568–575.

Piko L, Clegg K (1982): Quantitative changes in total RNA, total poly(A) and ribosomes in early mouse embryos. Dev Biol 89:362–378.

Propst F, Vande Woude GF (1985): Expression of *c-mos* proto-oncogene transcripts in mouse tissues. Nature 315:516–518.

Qui F, Ray P, Brown K, Barker PE, Jhanwar S, Ruddle FH, Besmer P (1988): Primary structure of c-kit: Relationship with the CSF-1/PDGF receptor kinase family—Oncogenic activation of *v-kit* involves deletion of extracellular domain and C terminus. EMBO J. 7:1003–1011.

Richter JD (1991): Translational control during early development. BioEssays 13:179–183.

Ringuette MJ, Chamberlin ME, Baur AW, Sobieski DA, Dean J (1988): Molecular analysis of cDNA coding for ZP3, a sperm binding protein of the mouse zona pellucida. Dev Biol 127:287–295.

Ringuette MJ, Sobieski DA, Chamow SM, Dean J (1986): Oocyte-specific gene expression: molecular characterization of a cDNA coding for *ZP3,* the sperm receptor of the mouse zona pellucida. Proc Natl Acad Sci USA 83:4341–4345.

Rodman TC, Bachvarova R (1976): RNA synthesis in preovulatory mouse oocytes. J Cell Biol 70:251–257.

Roller RJ, Kinloch RA, Hiraoka BY, Li SS-L, Wassarman PM (1989): Gene expression during mammalian oogenesis and early embryogenesis: quantification of three messenger RNAs abundant in fully-grown mouse oocytes. Development 106:251–261.

Rosenthal ET, Ruderman JV (1987): Widespread changes in the translation and adenylation of maternal messenger RNAs following fertilization of *Spisula* oocytes. Dev Biol 121:237–246.

Rosner MH, Vigano MA, Ozato K, Timmons PM, Poirier F, Rigby PWJ, Staudt LM (1990): A POU-domain transcription factor in early stem cells and germ cells of the mammalian embryo. Nature 345:686–692.

Rosner MJ, De Santo RJ, Arnheiter H, Staudt LM (1991): *oct-3* is a maternal factor required for the first mouse embryonic division. Cell 64:1103–1110.

Roy LM, Singh B, Gauthier J, Arlinghaus RB, Nordeen SK, Maller J (1990): The cyclin B2 component of MPF is a substrate for the *c-mos* protooncogene product. Cell 61:825–831.

Russell ES (1979): Hereditary anemias of the mouse: A review for geneticists. Adv Genet 20:357–459.

Ruvkun G, Finney M (1991): Regulation of transcription and cell identity by POU domain proteins. Cell 64:475–478.

Sagata N, Watanabe N, Vande Woude GF, Ikawa Y (1989): The *c-mos* proto-oncogene product is a cytostatic factor responsible for meiotic arrest in vertebrate eggs. Nature 342:512–518.

Salzmann GS, Greve JM, Roller RJ, Wassarman PM (1983): Biosynthesis of the sperm receptor during oogenesis in the mouse. EMBO J 2:1451–1456.

Schickler M, Lira SA, Kinloch RA, Wassarman PM (1992): A mouse oocyte-specific protein that binds to a region of mZP3 promoter responsible for oocyte-specific mZP3 gene expression. Mol Cell Biol 12:120–127.

Schöler HR, Dressler GR, Balling R, Rohdewohld H, Gruss P (1990b): oct-4: a germline-specific transcription factor mapping to the mouse t-complex. EMBO J 9:2185–2195.

Schöler HR, Hatzopoulos AK, Balling R, Suzuki N, Gruss P (1989a): A family of octamer-specific proteins present during mouse embryogenesis: evidence for germline-specific expression of an oct factor. EMBO J 8:2543–2550.

Schöler HR, Balling R, Hatzopoulos AK, Suzuki N, Gruss P (1989b): Octamer binding proteins confer transcriptional activity in early mouse embryogenesis. EMBO J 8:2551–2557.

Schöler HR, Ruppert S, Suzuki N, Chowdhury K, Gruss P (1990a): New type of POU domain in germ line-specific protein oct-4. Nature 344:435–439.

Schultz RM (1986): Molecular aspects of mammalian oocyte growth and maturation. In Rossant J, Pedersen RA (eds): Experimental Approaches to Mammalian Embryonic Development. Cambridge: Cambridge University Press, pp 195–237.

Schultz RM, LaMarca MJ, Wassarman PM (1978): Absolute rates of protein synthesis during meiotic maturation of mouse oocytes in vitro. Proc Natl Acad Sci USA 75:4160–4164.

Schultz RM, Letourneau GE, Wassarman PM (1979a): Program of early development in the mammal: Changes in patterns and absolute rates of tubulin and total protein synthesis during oogenesis and early embryogenesis in the mouse. Dev Biol 68:341–359.

Schultz RM, Letourneau GE, Wassarman PM (1979b): Program of early development in the mammal: Changes in patterns and absolute rates of tubulin and total protein synthesis during oocyte growth in the mouse. Dev Biol 73:120–133.

Schultz RM, Wassarman PM (1977): Biochemical studies of mammalian oogenesis: protein synthesis during oocyte growth and meiotic maturation in the mouse. J Cell Sci 24:167–194.

Schultz RM, Wassarman PM (1980): Efficient extraction and quantitative determination of nanogram amounts of cellular RNA. Anal Biochem 104:328–334.

Silvers WK (1979): The Coat Colors of Mice. New York: Springer-Verlag.

Sternlicht A, Schultz R (1981): Biochemical studies of mammalian oogenesis: kinetics of accumulation of total and poly(A)-containing RNA during growth of the mouse oocyte. J Exp Zool 215:191–200.

Strickland S (1980): Plasminogen activator in early development. In Johnson MH (ed): Development in Mammals. Amsterdam: Elsevier/North Holland, pp 81–100.

Strickland S, Huarte J, Belin D, Vassalli A, Rickles RJ, Vassalli J-D (1988): Antisense RNA directed against the 3' noncoding region prevents dormant mRNA activation in mouse oocytes. Science 241:680–684.

Vassalli J-D, Huarte H, Belin D, Gubler P, Vassalli A, O'Connell ML, Parton LA, Rickles RJ, Strickland S (1989): Regulated polyadenylation controls mRNA translation during meiotic maturation of mouse oocytes. Genes Dev 3:2163–2171.

Wassarman PM (1983): Oogenesis: Synthetic events in the developing mammalian egg. In Hartmann JF (ed): Mechanism and Control of Animal Fertilization. New York: Academic Press, pp 1–54.

Wassarman PM (1987a): The biology and chemistry of fertilization. Science 235:553–560.

Wassarman PM (1987b): Early events in mammalian fertilization. Annu Rev Cell Biol 3:109–142.

Wassarman PM (1988a): Zona pellucida glycoproteins. Annu Rev Biochem 57:415–442.

Wassarman PM (1988b): The mammalian ovum. In Knobil E, Neill JD (eds): The Physiology of Reproduction. New York: Raven Press, pp 69–102.

Wassarman PM (1990): Profile of a mammalian sperm receptor. Development 108:1–17.

Wassarman PM (ed) (1991): Elements of Mammalian Fertilization, Vols 1 and 2. Boca Raton, FL: CRC Press.

Wassarman PM, Bleil JD, Florman HM, Greve JM, Roller RJ, Salzmann GS, Samuels FG (1985): The mouse egg's receptor for sperm: What is it and how does it work? Cold Spring Harbor Symp Quant Biol 50:11–19.

Wassarman PM, Josefowicz WJ (1978): Oocyte development in the mouse: An ultrastructural comparison of oocytes isolated at various stages of growth and meiotic competence. J Morphol 156:209–236.

Wassarman PM, Letourneau GE (1976): RNA synthesis in fully grown mouse oocytes. Nature 261:73–74.

Yarden Y, Kuang W-J, Yang-Feng T, Coussens L, Munemitsu S, Dull TJ, Chen E, Schlessinger J, Francke U, Ullrich A (1987): Human proto-oncogene c-kit: A new cell surface receptor tyrosine kinase for an unidentified ligand. EMBO J 6:3341-3351.

Zhao X, Singh B, Batten BE (1990): The role of c-mos proto-oncoprotein in mammalian meiotic maturation. Oncogene 6:43–49.

Zsebo KM, Williams DA, Geissler EN, Broudy VC,

Martin FH, Atkins HL, Hsu R-Y, Birkett NC, Okino KH, Murdock DC, Jacobsen FW, Langley KE, Smith KA, Takeishi T, Cattanach BM, Galli SJ, Suggs S (1990b): Stem cell factor is encoded at the *Sl* locus of the mouse and is the ligand for the *c-kit* tyrosine kinase receptor. Cell 63:213–224.

Zsebo KM, Wypych J, McNiece IK, Lu HS, Smith KA, Karkare SB, Sachdev RK, Yuschenkoff VN, Birkett NC, Williams LR, Satyagal VN, Tung W, Bosselman RA, Mendiaz EA, Langley KE (1990a): Identification, purification and biological characterization of hematopoietic stem cell factor from Buffalo rat liver-conditioned medium. Cell 63:195–201.

Zuckerman S, Weir B (eds) (1977): The Ovary. New York: Academic Press.

ABOUT THE AUTHORS

ROSS A. KINLOCH is a Research Fellow in the Department of Cell and Developmental Biology at the Roche Institute of Molecular Biology, where he carries out research on mammalian sperm receptor genes and glycoproteins. Dr. Kinloch received both his B.Sc. and Ph.D. from the University of Glasgow, Scotland. His Ph.D. research on recombination in Adenovirus was carried out under the supervision of Dr. Vivian Mautner at the Medical Research Council Virology Unit in Glasgow. In 1985 Dr. Kinloch joined the laboratory of Dr. Paul M. Wassarman at the Roche Institute of Molecular Biology in order to pursue postdoctoral research in mammalian development. His research has led to a better understanding of the structural organization of mammalian sperm receptor genes, the control of expression of these genes during development, and the regulation of biosynthesis of mammalian sperm receptor glycoproteins. Dr. Kinloch's research papers have appeared in such journals as *Proceedings of the National Academy of Sciences of the United States of America, Journal of Cell Biology, Molecular and Cellular Biology, Development,* and *Developmental Biology.*

PAUL M. WASSARMAN is a staff member in the Department of Cell and Developmental Biology at the Roche Institute of Molecular Biology, where he and his colleagues carry out research on various aspects of mammalian gametogenesis, fertilization, and preimplantation development. Dr. Wassarman received his B.S. and M.S. from the University of Massachusetts, Amherst, and his Ph.D. in biochemistry, under the supervision of Dr. Nathan O. Kaplan, from Brandeis University. He pursued postdoctoral research in structural studies as a Helen Hay Whitney Fellow, under the sponsorship of Sir John C. Kendrew, at the Medical Research Council, Laboratory of Molecular Biology, Cambridge, England. In 1972, after three years as a staff member in the Department of Biological Sciences at Purdue University, Dr. Wassarman joined the staff of the Department of Biological Chemistry at Harvard Medical School, where he commenced his research on mammalian development. Dr. Wassarman moved to the Roche Institute of Molecular Biology in 1985; there he headed the Department of Cell and Developmental Biology from 1986 through 1991. His laboratory continues to focus its research on genes that are expressed during mammalian gametogenesis and that encode proteins which participate directly in the fertilization process.

Genes in Mammalian Reproduction: 45–71
© 1993 Wiley-Liss, Inc.

Genes Involved in Cleavage, Compaction, and Blastocyst Formation

Gerald M. Kidder

I. INTRODUCTION

For many years preimplantation development remained *terra incognita* with respect to the involvement of specific genes and the genetic control of morphogenetic events. While the tools of biochemistry and, later, molecular biology were being applied to the early embryos of many nonmammalian species, the difficulty of obtaining sufficient numbers of preimplantation mammalian conceptuses for analysis, even from the laboratory mouse, proved to be a real obstacle. While working with preimplantation stages of development is still not easy, investigators are now bringing the power of molecular genetic approaches to bear on some long-standing problems. Some very new techniques, most notably mRNA detection by means of the polymerase chain reaction, have increased analytical sensitivity to the point where experiments can be performed with tens instead of thousands of conceptuses.

The purpose of this chapter is to review both old and new information concerning the genetic program underlying preimplantation development, with particular attention to specific genes that are expressed during this period and whose functions in cleavage, compaction, and blastocyst formation are beginning to be understood. The emphasis is on known genes whose transcripts have been detected in preimplantation development and that are therefore zygotically expressed. Little attention will be devoted to the many gene products that have been identified in preimplantation development but have not yet been shown to result from transcription of the embryonic genome. Likewise, this review will not consider the many uncharacterized genes that, when mutated, deleted, or rearranged, manifest themselves as embryonic lethals having effects before implantation. For information on this latter topic the reader is referred to reviews by Magnuson [1986] and Magnuson and Epstein [1987].

II. CURRENT STATUS OF THE FIELD

A. The Genetic Program for Preimplantation Development

Despite their unique features, mammalian conceptuses employ the same basic strategy for the genetic programming of events after fertilization that has been demonstrated in a variety of nonmammalian species: the stockpiling, in the growing oocyte, of gene transcripts and their translation products that will be utilized later for the initiation of embryogenesis [reviewed by Schultz, 1986]. In all embryos, this set of genetic instructions is eventually superceded by the products of embryonic transcription. In mammals, the transition from oogenetic to embryonic control of development starts with ovulation, as oogenetic mRNAs begin to decline, and is essentially complete after one to three cleavage divisions, by which time embryonic transcription has begun [reviewed by Telford et al., 1990]. An important consequence of this is that all phases of morphogenesis leading to the blastocyst are dependent on expression of embryonic genes, as revealed by the sensitivity of preimplantation development to agents that disrupt transcription or protein synthesis [Kidder and McLachlin, 1985; Levy et al., 1986; Braude et al., 1988]. This is not to say, however, that oogenetic products have no role to play after the embryonic genome is activated. In the mouse there is evidence for the persistence of some oogenetic mRNAs and proteins through the oogenetic–embryonic transition, and it is possible that these products persist and are functional throughout preimplantation development [Taylor and Pikó, 1987; Barron et al., 1989; Brenner et al., 1989; West and Flockhart, 1989].

Most of the currently available information on the genetic program for preimplantation development comes from the mouse model, although enough data on the oogenetic-to-embryonic transition have been collected from other mammals to begin to give a broader picture. These findings have been reviewed recently [Telford et al., 1990] and will only be summarized here. It is clear that the timing of this transition is a species characteristic. It occurs in the two-cell stage in mice and is characterized by a precipitous decline in oogenetic mRNA content, activation of transcription of the embryonic genome, a corresponding shift in polypeptide synthesis pattern, and the onset of sensitivity to the transcriptional inhibitor α-amanitin. All but the first of these changes have been noted in human conceptuses between the four- and eight-cell stages. Similar but more limited data suggest that the transition occurs in the two-cell stage in hamsters [Seshagiri et al.,

1990], in the four-cell stage in pigs [Jarrell et al., 1991], and in the 8–16-cell stage of sheep and cows [Telford et al., 1990]. It may occur as early as the two-cell stage in goats, although this suggestion was based entirely on nucleolar morphology [Chartrain et al., 1987]. In rabbits, the transition appears to be more gradual, occurring between the 2- and 16-cell stages [Telford et al., 1990]. Given the obvious relevance of this kind of information for embryo transfer programs, we can expect a more complete picture to emerge for most of these species in the near future.

The rapidly expanding fund of information on early mouse development has provided a framework for understanding how specific embryonic genes operate to bring about morphogenesis and differentiation leading to formation of the blastocyst. Before the advent of molecular probe technology, the timing of expression of unknown genes involved in cleavage and morphogenesis was examined using inhibitors of RNA or protein synthesis. Protein synthesis is continuous throughout preimplantation development: it occurs at a low level through fertilization and the first cleavage, then accelerates thereafter to achieve a rate of amino acid incorporation in blastocysts that is about seven times greater than that in zygotes [Abreu and Brinster, 1978]. In general, mouse conceptuses will complete no more than one additional cleavage division when RNA or protein synthesis is inhibited; hence gene expression underlying this process must be renewed in each cell cycle [Kidder and McLachlin, 1985; Levy et al., 1986]. Although both major morphogenetic transitions of compaction and cavitation can be blocked by such inhibitors, the timing of some of the underlying gene expression events is specific to each transition. For compaction, which in the mouse occurs during the eight-cell stage, the necessary transcriptional events are completed (or at least sufficient mRNAs have accumulated) by the mid four-cell stage, at least 10 hours before the onset of cell flattening (the first visible sign of compaction). The required protein synthesis reaches a state of sufficiency at various points during the four-cell stage, depending on which component of compaction (cell flattening, cytoplasmic or surface polarization, or the establishment of intercellular junctional communication) is examined [McLachlin et al., 1983; Kidder and McLachlin, 1985; McLachlin and Kidder, 1986; Levy et al., 1986]. For cavitation the coupling between transcription and morphogenesis is more immediate: mRNA required for the onset of fluid transport does not reach a level of sufficiency until about 5 hours preceding this event, with protein synthesis being virtually concomitant with it. During cavitation, an increased rate of protein synthesis is maintained in part by increases in both the overall stability and the extent of utilization of newly synthesized mRNA [Kidder and Pedersen, 1982; Kidder and Conlon, 1985]. Once cavitation has been completed, development becomes less tightly coupled with transcription such that escape from the zona pellucida can occur in blastocysts cultured in the presence of α-amanitin for at least 14 hours [Kidder and McLachlin, 1985]. These results showing that transcription and morphogenesis are not always closely linked imply that gene expression during preimplantation development can be regulated at posttranscriptional levels, a suggestion that can only be evaluated by examining the behavior of specific mRNAs.

This general view of the programming of preimplantation mouse development is now becoming understood in molecular terms as the accumulation profiles of individual mRNAs are examined, as reviewed in the following sections. One particular study is worth mentioning here, as it examined 37 unidentified, rare to moderately abundant transcripts that were represented in a random cDNA library made from late two-cell stage RNA [Taylor and Pikó, 1987]. Many of the two-cell transcripts were not detected in ovulated oocytes and therefore were products of the embryonic genome, confirming that a major influx of new genetic information occurs in the cytoplasm by the late two-cell stage. Those transcripts that were oogenetic declined in abundance by about one half, on average, between the one- and two-cell stages; most of them were subsequently replenished as the result of embryonic transcription (Fig. 1). Nearly all of the transcripts accumulated continuously from the two-cell stage onward, achieving an average 15-fold increase in number of copies per embryo by the blastocyst stage. This observation implies that most genes, once activated in the two-cell stage, continue to be transcribed throughout preimplantation development, a conclusion substantiated by the accumulation profiles of a number of known mRNAs (see below). Continuous accumulation of mRNAs from the two-cell stage leaves unexplained the relatively tight coupling between transcription and cavitation that was revealed by the α-amanitin experiments. However, transcripts appearing for the first time after the two-cell stage would not have been detected in the study by Taylor and Pikó [1987]. The overall impression from analyses of polypeptide synthesis patterns during preimplantation mouse development is that the diversity of the cytoplasmic mRNA population does not change appreciably between the four-cell and blastocyst stages, but several new polypeptides undergo an amanitin-sensitive increase in synthesis rate during the morula-to-blastocyst transition [Levinson et al., 1978; Braude, 1979]. Qualitative constancy of the polypeptide synthesis pattern between the four-cell and blastocyst stages has also been noted in the hamster [Seshagiri et al., 1990].

The recent availability of human zygotes as "spares" from in vitro fertilization programs has made it possible to collect some information on the genetic program for preimplantation development in our own

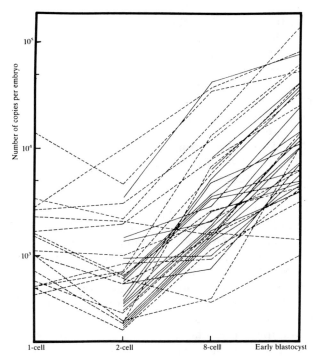

Fig. 1. *Semilogarithmic plot showing changes in the amounts per embryo of individual cloned, unidentified transcripts during mouse preimplantation development. The clones were obtained from a late two-cell cDNA library. Broken lines represent transcripts detected in unfertilized eggs (one-cell) as well as subsequent stages; solid lines represent transcripts not detected in eggs, but present in subsequent stages. The data illustrate the overall decline in transcript abundance between the one- and two-cell stages reflecting destruction of most (but not all) oogenetic mRNAs. Embryonic transcription then results in reaccumulation of the same mRNAs, most at an exponential rate. (Reproduced from Taylor and Pikó, 1987, with permission of the Company of Biologists, Ltd.)*

species, and preliminary indications are that there are important differences between the human and mouse. Autoradiography has been used to show that the first detectable extranucleolar RNA synthesis in the human occurs in the four-cell stage, with nucleolar labeling, indicative of transcription of rRNA genes, beginning after the next cleavage division [Tesařík et al., 1986a, b, 1987]. Subsequent to genomic activation, the incorporation of uridine into nucleoplasmic RNA (presumably mRNA precursors) increases, and this is accompanied by changes in the fractional volumes of membranous cytoplasmic organelles such as the Golgi complex, endoplasmic reticulum, and lysosomes [Tesařík et al., 1988]. However, these changes do not occur synchronously: Some blastomeres apparently fail to undergo the transition to a higher rate of nucleoplasmic RNA synthesis in the eight-cell stage, retaining an autoradiographic labeling pattern typical of four-cell stage nuclei, and the same blastomeres also fail to exhibit the cytoplasmic changes. The implication is that the postactivation increase in transcription in the human occurs asynchronously among the blastomeres and that the observed changes in cytoplasmic organelles depend on it. This could be explained if genes governing membrane biogenesis were among the first to be transcribed in human development. Other morphological changes, on the other hand, seem not to depend on increased transcription, since they occur even when some of the blastomeres involved retain the four-cell nuclear labeling pattern [Tesařík et al., 1986a; Tesařík, 1989]. These include cleavage to the 32-cell stage, regionalization of plasma membranes into microvillous and non-microvillous surfaces, compaction (involving cell flattening and the assembly of both adherens and putative gap junctions), and the segregation of distinct populations of inner and outer cells. Two of the latest events of human preimplantation development, formation of the trophectodermal tight junctions and development of the blastocele, do appear to depend on increased transcription: Tight junctions were only seen between blastomeres when both of them had undergone an increase in nuclear RNA labeling, and cavitation occurred only when the majority of blastomeres had done so [Tesařík, 1989]. It is tempting to conclude from these observations that the influence of oogenetic mRNA in the human extends well beyond the first few cleavages, rendering certain morphogenetic events independent of the increase in transcription that occurs in most blastomeres after the four-cell stage. However, it is also possible that the pattern of gene expression established at the time of genomic activation in the four-cell stage is necessary as well as qualitatively and quantitatively sufficient to support most structural changes occurring up to the morula stage, even when no further increase in RNA synthesis occurs in some

blastomeres. This issue could be clarified by studying the developmental profiles of individual mRNAs, as has been done in the mouse.

B. Genes Involved in Nuclear Functions

Construction of the zygotic nucleus, DNA replication, chromatin assembly, activation of embryonic transcription, and processing of primary transcripts are all functions that must be fully operational within the first one to three cleavage cycles of mammalian development. Recent research on the mouse has begun to supply some of the details concerning the development of these functions.

1. Transcription factors. It is now clear that there is a change in nuclear structure associated with formation of the diploid nuclei after the first cleavage. This is demonstrated by the finding that enhancers, DNA sequence elements that facilitate transcription of nearby genes by providing binding sites for transcriptional regulatory proteins, are required for transcription after the first cleavage but not before [Martinez-Salas et al., 1988, 1989]. Apparently, some component of nuclear structure arising during formation of the diploid nuclei acts to suppress transcription, an effect that must be alleviated through enhancer interaction with specific binding factors, known as transcription factors [Martinez-Salas et al., 1989]. Several such transcription factors and the sequence elements they recognize have been identified [Schöler et al., 1989a,b; Dooley et al., 1989]. One of these, termed *oct-3* by some investigators and *oct-4* by others (it belongs to a family of transcription factors that bind to a specific octamer of bases in DNA), is expressed only in the germ line and in totipotent or pluripotent stem cells, including the preimplantation conceptus. Transcripts of this gene have been detected in ovarian oocytes, fertilized eggs, morulae, and blastocysts of the mouse by in situ hybridization; in late blastocysts the transcripts become restricted to the inner cell mass [Rosner et al., 1990; Schöler et al., 1990]. The gene maps to mouse chromosome 17, to a region that also includes some of the *t*-complex mutants that affect preimplantation or early postimplantation development [Schöler et al., 1990]. It is possible that this transcription factor is itself one of the first products of embryonic gene expression, although the exact time of onset of its transcription has not yet been reported.

Since oct-3 is expressed in oocytes as well as in preimplantation stages, this transcription factor is a good candidate for an oogenetic gene product required for postfertilization development. Mutations in the maternal gene encoding oct-3 would be expected to impair transcription beginning in the two-cell stage of mouse development, thus causing cleavage arrest. This prediction has been tested by microinjection of antisense oligodeoxynucleotides complementary to oct-3 mRNA into mouse zygotes [Rosner et al., 1991]. The antisense oligos caused the destruction of oct-3 mRNA (presumably because the mRNA-oligo hybrids fell victim to endogenous RNase H), but the outcome was not what was expected: the injected zygotes failed to undergo even the first cleavage, although later injection led to arrest at the two-cell stage. Sense oligos did not cause cleavage arrest, and the effect of the antisense oligos could be partially alleviated by coinjection of synthetic oct-3 mRNA, demonstrating the specificity of the effect. In addition to its effect on oct-3 mRNA, antisense oligo injection also caused a severe reduction in DNA synthesis in zygotes. The most likely explanation for these results is that oct-3 is required not only for transcription beginning in the two-cell stage, but also for initiating DNA replication in the first and possibly subsequent cell cycles, as is known to be the case with viral systems [discussed by Rosner et al., 1991].

In addition to maternal factors, it is likely that some nuclear functions associated with the early cleavages in the mouse are encoded by genes activated in the two-cell stage. One such group of proteins may be the set of three (M_r 73,000, 70,000, and 68,000) that were identified as being among the most prominent products of genomic activation[Conover et al., 1991]. Because of the sensitivity of their synthesis to α-amanitin, they have been called the *transcription-requiring complex* (TRC). These proteins are related to each other and may represent products of an as yet unidentified gene family. Their synthesis is limited to the two-cell stage, they are partially associated with nuclei, and they have solubility properties similar to nuclear lamins (although they do not cross-react with lamin antibodies). It was suggested that the TRC may be products of a family of regulatory genes involved in the burst of transcription that occurs in the late two-cell stage.

2. Small nuclear RNAs. Another early postfertilization nuclear function that relies on oogenetic gene products is the processing (splicing) of primary transcripts. Since fully functional mRNAs resulting from embryonic transcription enter the cytoplasm beginning in the two-cell stage in the mouse [reviewed by Schultz, 1986], the nuclear RNA processing machinery must be operational from the moment embryonic transcription is activated. RNA processing is carried out by small nuclear ribonucleoproteins (snRNPs) that consist of a specialized class of small nuclear RNAs (snRNAs) and their associated proteins [reviewed by Maniatis and Reed, 1987]. Cloned DNA probes for the snRNAs and antibodies specific for their associated proteins have been used to determine the levels and localization of the various snRNPs during fertilization and preimplantation development [Lobo et al., 1988; Dean et al., 1989; Prather et al., 1990]. According to in situ hybridization analysis, the snRNAs are concentrated in the germinal vesicle of the primary oocyte. They disperse through the cytoplasm upon germinal vesicle breakdown, to become reincorporated in the pronuclei of the zygote and later the diploid nuclei after the first cleavage (Fig. 2). Immunocytochemical analyses have confirmed the same shifts in localization for the intact snRNPs. Quantitatively, the levels of snRNAs remain fairly constant during ovulation, fertilization, and the first cleavage, while the levels of most mRNAs are declining (these measurements have been made by ribonuclease protection assays as well as Northern blotting). After genomic activation the snRNAs increase continuously on a per-embryo basis, but by the second or third cleavage they reach a plateau on a per-cell basis. Thus while oogenetic snRNAs supply the processing machinery for the first transcripts of the embryonic genome, they are soon augmented (replaced?) by embryonic transcription such that a constant complement is maintained in the nuclei as cleavage progresses.

3. Histones. These proteins constitute another component of nuclei that must accumulate continuously throughout cleavage [for a review of the contribution of histones to chromatin structure, see Svaren and Chalkley, 1990]. The synthesis of all four types of nucleosomal core histone (H2A, H2B, H3, and H4) as well as H1 has been detected during preimplantation development of the mouse from fertilization onward, and measurement of the absolute rates of synthesis of three of them suggested a balance between histone and DNA synthesis in each cell cycle [Kaye and Wales, 1981]. Interestingly, this relationship does not appear to be due to direct coupling, at least in the early cleavages, since DNA synthesis can be inhibited without affecting histone synthesis [Kaye and Church, 1983]. The histone genes were among the first to be studied in preimplantation development using recombinant probes [Giebelhaus et al., 1983]: Not

surprisingly, given the need to double the histone content with each doubling of DNA, histone mRNAs reach high levels during cleavage and are therefore relatively easy to quantify. The mRNA levels for histones H3, H2A, and H2B have been measured by dot-blotting [Graves et al., 1985]. The three types of mRNA are abundant and roughly equal in number in ovulated, unfertilized oocytes, where mRNA for core histones has been estimated to make up about 3% of the total mRNA pool. These mRNAs decline by 80%–90% after fertilization, reaching a low point in the late two-cell stage. They subsequently increase again (see Fig. 3), maintaining a constant number of mRNAs per cell between the four-cell and blastocyst stages and reaching a total core histone mRNA content equivalent to about 1.6% of the total pool. The implication of this finding is that the balance between histone synthesis and DNA content is controlled by regulation of the cytoplasmic mRNA content, perhaps by maintaining a constant rate of histone gene transcription per nucleus. When ribonuclease protection assays were performed to distinguish between mRNAs encoding various subtypes (variants) of these three histones, it was found that their proportions relative to total histone mRNA in blastocysts are not the same as in ovulated oocytes [Graves et al., 1985]. Apparently, then, the pattern of expression of histone subtype genes shifts during preimplantation development, with the result that blastocyst chromatin differs in histone composition from that of the oocyte. It remains to be determined when this shift occurs and what its developmental significance might be.

C. Genes Encoding Cytoskeletal Elements

Genes encoding the subunits of cytoskeletal elements are of interest because of the involvement of these elements in cytoarchitectural changes associated with fertilization, compaction, and cavita-

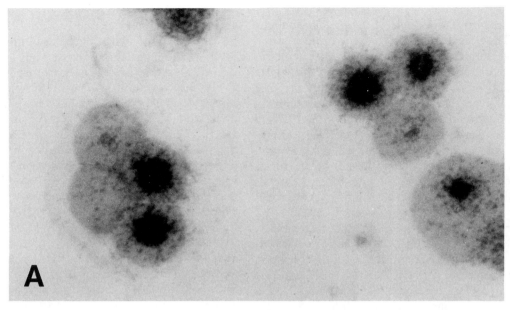

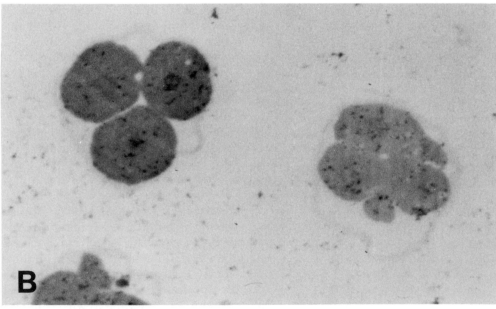

Fig. 2. *Nuclear localization of U1b gene transcripts in four- to eight-cell stage mouse embryos visualized by in situ hybridization (unpublished results of Andrew J. Watson). U1b RNA is representative of the small nuclear RNAs that participate in the splicing of pre-mRNAs resulting from embryonic transcription. Embryos were embedded in polyethylene glycol [Watson and Kidder, 1988], sectioned at 1 μm thickness, and the sections hybridized with ^{35}S-labeled riboprobes transcribed from U1b DNA cloned in the antisense (part **A**) or sense (part **B**) orientation. Hybridization was carried out overnight at 50°C in 4× SSPE, 50% formamide, 20 mM dithiothreitol, 1× Denhardt's solution, with 40 μg/ml tRNA. The sections were washed with 2× SSC at 50°C for 30 minutes before treatment with RNase A (20μg/ml) for 30 minutes at 37°C followed by several more rinses at room temperature. After development of the autoradiograms the sections were lightly stained with Geimsa. The lack of hybridization with the sense riboprobe (B) illustrates the specificity of the detection of U1b RNA.*

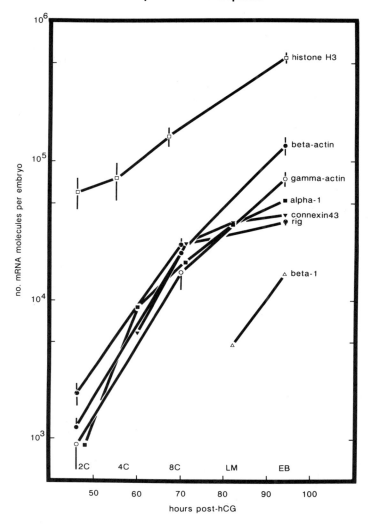

Fig. 3. *Semilogarithmic plot showing changes in the amounts per embryo of several known mRNAs during mouse preimplantation development. Data for histone H3 mRNA [Graves et al., 1985], β- and γ-actin mRNAs [Taylor and Pikó, 1990], and rig mRNA (which encodes a highly conserved nuclear protein of unknown function; [Taylor and Pikó, 1991] were obtained by quantitative dot or slot blotting. Data for the mRNAs encoding α1- and β1-subunits of Na^+,K^+-ATPase [Watson et al., 1990b] and connexin43 [Valdimarsson et al., 1991] were obtained from Northern blots such as those shown in Figure 4, in which each hybridization signal, as measured by densitometric scanning of an appropriately exposed autoradiogram, was expressed as a fraction of the histone signal in the same lane. The fractions were converted to transcript numbers by reference to the histone H3 data of Graves et al. [1985], taking into account differences in probe length and specific activity.*

tion [reviewed by Fleming and Johnson, 1988]. Some of these proteins, particularly actins and tubulins (the subunits of microfilaments and microtubules, respectively), are also among the most abundant proteins in the conceptus, facilitating analyses at the molecular level. It has been estimated, for example, that actins account for almost 6% of total protein synthesis in mouse blastocysts, while tubulins account for about 2% [Abreu and Brinster, 1978]. The synthesis of both of these can already be detected at low levels in unfertilized oocytes and zygotes, with the rate of amino acid incorporation into them increasing continuously from the two-cell stage to the blastocyst: the extent of increase is about 77-fold for actins and 12-fold for tubulins [Abreu and Brinster, 1978]. The actins are prominent, newly synthesized proteins in rabbit conceptuses as well, constituting about 2% of total protein synthesis in blastocysts [Overström et al., 1989]. Other cytoskeletal proteins whose synthesis has been detected (but not quantified) in preimplantation development include cytokeratins, one of the classes of protein that forms intermediate filaments [reviewed by Chisholm and Houliston, 1987], and α- and β-spectrin, constituents of the plasma membrane–cytoskeletal complex [Sobel and Goldstein, 1988].

1. Actins. In terms of understanding the program of expression of cytoskeletal genes, only the actins have been studied extensively. These constitute a multigene family, with six isoforms having been characterized in mammals: two cytoplasmic isoforms (β and γ) that are present in most if not all cell types, two sarcomeric (α) isoforms unique to skeletal or cardiac muscle, and two smooth muscle isoforms [summarized by Taylor and Pikó, 1990]. The mRNAs encoding these isoforms share extensive sequence identity in their coding regions but can be distinguished by probes specific for their 3' untranslated regions. Although the synthesis of α, β, and

γ isoforms had been reported in preimplantation stages of the mouse [Abreu and Brinster, 1978], only β and γ mRNAs could be detected by quantitative hybridization with isoform-specific riboprobes [Taylor and Pikó, 1990]. Furthermore, the total actin mRNA content, as indicated by hybridization with a coding sequence probe, approximated the sum of the β and γ mRNAs. These two mRNAs accumulate during oocyte growth and then decline in parallel after ovulation, reaching their lowest point (about 17% of the amount in ovulated oocytes) in the late two-cell stage [Bachvarova et al., 1989; Taylor and Pikó, 1990]. The two then reaccumulate, with the rate of increase in β-actin mRNA slightly exceeding that of γ-actin mRNA [Taylor and Pikó, 1990] (data plotted in Fig. 3). By the early blastocyst stage, the total actin mRNA content of the conceptus has increased 60-fold over the amount in the two-cell stage, and these mRNAs make up an estimated 0.7% of poly(A)-containing mRNA. Thus the kinetics of actin mRNA accumulation reflect the steadily increasing rate of actin synthesis, suggesting that the production of actins in preimplantation development is primarily a function of mRNA content. Unlike the histone mRNAs, whose steady-state levels increase in proportion to cell number during cleavage [Graves et al., 1985], the actin mRNAs increase in excess of cell number. This presumably reflects the importance of the actin microfilament system, and the need for accelerated actin synthesis, in the morphogenetic events leading to blastocyst formation. As expected, in situ hybridization revealed that cytoskeletal actin mRNAs accumulate throughout the conceptus [Nisson et al., 1989].

2. Tubulins. There is much less information on the accumulation of mRNAs encoding other cytoskeletal proteins in the mouse. Both α- and β-tubulins are synthesized in ovulated oocytes and through preimplantation development [Abreu and

Brinster, 1978], but only α mRNA has been analyzed by hybridization with a cDNA. By the late two-cell stage this mRNA has declined to 14% of its level in the unfertilized oocytes; it then increases through cleavage to reach a maximum in blastocysts [Paynton et al., 1988]. Based on the frequency of clones in a random-primed cDNA library, the abundance of β-tubulin mRNA in early blastocysts was estimated to be 0.3% of total mRNA [Weng et al., 1989], a value that is surprisingly low considering the proportion of protein synthesis (2%) reported to be contributed by the tubulins at that stage [Abreu and Brinster, 1978].

3. Cytokeratins. Two intermediate filament proteins, cytokeratins Endo A and Endo B, have been shown to be synthesized and assembled into filaments by the eight-cell stage [Oshima et al., 1983; Emerson, 1988], a finding that correlates with the detection of Endo A mRNA as early as the eight-cell stage by means of a nuclease protection assay [Duprey et al., 1985]. The amount of this mRNA then increases approximately 10-fold by the blastocyst stage. Endo B mRNA probably behaves similarly, since the two mRNAs are about equally abundant (0.2%–0.3%) in blastocysts, based on cDNA clone frequencies in a blastocyst library [Weng et al., 1989]. Cytokeratin filaments and the synthesis of cytokeratins are markers of trophectoderm differentiation [Chisholm and Houliston, 1987], and Endo A mRNA accumulates preferentially in trophectoderm [Duprey et al., 1985; Nisson et al., 1989]. Since this gene is first transcribed before the establishment of separate inside and outside cell populations, it follows that production of the message must somehow be downregulated in progenitors of the inner cell mass. Despite their association with trophectoderm differentiation, expression of cytokeratin genes does not appear to play an essential role in blastocyst formation since inhibi-

tion of cytokeratin assembly by antibody microinjection failed to perturb development [Emerson, 1988].

D. Genes Encoding Membrane Channels and Ion Transporters

Several plasma membrane functions are known to play important roles in specific morphogenetic events in preimplantation development, and in the past few years the genes encoding these functions have been identified and their patterns of expression elucidated. These include genes encoding connexins (the protein subunits of gap junction channels) and the subunits of Na^+,K^+-ATPase (the plasma membrane sodium pump). It is this category of genes that has provided the clearest picture thus far of the relation between specific gene expression and morphogenesis. Most of this research has been done with mice, although rabbits and pigs have also provided information on sodium pump expression.

1. Connexins. Gap junctions are aggregations of intramembranous particles (connexons) that can be found in regions of closely apposed plasma membranes in many tissues; they are morphological manifestations of intercellular communication. Each connexon is a hexameric assembly of protein subunits called *connexins,* arranged in a cylindrical array forming a hydrophilic channel. When connexons in apposed plasma membranes align, intercellular channels are created that allow the direct passage of small molecules (less than about 1,000 Da) between cells. Gap junctional intercellular communication has been implicated in metabolic coupling, cell growth control, and cell patterning [for reviews of the structure and function of gap junctions, see Guthrie and Gilula, 1989; Beyer et al., 1990; Bennett et al., 1991].

Gap junctions are assembled de novo during the eight-cell stage of mouse development, during compaction, providing channels for embryo-wide intercellular coupling that is maintained for the remainder of the preimplantation period [reviewed by Kidder, 1987]. Three independent lines of evidence have indicated that gap junctional communication between the blastomeres is essential for maintaining the compacted state and thus for subsequent blastocyst formation. Two of these have involved perturbing the formation or function of embryonic gap junctions by injecting antibodies or antisense RNA specific for connexins [Lee et al., 1987; Bevilacqua et al., 1989]. In both cases, the injected blastomeres uncoupled from their neighbors and underwent decompaction. The third line of evidence comes from an analysis of intercellular coupling in conceptuses derived from the lethal DDK × C3H interstrain cross [Buehr et al., 1987]. These suffer reduced gap junctional coupling and eventually decompact, but this can be prevented by treatment with a weak organic base to raise intracellular pH and restore a normal level of coupling. Thus gap junctions play a direct (although unspecified) role in compaction in the mouse, so that knowledge of the program of expression of the gene(s) encoding the embryonic connexins should provide insight into the genetic control of compaction itself.

The connexins constitute a family of related proteins, each of which has a particular tissue distribution in adults [Beyer et al., 1990; Bennett et al., 1991; Hoh et al., 1991]. The most common nomenclature uses the predicted molecular mass of each polypeptide to distinguish it: Connexin32 is 32 kD, connexin43 is 43 kD, and so on. Five rodent connexins (26, 31, 32, 43, and 46) have been cloned and their sequences published as of this writing, and at least five more have been identified (D.L. Paul, personal communication). The single copy genes encoding the connexins are scattered among several chromosomes in both mouse and human genomes [Hsieh et al., 1991; Willecke et al., 1990]. At least two of the known rodent connexins are present

in preimplantation mouse development. Connexin32 appears to be a persistent oogenetic product, because a polypeptide of the same size and immunoreactivity can be detected at roughly the same level in all preimplantation stages. Connexin32 mRNA, however, could not be detected in any stage of preimplantation development [Barron et al., 1989]. Transcripts of connexin26 were not detected either. Connexin43 has also been detected by immunoblotting in all preimplantation stages, but in this case it clearly increases after the second cleavage. The accumulation of connexin43 is driven by embryonic transcription, since its mRNA increases from an undetectable (by Northern blotting) level in the two-cell stage to a maximum in blastocysts (Fig. 4A). Overall, connexin43 transcripts accumulate at about the same rate as histone H3 transcripts from the four-cell stage to the late morula, although the latter are much more abundant (Fig. 3) [Valdimarsson et al., 1991]. Immunocytochemistry has revealed that connexin43, but not connexins32 or -26, is incorporated into gap junctions beginning with compaction, when intercellular coupling is established [Valdimarsson et al., 1991; Nishi et al., 1991; De Sousa and Kidder, in preparation]. It can be concluded, then, that the connexin43 gene supplies subunits for gap junction assembly beginning in the eight-cell stage, and this member of the connexin gene family (perhaps along with others) plays an important role in preimplantation morphogenesis. The function of the maternally inherited connexin32, if any, is yet to be discovered.

2. Na$^+$,K$^+$-ATPase. The plasma membrane sodium pump has long been assumed to be a principal mediator of transtrophectodermal fluid transport during cavitation [reviewed by Benos and Balaban, 1990]. In blastocysts of rabbits, pigs, and mice fluid transport has been shown to be sensitive to ouabain, a specific

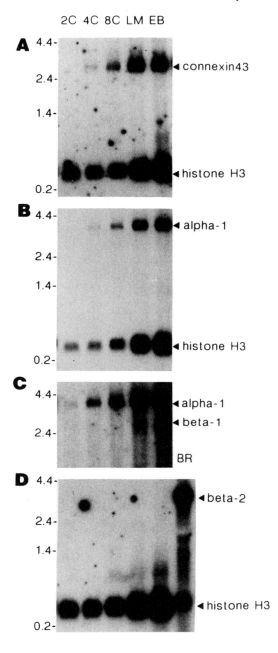

2C 4C 8C LM EB

A
4.4-
2.4-
1.4-
0.2-
◄ connexin43
◄ histone H3

B
4.4-
2.4-
1.4-
0.2-
◄ alpha-1
◄ histone H3

C
4.4-
2.4-
◄ alpha-1
◄ beta-1
BR

D
4.4-
2.4-
1.4-
0.2-
◄ beta-2
◄ histone H3

Fig. 4. *Autoradiograms obtained by double-probe hybridizations of Northern blots illustrating the accumulation of several mRNAs during mouse preimplantation development [for technical details see Watson et al., 1990b; Valdimarsson et al., 1991]. Autoradiograms such as these were used to estimate the numbers of transcripts encoding connexin43 (A) and the α1- and β1-subunits of Na+,K+-ATPase (B, C) shown in Figure 3. The number of embryos per lane (1,500–2,500) was kept constant for each blot. D documents the failure to detect transcripts encoding the Na+,K+-ATPase β2 isozyme in any preimplantation stage, although the expected 3.4 kb transcript [Martin-Vasallo et al., 1989] was detected in an RNA sample from adult brain (BR). The positions of RNA length markers are indicated along the left margins.*

sponding flow of water across the trophectoderm into the blastocele. Results showing that Na+,K+-ATPase is localized in membranes lining the blastocele cavity (i.e., in the basolateral plasma membranes of trophectoderm cells) are consistent with this view [Benos and Balaban, 1990]. According to immunocytochemical evidence, Na+,K+-ATPase in the mouse is concentrated in the basolateral plasma membranes of mural trophectoderm, including its extensions covering the inner cell mass [Watson and Kidder, 1988; Kidder and Watson, 1990; Watson et al., 1990a]. There is also evidence for a variety of apical Na+ transport systems in the mouse that provide routes of Na+ entry into trophectoderm cells [Manejwala et al., 1989]. One such system, related to an Na+ glucose cotransporter, was identified in preimplantation stages by immunoblotting and immunocytochemistry; it is not yet known if this polypeptide is a product of embryonic gene expression [Wiley et al., 1991].

Na+,K+-ATPase consists of two types of subunit, a catalytic (α) subunit and a noncatalytic, glycosylated (β) subunit, the function of which is still under investigation [reviewed by Geering, 1990]. Both

inhibitor of Na+,K+-ATPase, and to depend on a transtrophectodermal sodium flux [Benos and Biggers, 1981; Overström, 1987; Manejwala et al., 1989]. The currently favored hypothesis is that this sodium flux, driven by Na+,K+-ATPase, drives a corre-

subunits exist in multiple isozymic forms that are differentially distributed among adult tissues, reflecting the fact that both are encoded by families of related genes. In the mouse there are three α-subunit genes, producing the $\alpha1$, $\alpha2$, and $\alpha3$ isozymes; these reside on three different chromosomes [Kent et al., 1987]. The two β-subunit genes that have been cloned thus far (encoding the $\beta1$ and $\beta2$ isozymes) also reside on different chromosomes [Kent et al., 1987; Malo et al., 1990]. Northern blotting with isoform-specific cDNAs was used to show that only the $\alpha1$ and $\beta1$ genes are transcribed during cavitation in the mouse [Watson et al., 1990b] (Fig. 4D). The $\alpha1$ transcript was detectable from the two-cell stage onward, accumulating steadily at about the same rate as histone H3 transcripts (but less abundant than the latter; see Figs. 3, 4B). The $\beta1$ transcript, on the other hand, could not be detected on Northern blots until the late morula stage, after which its level increases rapidly (Figs. 3, 4C). The Na$^+$,K$^+$-ATPase $\beta1$ subunit gene thus provides an exception to the pattern of expression exhibited by most genes transcribed in preimplantation development in that the accumulation of its mRNA does not proceed steadily after genomic activation in the two-cell stage (more recent measurements in this laboratory, using the much more sensitive polymerase chain reaction technique described in the next section, have indicated that $\beta1$ transcripts are present, although in very low abundance, during early cleavage).

The subunits of Na$^+$,K$^+$-ATPase illustrate an important principle, namely, that knowledge of the accumulation profile of a transcript in early development is sometimes not sufficient to explain the timing of accumulation of the final gene product. Although $\alpha1$ mRNA accumulates steadily after the two-cell stage, reaching about 60%–70% of its blastocyst level by the completion of compaction [Watson et al.,

1990b], the α-subunit itself could not be detected by immunocytochemistry until the late morula stage, just hours before the onset of fluid transport [Watson and Kidder, 1988]. This suggests that $\alpha1$ mRNA is not translated efficiently prior to this stage or that newly synthesized α-subunits are unstable or incompletely processed so as to be unrecognizable by antibodies. Furthermore, neither the total α-subunit content, as measured by Western blotting, nor the total Na$^+$, K$^+$-ATPase activity, as indicated by ouabain-sensitive Rb$^+$ uptake, changes appreciably (on a per-embryo basis) throughout preimplantation development [Gardiner et al., 1990b; Van Winkle and Campione, 1991]. One explanation for this might be that much of the α-subunit content and enzymatic activity of the conceptus during this period is of oogenetic origin. The β-subunit content, however, does increase markedly during the blastocyst stage following the rapid increase in $\beta1$ mRNA content [Gardiner et al., 1990b; Watson et al., 1990b]. The β-subunit is known to be required for processing, maturation, and transport of the α-subunit [reviewed by McDonough et al., 1990] and may be acting during preimplantation development to trigger an influx of new (embryonically coded) Na$^+$,K$^+$-ATPase molecules during cavitation.

The situation may be less complicated in the rabbit. Recent studies have shown that both the α-subunit of Na$^+$,K$^+$-ATPase and its mRNA increase markedly during the period of maximal blastocele expansion, as does the rate of synthesis of the enzyme [Gardiner et al., 1990a; Overström et al., 1989].

E. Genes Encoding Cell Surface and Secreted Proteins

Proteins expressed externally on the surfaces of embryonic cells or secreted into extracellular spaces are good candidates for agents involved in interactions between blastomeres or signalling between the conceptus and the uterus. The expression of a number of genes encoding such proteins has been studied in the mouse, and some of them are discussed in this section. Many of the proteins in this category belong to several families of growth factors and their receptors, which are treated separately in the next section.

1. Alkaline phosphatase. The enzyme alkaline phosphatase is a prominent cell surface protein, one that has received recent attention. It is expressed on the plasma membranes in regions of close cellular apposition and is thus a marker of membrane regionalization [Izquierdo et al., 1980; Sepúlveda and Izquierdo, 1990]. Its activity increases continuously from the two-cell stage onward, eventually disappearing from trophectoderm and becoming restricted to the inner cell mass in the late blastocyst [Barron et al., 1989; Hahnel et al., 1990; Mulnard and Huygens, 1978]. Its function in preimplantation development is unknown, although there is good evidence from another developmental system, the axolotl embryo, that it is involved in cell–cell interactions that guide cell migration [Zackson and Steinberg, 1989]. Mammalian alkaline phosphatase comprises a group of isozymes, and cDNAs representing four of them have been cloned [summarized by Hahnel et al., 1990]. Using sequence information from these cDNAs, Hahnel et al. [1990] constructed oligonucleotide primers for use in the polymerase chain reaction (PCR). This procedure uses a thermostable DNA polymerase and sequence-specific primers to amplify a defined segment of DNA many thousands of times by repeated cycles of denaturation, primer annealing, and primer extension; coupled with a prior reverse transcription step, it can be used to amplify a particular segment of mRNA, allowing the "mRNA phenotype" of a single cell to be established [Rappolee et al., 1989]. Using this approach, transcripts of two different alkaline

phosphatase genes were detected and found to increase from the two-cell stage to the blastocyst, with one maintaining a 5–10-fold excess over the other (Fig. 5). The correspondence between the continuously increasing amounts of mRNA and of enzyme activity from the two-cell stage implies that the synthesis of these isozymes begins with genomic activation and is governed by the rates of accumulation of their respective mRNAs.

2. Extracellular matrix components. Proteinaceous components of extracellular matrix constitute another class of gene products that are involved in mediating cell interactions. Fibronectin is one such protein that has been studied in preimplantation development. Fibronectin, a glycoprotein dimer, acts as a ligand between cells and other extracellular matrix components such as heparin or collagen, and it has been demonstrated to be involved in cell adhesion and migration during amphibian and chick development [reviewed by Hynes, 1985; Thiery et al., 1985]. When examined by indirect immunofluorescence, fibronectin was detected for the first time in mouse embryogenesis between the inner cell mass (ICM) cells of the early blastocyst, and its level increased in late blastocysts. It could not be detected in the trophectoderm layer [Zetter and Martin, 1978]. Thus fibronectin's role would appear to be restricted to mediating cell interactions within the primitive ectoderm lineage or between this group of cells and the primitive endoderm layer. Given these findings, one would predict that fibronectin mRNA is restricted to the ICM; surprisingly, this is not the case. Nisson et al. [1989] used in situ hybridization to show that fibronectin transcripts are present in all blastomeres at the eight-cell stage and in both ICM and trophectoderm cells of blastocysts. The hybridization signal was greatest over the ICM. The presence of fibronectin transcripts in trophectoderm cells may indicate

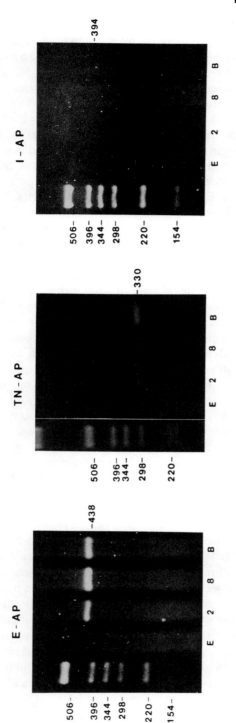

that the restriction of fibronectin expression to the ICM is brought about primarily by posttranscriptional regulation, although it is possible that a low level of the protein in trophectoderm had gone undetected in the immunofluorescence analysis. By the same token, it is possible that fibronectin synthesis begins earlier than was previously thought.

Fibronectin is just one of several extracellular matrix components identified in preimplantation stages of the mouse; the list includes laminin, nidogen, thrombospondin, collagen, and heparan sulfate proteoglycan [reviewed by O'Shea et al., 1990]. Laminin is an adhesive glycoprotein typically found in basement membranes [reviewed by Mercurio, 1990]. Its three different types of subunit (A, B1, and B2) bind to cells as well as other extracellular matrix components, hence the function of laminin in mediating cell interactions is similar to that of fibronectin. Laminin is present on the surface of blastomeres at least as early as the two-cell stage and continues to be expressed throughout preimplantation development [Cooper and MacQueen, 1983; Dziadek and Timpl, 1985]. By the blastocyst stage, this expression is driven by embryonic transcription

Fig. 5. *Detection of transcripts encoding alkaline phosphatase isozymes in mouse preimplantation development by means of RT/PCR. Total RNA from 21 unfertilized eggs (E), 31 two-cell embryos (2), 17 eight-cell embryos (8), and 14 early blastocysts (B) was reverse transcribed and specific segments of the resulting cDNA amplified using primer pairs specific for the embryonic (E-AP), tissue nonspecific (TN-AP), or intestinal (I-AP) mRNA isoforms. The amplification products were separated on agarose gels along with DNA size markers, stained with ethidium bromide, and photographed on a UV transilluminator. Amplified segments of the expected sizes for E-AP (438 bp) and TN-AP (330 bp) mRNAs were obtained, but the I-AP gene appears not to be transcribed during this period. (Reproduced from Hahnel et al., 1990, with permission of the Company of Biologists, Ltd.)*

since the transcript encoding the B1 sub-unit has been detected by quantitative slot-blotting [Kang et al., 1990]. The level of B1 transcripts was found to increase during blastocyst expansion in parallel with blastocele volume.

3. Secreted proteinases. In contrast to their role in mediating cell interactions, extracellular matrix proteins in some situations pose barriers to cell migration or morphogenesis. Thus they sometimes serve as substrates for extracellular proteinases. A family of metalloproteinases has been implicated as key agents in extracellular matrix degradation [discussed by Brenner et al., 1989], and the timing of expression of the genes encoding two of these has been investigated in preimplantation mouse development. Polymerase chain reaction amplification of reverse-transcribed RNA preparations revealed that transcripts of the stromelysin and collagenase genes are present in unfertilized eggs and in all pre-implantation stages, increasing quantitatively through time. Transcripts encoding the natural metalloproteinase inhibitor TIMP are also present in the same stages. Interestingly, these transcripts do not appear to suffer the precipitous decline that most oogenetic mRNAs undergo between ovulation and the late two-cell stage. Secretion of the enzymes themselves, analyzed using zymography, occurred in unfertilized eggs and through the first cleavage but then declined, later to increase again in the blastocyst stage and during trophoblast outgrowth [Brenner et al., 1989]. This is undoubtedly an example of a group of genes expressed in association with cavitation but whose functions are primarily related to the initial stages of implantation, as the trophoblast invades the uterine lining.

F. Genes Encoding Growth Factors and Receptors

Polypeptide growth factors have attracted considerable attention recently as mediators of cell interactions in a wide range of animal embryos, including mammalian preimplantation conceptuses. Analyses of the structure of growth factors and their genes have made it clear that they can be assigned to several superfamilies of related polypeptides [reviewed by Mercola and Stiles, 1988]. It had been suspected that growth factors play important roles in preimplantation development because of the stimulatory effects of serum added to embryo culture media, and the finding that similar enhancement of development can be obtained without serum if conceptuses are cultured at high embryo-to-volume ratios [Wiley et al., 1986; Rizzino, 1987; Paria and Dey, 1990]. Now, with the advent of reverse transcription/polymerase chain reaction (RT/PCR) technology, it has become possible to assay for the transcripts of genes encoding both specific growth factors and their receptors in preimplantation development.

Information concerning the expression of growth factor and cognate receptor genes has been useful in distinguishing between the probable autocrine vs. paracrine roles of such agents: In the former case, the conceptus produces both the growth factor and its receptor, whereas in the latter the conceptus produces either one but not the other. In the mouse, several growth factors, including transforming growth factor α (TGF-α), transforming growth factor $\beta1$ (TGF-$\beta1$), and insulin-like growth factor II (IGF-II), are likely to play autocrine roles, since the conceptus transcribes all three genes and responds to the addition of these growth factors to the culture medium (as evidenced by enhancement of amino acid incorporation, success of blastocyst development, or increase in cell number [Rappolee et al., 1988, 1990; Werb, 1990; Paria and Dey, 1990]). In the case of IGF-II, mRNA encoding the type 2 receptor was reported to be present from the two-cell stage onward [Rappolee et al., 1990; Werb, 1990], and the receptor itself was detected on two-cell and later stages by immunocytochemistry [Harvey and Kaye, 1991]. The situation may be different in the human, however, since in situ hybridization failed to detect IGF-II transcripts in blastocysts [Brice et al., 1989; Ohlsson et al., 1989]. Regarding paracrine interactions, the assumption has been that they involve signalling between the conceptus and the uterus. In the mouse, insulin, insulin-like growth factor-I (IGF-I), and epidermal growth factor (EGF) probably fall into this category. No transcripts of any of these genes have been detected by RT/PCR [Rappolee et al., 1988; Werb, 1990], yet there is evidence that each of the three receptors is expressed from the eight-cell stage onward: Transcripts encoding receptors for insulin and IGF-I have been detected by RT/PCR [Schultz et al., 1990; Rappolee et al., 1988; Werb, 1990], specific binding of labeled insulin, IGF-I, and EGF to postcompaction conceptuses has been demonstrated [Mattson et al., 1988; Paria and Dey, 1990], and insulin and EGF have been shown to stimulate metabolism, enhance blastocyst development, and/or cause an increase in cell number [Rao et al., 1988; Harvey and Kaye, 1988, 1990; Heyner et al., 1989; Wood and Kaye, 1989; Paria and Dey, 1990; Gardner and Kaye, 1991]. The source of these growth factors is most likely the maternal reproductive tract [Rao et al., 1988; Huet-Hudson et al., 1988; Heyner et al., 1989].

Although it has been difficult to quantify accurately the amounts of specific transcripts in different stages using RT/PCR, it is possible to discern differences in the accumulation profiles of the various growth factor and receptor mRNAs that have been detected in preimplantation development of the mouse [summarized by Rappolee et al., 1988; Werb, 1990] (see Table I). Some, such as TGF-α and platelet-derived growth

TABLE I. Temporal Patterns of Expression of Known Genes in Preimplantation Development of the Mouse

Category	Gene	Stage by which accumulation of mRNA begins	Method of detection	Pattern of accumulation	References
Nuclear functions	Histone H3	4-Cell	Dot blot	Continuous (constant/cell)	Graves et al. [1985]
	rig Protein	8-Cell*	Slot blot	Continuous (declining /cell)	Taylor and Pikó [1991]
Cytoskeletal proteins	β,γ-Actin	4–8-Cell	Northern blot, slot blot	Continuous (increasing/cell)	Paynton et al. [1988] Taylor and Pikó [1990]
	α-Tubulin	4–8-Cell	Northern blot	Continuous	Paynton et al. [1988]
Membrane channels, transporters	Connexin43	4-Cell	Northern blot	Continuous (declining /cell)	Valdimarsson et al. [1991]
	Na⁺,K⁺-ATPase α1 subunit;	4-cell	Northern blot	Continuous	Watson et al. [1990b]
	Na⁺,K⁺-ATPase β1 subunit	Late morula	Northern blot	Continuous	Watson et al. [1990b]

(Continued)

TABLE I. Temporal Patterns of Expression of Known Genes in Preimplantation Development of the Mouse (Continued)

Category	Gene	Stage by which accumulation of mRNA begins	Method of detection	Pattern of accumulation	References
Surface, secreted proteins	Alkaline phospha- tases	2-Cell	RT/PCR	Continuous	Hahnel et al. [1990]
	Stromelysin	Blastocyst**	RT/PCR	Unable to assess**	Brenner et al. [1989]
	Collagenase	Blastocyst**	RT/PCR	Unable to assess**	Brenner et al. [1989]
	'TIMP	Blastocyst**	RT/PCR	Unable to assess**	Brenner et al. [1989]
Growth factors, receptors	TGF-α	4-Cell	RT/PCR	Continuous	Werb [1990]
	TGF-β1	2-Cell	RT/PCR	Continuous	Werb [1990]
	PDGF-A	4-Cell	RT/PCR	Continuous	Werb [1990]
	kFGF	4-Cell[†]	RT/PCR	Continuous	Werb [1990]
	IGF-II	2-Cell	RT/PCR	Continuous	Werb [1990]
	IGF-II receptor	2-Cell	RT/PCR	Continuous	Werb [1990]
	Insulin receptor	8-Cell	RT/PCR	Continuous	Werb [1990]
	IGF-I receptor	8-Cell	RT/PCR	Continuous	Werb [1990]
House- keeping functions	β-Glucur- onidase	8-Cell	Dot blot	Continuous	Bevilacqua et al. [1988]
	HPRT	4–8-Cell	Northern blot	Continuous	Paynton et al. [1988]
	MT-I, -II	2-Cell	RT/PCR	Continuous	Andrews et al. [1991]
Retrovirus	IAP-I, -II	8-Cell*	Slot blot, RT/PCR	Continuous	Poznanski and Calar- co [1991]

*Transcript presence was not examined in the four-cell stage.
**The temporal pattern of embryonic gene expression could not be assessed because of apparent persistence of oogenetic mRNA.
[†]Transcript presence was not examined in the two-cell stage.

factor A chain (PDGF-A) transcripts, are present in the unfertilized egg and then are lost by the two-cell stage, to reaccumulate from the four-cell stage onward. Others, including transcripts encoding IGF-II and its receptor, TGF-β1, and Kaposi's sarcoma-type fibroblast growth factor (kFGF) appear for the first time in the two- or four-cell stage and accumulate continuously thereafter. Transcripts encoding insulin and IGF-I receptors have not been de-

tected before the eight-cell stage. Whether this finding truly reflects a delay in the onset of transcription of those genes remains to be determined.

G. Genes Encoding "Housekeeping" Functions

1. Metabolic enzymes. Despite the importance of intermediary metabolism in governing viability of the conceptus and perhaps its rate of development, relatively little attention has been devoted to genes encoding metabolic enzymes. Some investigators have taken advantage of allelic variants of enzymes to determine the time of activation of the locus on the paternal chromosome of the persistence of the oogenetic gene product through development. For example, oocyte-encoded glucose phosphate isomerase (GPI-1) activity persists through the first 4–5 days of development in the mouse, depending on the amount of enzyme present in the unfertilized egg in different strains, and embryo-encoded enzyme can first be detected after 3.5 days (late morulae) [West and Flockhart, 1989]. Because of their ample oogenetic store of GPI, zygotes homozygous for a null allele of *Gpi-1* were able to develop beyond implantation [West et al., 1990]. Another gene activated late in preimplantation development is *Pgk*-1, which encodes phosphoglycerate kinase (PGK-1). The maternally inherited *Pgk*-1 gene is activated in the blastocyst stage, whereas the paternally encoded allozyme was not detected until day 6, just after implantation [Krietsch et al., 1982]. On the other hand, the limited data available show that some genes encoding metabolic enzymes are transcribed from the early cleavages onward. Hybridization of total RNA with an antisense riboprobe complementary to β-glucuronidase mRNA was used to determine that the amount of this transcript increases steadily after reaching a low point in the four-cell stage. The activity of the enzyme also increases during this period [Bevilacqua et al., 1988]. A similar accumulation profile was found for transcripts of the HPRT (hypoxanthine phosphoribosyltransferase) gene [Paynton et al., 1988]. In this case, there is evidence that both oogenetic and embryonic transcripts are translated in the early cleavages [see discussion by Paynton et al., 1988].

2. Metallothioneins. The metallothioneins are another group of "housekeeping" genes that have come under recent scrutiny [Andrews et al., 1991]. The metallothioneins are small metal-binding proteins that are considered to be important for buffering cells against toxic levels of metals such as zinc, copper, or cadmium. RT/PCR was used to show that two metallothionein genes, *MT-I* and *MT-II,* are constitutively transcribed at low levels throughout preimplantation development of the mouse. From the eight-cell stage onward the number of transcripts of *MT-I* could be increased by exposure to zinc or cadmium. Surprisingly, the acquisition of metal inducibility of metallothionein expression was correlated with a reduction in sensitivity to the toxic effects of zinc, but not cadmium, suggesting that these two metals exert their toxic effects in different ways. At present the role of metallothionein genes in preimplantation development remains unclear.

H. Genes Associated With Endogenous Retroviruses

Endogenous retroviruses, known as *intracisternal A particles* (IAP) to reflect the fact that the viruses form by budding an RNA nucleoid core into the cisternae of the endoplasmic reticulum, account for a significant fraction of the mRNA present in the cytoplasm during preimplantation mouse development (10^3–10^5 copies per embryo) [Pikó et al., 1984]. Two subfamilies of IAP genes, type I and type II, have been distinguished, and both are transcribed throughout this period [Poznanski and Calarco, 1991]. These transcripts exhibit the common profile of decline after ovulation followed by reaccumulation from the two-cell stage onward, with both the rate of synthesis of the mRNAs and their total accumulated mass during cleavage and blastocyst formation keeping in step with cell number [Pikó et al., 1984; Poznanski and Calarco, 1991]. Hence these genes must be under the same regulatory influences that govern the rates of accumulation of most mRNAs in preimplantation development, ensuring a roughly constant mRNA content per cell. Immunoprecipitation of IAP proteins after labeling with ^{35}S-methionine demonstrated that synthesis of retroviral proteins begins in the two-cell stage and continues at least into the blastocyst stage [Poznanski and Calarco, 1991]. It is not known if IAP genes or the retroviruses themselves play any role in development.

III. CONCLUSIONS AND FUTURE DIRECTIONS

There is now a sufficiently large body of information about the temporal pattern of expression of individual genes in preimplantation development, at least in the mouse, to provide a reasonably accurate view of the genetic program underlying cleavage, compaction, and blastocyst formation. The information reviewed in this chapter concerning the time of onset of mRNA accumulation for known genes (where transcript presence has been examined throughout preimplantation development) is summarized in Table I. This information, along with the accumulation profiles of the numerous unidentified mRNAs studied by Taylor and Pikó [1987] (Fig. 1), leads to several conclusions. First, in the large majority of cases in which transcript level was examined in the two- and four-cell stages, accumulation of the mRNA was clearly occurring by this time. Most of the genes examined to date, then, begin transcription by the four-cell stage (some as early as the two-cell stage). There are obvious exceptions such as the insulin and IGF receptor mRNAs, which were not detected before the eight-cell stage, and the Na$^+$,K$^+$-ATPase β-subunit mRNA, which accumulates rapidly beginning in the morula stage. However, even in these cases it is still possible that transcription is activated earlier, but the mRNA fails to accumulate because of a very short half-life.

A second conclusion is that, once acti-

vated, a gene continues to be transcribed at least into the blastocyst stage, resulting in continuous mRNA accumulation. In many cases the rate of accumulation is exponential such that the mRNA content per cell remains roughly constant, whereas in other cases the rate of accumulation either falls behind or exceeds the rate of increase in cell number. No mRNA has yet been identified that, once present, disappears at a later stage of preimplantation development. This finding explains the relative constancy of polypeptide synthesis patterns through preimplantation stages and implies that the few qualitative changes that have been noted are probably due to translational regulation. It also explains why compaction is relatively insensitive to transcriptional blockade (mRNAs necessary for this event probably begin to accumulate in the two- and four-cell stages), but does not explain why cavitation should be more sensitive. It is tempting to speculate that one or a few key mRNAs required for blastocele formation reach a critical level in the late morula because of transcriptional activation or transcript stabilization, and this triggers cavitation. The Na$^+$,K$^+$-ATPase β-subunit mRNA is a likely candidate for this role.

To what extent does this view of the genetic program for preimplantation development apply to other mammals? This question must go largely unanswered at present because of the dearth of information from other species. It is definitely not safe to *assume* that what is true for the mouse is true for other mammals. The timing of genomic activation is one important feature that is known to vary, and there are likely to be others. To cite just one example, the importance of connexin gene expression for compaction seems well established in the mouse, yet in the human there is some evidence that functional gap junctions may not appear until the blastocyst stage, well after compaction [Dale et al., 1991]. Obtaining data on the expression of specific genes during preimplantation development of human and other species must be a major focus of future research. Knowledge of the activity of genes in early mouse development can provide a guide for planning such explorations, and new techniques such as RT/PCR make such research feasible. The coming years are likely to bring a much broader understanding of preimplantation development as more mammals come under the intense scrutiny that the mouse now enjoys.

The task now facing researchers working with the mouse model is to move beyond the identification of genes involved in preimplantation events to an understanding of how the expression of those genes is regulated and what precise roles they play. The seeds are already sown for advances on both of these fronts. The discovery of a transcription factor that operates during and after genomic activation is a first step in unravelling the complex array of gene regulatory interactions that must govern transcription in this as in other periods of development. Regarding the elucidation of gene function, several approaches are available, and some are already beginning to be applied to preimplantation development. The use of microinjected antibodies, antisense RNAs, or oligodeoxynucleotides has been discussed; these approaches suffer from a lack of precise understanding of their modes of action and the difficulty of introducing macromolecules into more than a few blastomeres in the eight-cell stage and beyond. Future experiments will involve the construction of transgenic lines expressing antisense constructs under the control of inducible promoters or the use of gene-targeting technology to generate mice heterozygous for null mutations, which can then be mated to produce zygotes lacking even one functional copy of a gene of interest. Eventually it should be possible to produce mice with one copy of a gene replaced with a modified version so that the effects of subtle changes in the structure of the

gene product can be assessed. The laboratory mouse will thus remain at the forefront of research in preimplantation development for many years to come. However, applying this knowledge to understanding the involvement of genes in preimplantation development of other mammals, particularly our own species, remains the greatest challenge.

ACKNOWLEDGMENTS

Thanks are due to Dr. Andy Watson for supplying the unpublished micrographs shown in Figure 2 and to Drs. Gil Schultz and Lajos Pikó for allowing the reuse of their published data. The work from my laboratory reviewed in this chapter was supported by grants from the Natural Sciences and Engineering Research Council of Canada and the National Institutes of Health.

NOTE ADDED IN PROOF

Readers are alerted to the fact that the study by Rosner et al. [1991] cited on p. 50 of this chapter has been retracted in its entirety. The retraction can be found in Cell 69:724, 1992.

REFERENCES

Abreu SL, Brinster RL (1978): Synthesis of tubulin and actin during the preimplantation development of the mouse. Exp Cell Res 114:135–141.

Andrews GK, Huet-Hudson YM, Paria BC, McMaster MT, De SK, Dey SK (1991): Metallothionein gene expression and metal regulation during preimplantation mouse embryo development. Dev Biol 145:13–27.

Bachvarova R, Cohen EM, De Leon V, Tokunaga K, Sakiyama S, Paynton BV (1989): Amounts and modulation of actin mRNAs in mouse oocytes and embryos. Development 106:561–565.

Barron DJ, Valdimarsson G, Paul DL, Kidder GM (1989): Connexin32, a gap junction protein, is a persistent oogenetic product through preimplantation development of the mouse. Dev Genet 10:318–323.

Bennett MVL, Barrio LC, Bargiello TA, Spray DC, Hertzberg E, Sáez JC (1991): Gap junctions: new tools, new answers, new questions. Neuron 6:305–320.

Benos DJ, Balaban RS (1990): Transport mechanisms in preimplantation mammalian embryos. Placenta 11:373–380.

Benos DJ, Biggers JD (1981): Blastocyst fluid formation. In Mastroianni L, Biggers JD (eds): Fertilization and Embryonic Development In Vitro. New York: Plenum Press, pp 283–297.

Bevilacqua A, Erickson RP, Hieber V (1988): Antisense RNA inhibits endogenous gene expression in mouse preimplantation embryos: lack of double-stranded RNA "melting" activity. Proc Natl Acad Sci USA 85:831–835.

Bevilacqua A, Loch-Caruso R, Erickson RP (1989): Abnormal development and dye coupling produced by antisense RNA to gap junction protein in mouse preimplantation embryos. Proc Natl Acad Sci USA 86:5444–5448.

Beyer EC, Paul DL, Goodenough DA (1990): Connexin family of gap junction proteins. J Membrane Biol 116:187–194.

Braude PR (1979): Control of protein synthesis during blastocyst formation in the mouse. Dev Biol 68:440–452.

Braude PR, Bolton V, Moore S (1988): Human gene expression first occurs between the four- and eight-cell stages of preimplantation development. Nature 332:459–461.

Brenner CA, Adler RR, Rappolee DA, Pedersen RA, Werb Z (1989): Genes for extracellular matrix–degrading metalloproteinases and their inhibitor, TIMP, are expressed during early mammalian development. Genes Dev 3:848–859.

Brice AL, Cheetham JE, Bolton VN, Hill NCW, Schofield PN (1989): Temporal changes in the expression of the insulin-like growth factor II gene associated with tissue maturation in the human fetus. Development 106:543–554.

Buehr M, Lee S, McLaren A, Warner A (1987): Reduced gap junctional communication is associated with the lethal condition characteristic of DDK mouse eggs fertilized by foreign sperm. Development 101:449–459.

Chartrain I, Niar A, King WA, Picard L, St-Pierre H (1987): Development of the nucleolus in early goat embryos. Gamete Res 18:201–213.

Chisholm JC, Houliston E (1987): Cytokeratin filament assembly in the preimplantation mouse embryo. Development 101:565–582.

Conover JC, Temeles GL, Zimmermann JW, Burke B, Schultz RM (1991): Stage-specific expression of a family of proteins that are major products of zygotic gene activation in the mouse embryo. Dev Biol 144:392–404.

Cooper AR, MacQueen HA (1983): Subunits of laminin are differentially synthesized in mouse eggs and early embryos. Dev Biol 96:467–471.

Dale B, Gualtieri R, Talevi R, Tosti E, Santella L, Elder K (1991): Intercellular communication in the early human embryo. Mol Reprod Dev 29:22–28.

Dean WL, Seufert AC, Schultz GA, Prather RS, Simerly C, Schatten G, Pilch DR, Marzluff WF (1989): The small nuclear RNAs for pre-mRNA splicing are coordinately regulated during oocyte maturation and early embryogenesis in the mouse. Development 106:325–334.

Dooley TP, Miranda M, Jones NC, DePamphilis ML (1989): Transactivation of the adenovirus EIIa promoter in the absence of adenovirus E1a protein is restricted to mouse oocytes and preimplantation embryos. Development 107:945–956.

Duprey P, Morello D, Vasseur M, Babinet C, Condamine H, Brûlet P, Jacob F (1985): Expression of the cytokeratin endo A gene during early mouse embryogenesis. Proc Natl Acad Sci USA 82:8535–8539.

Dziadek M, Timpl R (1985): Expression of nidogen and laminin in basement membranes during mouse embryogenesis and in teratocarcinoma cells. Dev Biol 111:372–382.

Emerson JA (1988): Disruption of the cytokeratin filament network in the preimplantation mouse embryo. Development 104:219–234.

Fleming TP, Johnson MH (1988): From egg to epithelium. Annu Rev Cell Biol 4:459–485.

Gardiner CS, Grobner MA, Menino AR Jr (1990a): Sodium/potassium adenosine triphosphatase α-subunit and α-subunit mRNA levels in early rabbit embryos. Biol Reprod 42:539–544.

Gardiner CS, Williams JS, Mesino AR Jr (1990b): Sodium/potassium adenosine triphosphatase α- and β-subunit and α-subunit mRNA levels during mouse embryo development in vitro. Biol Reprod 43:788–794.

Gardner HG, Kaye PL (1991): Insulin increases cell numbers and morphological development in mouse pre-implantation embryos in vitro. Reprod Fertil Dev 3:79–91.

Geering K (1990): Subunit assembly and functional maturation of Na,K-ATPase. J Membrane Biol 115:109–121.

Giebelhaus DH, Heikkila JJ, Schultz GA (1983): Changes in the quantity of histone and actin messenger RNA during the development of preimplantation mouse embryos. Dev Biol 98:148–154.

Graves RA, Marzluff WF, Giebelhaus DH, Schultz GA (1985): Quantitative and qualitative changes in histone gene expression during early mouse embryo development. Proc Natl Acad Sci USA 82:5685–5689.

Guthrie SC, Gilula NB (1989): Gap junctional communication and development. Trends Neurosci 12:12–16.

Hahnel AC, Rappolee DA, Millan JL, Manes T, Ziomek CA, Theodosiou NG, Werb Z, Pedersen RA, Schultz GA (1990): Two alkaline phosphatase genes are expressed during early development in the mouse embryo. Development 110:555–564.

Harvey MB, Kaye PL (1988): Insulin stimulates protein synthesis in compacted mouse embryos. Endocrinology 122:1182–1184.

Harvey MB, Kaye PL (1990): Insulin increases the cell number of the inner cell mass and stimulates morphological development of mouse blastocysts in vitro. Development 110:963–967.

Harvey MB, Kaye PL (1991): IGF-2 receptors are first expressed at the 2-cell stage of mouse development. Development 111:1057–1060.

Heyner S, Smith RM, Schultz GA (1989): Temporally regulated expression of insulin and insulin-like growth factors and their receptors in early mammalian development. BioEssays 11:171–176.

Hoh JH, John SA, Revel J-P (1991): Molecular cloning and characterization of a new member of the gap junction gene family, connexin-31. J Biol Chem 266: 6524–6531.

Hsieh C-L, Kumar NM, Gilula NB, Francke U (1991): Distribution of genes for gap junction membrane channel proteins on human and mouse chromosomes. Somatic Cell Mol Genet 17:191–200.

Huet-Hudson YM, Andrews GK, Dey SK (1990): Epidermal growth factor and pregnancy in the mouse. In Heyner S, Wiley LM (eds): Early Embryo Development and Paracrine Relationships. New York: Wiley-Liss, pp 125–136.

Hynes R (1985): Molecular biology of fibronectin. Annu Rev Cell Biol 1:67–90.

Izquierdo L, Lopez T, Marticorena P (1980): Cell membrane regions in preimplantation mouse embryos. J Embryol Exp Morphol 59:89–102.

Jarrell VL, Day BN, Prather RS (1991): The transition from maternal to zygotic control of development occurs during the 4-cell stage in the domestic pig, Sus scrofa: quantitative and qualitative aspects of protein synthesis. Biol Reprod 44:62–68.

Kang HM, Kim K, Kwon HB, Cho WK (1990): Regulation of laminin gene expression in the expansion of mouse blastocysts. Mol Reprod Dev 27:191–199.

Kaye PL, Church RB (1983): Uncoordinated synthesis of histones and DNA by mouse eggs and preimplantation embryos. J Exp Zool 226:231–237.

Kaye PL, Wales RG (1981): Histone synthesis in pre-

implantation mouse embryos. J Exp Zool 216:453–459.

Kent RB, Fallows DA, Geissler E, Glaser T, Emanuel JR, Lalley PA, Levenson R, Housman DE (1987): Genes encoding α and β subunits of Na, K-ATPase are located on three different chromosomes in the mouse. Proc Natl Acad Sci USA 84:5369–5373.

Kidder GM (1987): Intercellular communication during mouse embryogenesis. In Bavister BD (ed): The Mammalian Preimplantation Embryo: Regulation of Growth and Differentiation In Vitro. New York: Plenum Press, pp 43–64.

Kidder GM, Conlon RA (1985): Utilization of cytoplasmic poly (A)$^+$ RNA for protein synthesis in preimplantation mouse embryos. J Embryol Exp Morphol 89:223–234.

Kidder GM, McLachlin JR (1985): Timing of transcription and protein synthesis underlying morphogenesis in preimplantation mouse embryos. Dev Biol 112:265–275.

Kidder GM, Pedersen RA (1982): Turnover of embryonic messenger RNA in preimplantation mouse embryos. J Embryol Exp Morphol 67:37–49.

Kidder GM, Watson AJ (1990): Gene expression required for blastocoel formation in the mouse. In Heyner S, Wiley LM (eds): Early Embryo Development and Paracrine Relationships. New York: Wiley-Liss, pp 97–107.

Krietsch WKG, Fundele R, Kuntz GWK, Fehlau M, Bürki K, Illmensee K (1982): The expression of X-linked phosphoglycerate kinase in the early mouse embryo. Differentiation 23:141–144.

Lee S, Gilula NB, Warner AE (1987): Gap junctional communication and compaction during preimplantation stages of mouse development. Cell 51:851–860.

Levinson J, Goodfellow P, Vadeboncoeur M, McDevitt H (1978): Identification of stage-specific polypeptides synthesized during murine preimplantation development. Proc Natl Acad Sci USA 75:3332–3336.

Levy JB, Johnson MH, Goodall H, Maro B (1986): The timing of compaction: Control of a major developmental transition in mouse early embryogenesis. J Embryol Exp Morphol 95:213–237.

Lobo SM, Marzluff WF, Seufert AC, Dean WL, Schultz GA, Simerly C, Schatten G (1988): Localization and expression of U1 RNA in early mouse embryo development. Dev Biol 127:349–361.

Magnuson T (1986): Mutations and chromosomal abnormalities: How are they useful for studying genetic control of early mammalian development? In Rossant J, Pedersen RA (eds): Experimental Approaches to Mammalian Embryonic Development. New York: Cambridge University Press, pp 437–474.

Magnuson T, Epstein, CJ (1987): Genetic expression during early mouse development. In Bavister BD (ed): The Mammalian Preimplantation Embryo: Regulation of Growth and Differentiation In Vitro. New York: Plenum Press, pp 133–150.

Malo D, Schurr E, Levenson R, Gros P (1990): Assignment of Na, K-ATPase β2-subunit gene (*Atpb*-2) to mouse chromosome 11. Genomics 6:697–699.

Manejwala FM, Cragoe EJ Jr, Schultz RM (1989): Blastocoel expansion in the preimplantation mouse embryo: role of extracellular sodium and chloride and possible apical routes of their entry. Dev Biol 133:210–220.

Maniatis T, Reed R (1987): The role of small nuclear ribonucleoprotein particles in pre-mRNA splicing. Nature 325:673–678.

Martin-Vasallo P, Dackowski W, Emanuel JR, Levenson R (1989): Identification of a putative isoform of the Na, K-ATPase β subunit. Primary structure and tissue-specific expression. J Biol Chem 264:4613–4618.

Martinez-Salas E, Cupo DY, DePamphilis ML (1988): The need for enhancers is acquired upon formation of a diploid nucleus during early mouse development. Genes Dev 2:1115–1126.

Martinez-Salas E, Linney E, Hassell J, DePamphilis ML (1989): The need for enhancers in gene expression first appears during mouse development with formation of the zygotic nucleus. Genes Dev 3:1493–1506.

Mattson BA, Rosenblum IY, Smith RM, Heyner S (1988): Autoradiographic evidence for insulin and insulin-like growth factor binding to early mouse embryos. Diabetes 37:585–589.

McDonough AA, Geering K, Farley RA (1990): The sodium pump needs its β subunit. FASEB J 4:1598–1605.

McLachlin JR, Caveney S, Kidder GM (1983): Control of gap junction formation in early mouse embryos. Dev Biol 98:155–164.

McLachlin JR, Kidder GM (1986): Intercellular junctional coupling in preimplantation mouse embryos: effect of blocking transcription or translation. Dev Biol 117:146–155.

Mercola M, Stiles CD (1988): Growth factor superfamilies and mammalian embryogenesis. Development 102:451–460.

Mercurio AM (1990): Laminin: multiple forms, multiple receptors. Curr Opin Cell Biol 2:845–849.

Mulnard J, Huygens R (1978): Ultrastructural localization of non-specific alkaline phosphatase during cleavage and blastocyst formation in the mouse. J Embryol Exp Morphol 44:121–131.

Nishi M, Kumar NM, Gilula NB (1991): Developmental regulation of gap junction gene expression during mouse embryonic development. Dev Biol 146:117–130.

Nisson PE, Francis S, Crain WR (1989): Spatial patterns of gene expression in preimplantation mouse embryos. Mol Reprod Dev 1:254–263.

Ohlsson R, Larsson E, Nilsson O, Wahlstrom T, Sundstrom P (1989): Blastocyst implantation precedes induction of insulin-like growth factor II gene expression in human trophoblasts. Development 106:555–559.

O'Shea KS, Liu L-HJ, Kinnunen LH, Dixit VM (1990): Role of the extracellular matrix protein thrombospondin in the early development of the mouse embryo. J Cell Biol 111:2713–2723.

Oshima RG, Howe WE, Klier FG, Adamson ED, Shevinsky LH (1983): Intermediate filament protein synthesis in preimplantation murine embryos. Dev Biol 99:447–455.

Overström EW (1987): In vitro assessment of blastocyst differentiation. In Bavister BD (ed): The Mammalian Preimplantation Embryo: Regulation of Growth and Differentiation In Vitro. New York: Plenum Press, pp 95–116.

Overström EW, Benos DJ, Biggers JD (1989): Synthesis of Na^+/K^+ ATPase by the preimplantation rabbit blastocyst. J Reprod Fertil 85:283–295.

Paria BC, Dey SK (1990): Preimplantation embryo development in vitro: Cooperative interactions among embryos and role of growth factors. Proc Natl Acad Sci USA 87:4756–4760.

Paynton BV, Rempel R, Bachvarova R (1988): Changes in state of adenylation and time course of degradation of maternal mRNAs during oocyte maturation and early embryonic development in the mouse. Dev Biol 129:304–314.

Pikó L, Hammons MD, Taylor KD (1984): Amounts, synthesis, and some properties of intracisternal A particle-related RNA in early mouse embryos. Proc Natl Acad Sci USA 81:488–492.

Poznanski AA, Calarco PG (1991): The expression of intracisternal A particle genes in the preimplantation mouse embryo. Dev Biol 143:271–281.

Prather R, Simerly C, Schatten G, Pilch DR, Lobo SM, Marzluff WF, Dean WL, Schultz GA (1990): U3 snRNPs and nucleolar development during oocyte maturation, fertilization and early embryogenesis in the mouse: U3 snRNA and snRNPs are not regulated coordinate with other snRNAs and snRNPs. Dev Biol 138:247–255.

Rao LV, Farber M, Smith RM, Heyner S (1990): The role of insulin in preimplantation mouse development. In Heyner S, Wiley LM (eds): Early Embryo Development and Paracrine Relationships. New York: Wiley-Liss, pp 109–124.

Rappolee DA, Brenner CA, Schultz R, Mark D, Werb

Z (1988): Developmental expression of PDGF, TGF-α, and TGF-β genes in preimplantation mouse embryos. Science 241:1823–1825.

Rappolee DA, Sturm KS, Schultz GA, Pedersen RA, Werb Z (1990): The expression of growth factor ligands and receptors in preimplantation mouse embryos. In Heyner S, Wiley LM (eds): Early Embryo Development and Paracrine Relationships. New York: Wiley-Liss, pp 11–25.

Rappolee DA, Wang A, Mark D, Werb Z (1989): Novel method for studying mRNA phenotypes in single or small numbers of cells. J Cell Biochem 39:1–11.

Rizzino A (1987): Defining the roles of growth factors during early mammalian development. In Bavister BD (ed): The Mammalian Preimplantation Embryo: Regulation of Growth and Differentiation In Vitro. New York: Plenum Press, pp 151–174.

Rosner MH, De Santo RJ, Arnheiter H, Staudt LM (1991): Oct-3 is a maternal factor required for the first mouse embryonic division. Cell 64:1103–1110.

Rosner MH, Vigano MA, Ozato K, Timmons PM, Poirer F, Rigby PWJ, Staudt LM (1990): A POU-domain transcription factor in early stem cells and germ cells of the mammalian embryo. Nature 345:686–692.

Schöler HR, Balling R, Hatzopoulos AK, Suzuki N, Gruss P (1989a): Octamer binding proteins confer transcriptional activity in early mouse embryogenesis. EMBO J 8:2551–2557.

Schöler HR, Dressler GR, Balling R, Rohdewohld H, Gruss P (1990): Oct-4: A germline-specific transcription factor mapping to the mouse t-complex. EMBO J 9:2185–2195.

Schöler HR, Hatzopoulos AK, Balling R, Suzuki N, Gruss P (1989b): A family of octamer-specific proteins present during mouse embryogenesis: evidence for germline-specific expression of an Oct factor. EMBO J 8:2543–2550.

Schultz GA (1986): Utilization of genetic information in the preimplantation mouse embryo. In Rossant J, Pedersen RA (eds): Experimental Approaches to Mammalian Embryonic Development. New York: Cambridge University Press, pp 239–265.

Schultz GA, Dean W, Hahnel A, Telford N, Rappolee D, Werb Z, Pedersen R (1990): Changes in RNA and protein synthesis during development of the preimplantation mouse embryo. In Heyner S, Wiley LM (eds): Early Embryo Development and Paracrine Relationships. New York: Wiley-Liss, pp 27–46.

Sepulveda S, Izquierdo L (1990): Effect of cell contact on regionalization of mouse embryos. Dev Biol 139:363–369.

Seshagiri PB, Bavister BD, Williamson JL, Aiken JM (1990): Qualitative comparison of protein production at different stages of hamster preimplantation embryo development. Cell Differ Dev 31:161–168.

Sobel JS, Goldstein EG (1988): Spectrin synthesis in the preimplantation mouse embryo. Dev Biol 128:284–289.

Svaren J, Chalkley R (1990): The structure and assembly of active chromatin. Trends Genet 6:52–56.

Taylor KD, Pikó L (1987): Patterns of mRNA prevalence and expression of B1 and B2 transcripts in early mouse embryos. Development 101:877–892.

Taylor KD, Pikó L (1990): Quantitative changes in cytoskeletal β- and γ-actin mRNAs and apparent absence of sarcomeric actin gene transcripts in early mouse embryos. Mol Reprod Dev 26:111–121.

Taylor KD, Pikó L (1991): Expression of the rig gene in mouse oocytes and early embryos. Mol Reprod Dev 28:319–324.

Telford NA, Watson AJ, Schultz GA (1990): Transition from maternal to embryonic control in early mammalian development: A comparison of several species. Mol Reprod Dev 26:90–100.

Tesařík J (1989): Involvement of oocyte-coded message in cell differentiation control of early human embryos. Development 105:317–322.

Tesařík J, Kopečný V, Plachot M, Mandelbaum J (1986a): Activation of nucleolar and extranucleolar RNA synthesis and changes in the ribosomal content of human embryos developing in vitro. J Reprod Fertil 78:463–470.

Tesařík J, Kopečný V, Plachot M, Mandelbaum J, Da Lage C, Fléchon J-E (1986b): Nucleologenesis in the human embryo developing in vitro: ultrastructural and autoradiographic analysis. Dev Biol 115:193–203.

Tesařík J, Kopečný V, Plachot M, Mandelbaum J (1987): High-resolution autoradiographic localization of DNA-containing sites and RNA synthesis in developing nucleoli of human preimplantation embryos: A new concept of embryonic nucleologenesis. Development 101:777–791.

Tesařík J, Kopečný V, Plachot M, Mandelbaum J (1988): Early morphological signs of embryonic genome expression in human preimplantation development as revealed by quantitative electron microscopy. Dev Biol 128:15–20.

Thiery JP, Duband JL, Tucker GC (1985): Cell migration in the vertebrate embryo: Role of cell adhesion and tissue environment in pattern formation. Annu Rev Cell Biol 1:91–113.

Valdimarsson G, DeSousa PA, Beyer EC, Paul DL, Kidder GM (1991): Zygotic expression of the connexin43 gene supplies subunits for gap junction assembly during mouse preimplantation development. Mol Reprod Dev 30:18–26.

Van Winkle LJ, Campione AL (1991): Ouabain-sensitive Rb$^+$ uptake in mouse eggs and preimplantation conceptuses. Dev Biol 146:158–166.

Watson AJ, Damsky CH, Kidder GM (1990a): Differentiation of an epithelium: Factors affecting the polarized distribution of Na$^+$,K$^+$-ATPase in mouse trophectoderm. Dev Biol 141:104–114.

Watson AJ, Kidder GM (1988): Immunofluorescence assessment of the timing of appearance and cellular distribution of Na/K-ATPase during mouse embryogenesis. Dev Biol 126:80–90.

Watson AJ, Pape C, Emanuel JR, Levenson R, Kidder GM (1990b): Expression of Na, K-ATPase α and β subunit genes during preimplantation development of the mouse. Dev Genet 11:41–48.

Weng DE, Morgan RA, Gearhart JD (1989): Estimates of mRNA abundance in the mouse blastocyst based on cDNA library analysis. Mol Reprod Dev 1:233–241.

Werb Z (1990): Expression of EGF and TGF-α genes in early mammalian development. Molec Reprod Dev 27:10–15.

West JD, Flockhart JH (1989): Genetic differences in glucose phosphate isomerase activity among mouse embryos. Development 107:465–472.

West JD, Flockhart JH, Peters J, Ball ST (1990): Death of mouse embryos that lack a functional gene for glucose phosphate isomerase. Genet Res 56:223–236.

Wiley LM, Lever JE, Pape C, Kidder GM (1991): Antibodies to a renal Na$^+$/glucose cotransport system localize to the apical plasma membrane domain of polar mouse embryo blastomeres. Dev Biol 143:149–161.

Wiley LM, Yamami S, Van Muyden D (1986): Effect of potassium concentration, type of protein supplement, and embryo density on mouse preimplantation development in vitro. Fertil Steril 45:111–119.

Willecke K, Jungbluth S, Dahl E, Hennemann H, Heynkes R, Grzeschik K-H (1990): Six genes of the human connexin gene family coding for gap junction proteins are assigned to four different human chromosomes. Eur J Cell Biol 53:275–280.

Wood SA, Kaye PL (1989): Effects of epidermal growth factor on preimplantation mouse embryos. J Reprod Fertil 85:575–582.

Zackson SL, Steinberg MS (1989): Axolotl pronephric duct cell migration is sensitive to phosphatidylinositol-specific phospholipase C. Development 105:1–7.

Zetter BR, Martin GR (1978): Expression of a high molecular weight cell surface glycoprotein (LETS protein) by preimplantation mouse embryos and teratocarcinoma stem cells. Proc Natl Acad Sci USA 75:2324–2328.

ABOUT THE AUTHOR

GERALD M. KIDDER is Professor of Zoology at the University of Western Ontario in London, Canada, where he teaches undergraduate and graduate courses dealing with topics in developmental biology, genetics, and developmental genetics. After receiving a B.A. *summa cum laude* from Hiram College in Ohio, he pursued doctoral research at Yale University in the laboratory of Clement Market, where he earned a Ph.D. in 1971 for work on the molecular biology of early molluscan development. This was followed by postdoctoral research at Reed College in Oregon, where he worked with Laurens Ruben on the ontogeny of the immune response in *Xenopus*. Since 1979 Dr. Kidder's research has involved explorations of the genetic and molecular controls underlying preimplantation morphogenesis in the mouse and other mammals. He has held Visiting Scientist positions at the University of California, San Francisco, and the Massachusetts Institute of Technology. In 1990 he was awarded a Distinguished Research Professorship in the Faculty of Science at the University of Western Ontario. He is an Associate Editor of *Molecular Reproduction and Development* and in 1992 became Editor-in-Chief of *Developmental Genetics*.

Genes in Mammalian Reproduction: 73–130
© 1993 Wiley-Liss, Inc.

Prospects for Preimplantation Diagnosis of Genetic Diseases

John R. Gosden and John D. West

I. INTRODUCTION

In classical times (and in some societies today) consulting the auguries to determine the most propitious time and conditions for conception was not uncommon and might be considered the earliest (in two senses) form of prenatal diagnosis. However, prenatal diagnosis in the accepted sense only became practicable in the mid-1950s, with the discovery that fetal cells were present in the amniotic fluid and that fetal sex could be determined by sampling the fluid in the second trimester of pregnancy and examining the amniocytes to determine the sex chromatin status [Serr et al., 1955; Fuchs and Riis, 1956]. Although Fuchs and Riis thought the procedure would be of limited value in man, Edwards [1956] was sufficiently far-sighted to realize that it could be of more general diagnostic value, and by 1960 Riis and Fuchs, working under the more liberal abortion laws of Denmark, were convinced of its importance in the prevention of hereditary disease.

With the suggestion that amniotic fluid cells could be cultured and metaphase chromosomes prepared [Fuchs and Philip, 1963] and the reports that this had been achieved [Steele and Breg, 1966; Jacobson and Barter, 1967] came the first report of a chromosomal abnormality (trisomy 21, Down's syndrome) being successfully iden-

tified in utero [Valenti et al., 1968]. Later reports identified conditions that could be diagnosed biochemically in amniotic fluid or cells, and more recently the proliferation of DNA-based diagnoses using restriction fragment length polymorphisms (RFLPs) has added to the range of conditions susceptible to prenatal diagnosis. However, this was still largely based on amniotic fluid sampled in the third trimester of pregnancy at about weeks 16–18 of gestation, and, in the event of a major abnormality being detected, the most common solution was to offer the possibility of a therapeutic abortion, with all the consequent risk of psychological trauma.

In the 1970s the first reports of successful sampling of chorionic villi during the first trimester of pregnancy were published [Kullander and Sandahl, 1973; Hahnemann, 1974], bringing diagnosis forward to weeks 8–12 of gestation and reducing the risk of trauma due to late termination. However, it is only with the recent development of successful in vitro fertilization (IVF) and the polymerase chain reaction (PCR) to amplify specific DNA sequences for analysis that diagnosis prior to implantation of the embryo has become feasible.

Progress toward making preimplantation diagnosis feasible has been rapid. In 1985, McLaren published the first review to consider the prospects for preimplantation diagnosis. At that time (only 5 years ago) the problems appeared enormous and the outlook was pessimistic. Two years later, several significant advances had been made and, in their review of the subject, Penketh and McLaren [1987] were much more optimistic. Five years after McLaren's initial review, the first clinical use of preimplantation diagnosis was reported by Handyside et al. [1990], and this has culminated in the birth of twin girls (A Handyside, personal communication.). In this chapter we outline the goals of preimplantation diagnosis, describe some of the technical advances that have made these goals feasible, and consider some possible future trends.

II. WHAT IS PREIMPLANTATION DIAGNOSIS?

Preimplantation diagnosis is prenatal diagnosis of genetic disorders in the very early (preimplantation) stages of development. The advantage of this approach is that conceptuses are tested for the genetic disorder before being returned to the mother to implant and continue development, so it is possible to select for transfer only those that are free from the disorder. This avoids the implantation of affected conceptuses and thus the necessity for prenatal diagnosis, and possibly a therapeutic abortion, later in gestation. For a couple at risk for a particular genetic disorder, preimplantation diagnosis offers almost the only hope of starting a pregnancy that is known to be unaffected by that genetic disorder. (The only other hope would be artificial insemination with X-bearing spermatozoa to avoid conception of males affected by X-linked disorders. This approach would require a reliable method of separating X- and Y-bearing spermatozoa, but this has not yet been achieved.)

A. Preimplantation Development

Just before ovulation, the primary oocyte (encapsulated by the acellular glycoprotein zona pellucida) completes the first meiotic division to produce a large secondary oocyte and a much smaller first polar body. Both of these are haploid (with 2c amounts of DNA) but the first polar body has no function and often degenerates. At ovulation the second meiotic division begins but arrests in second metaphase until fertilization. Completion of the second meiotic division, after fertilization, produces a large fertilized ovum (egg) or zygote and a small, unfertilized second polar body. The sperm nucleus enlarges to become the male pronucleus. Male and female pronuclei do not fuse, but the membranes break down and the maternal and paternal sets of chromosomes come together (karyogamy) on the first mitotic spindle as the one-cell zygote divides to become a two-cell conceptus, embryo, or pre-embryo. (Until the primitive streak is formed at around 15–18 days the embryonic part of the conceptus is not dissociable from the extraembryonic part, and the conceptus may produce monozygotic twin embryos. For these reasons the term *pre-embryo* has been recommended in favor of *embryo* for these early stages [McLaren 1986; Leach, 1988] and is used in this chapter. This term is also favored by the American Fertility Society and, in the United Kingdom, by the Human Fertilisation and Embryology Authority (HFEA; formerly the Interim Licensing Authority [ILA] and the Royal College of Obstetricians and Gynaecologists).

The two-cell pre-embryo undergoes further cleavage divisions to produce 4-, 8-, and 16-cell stages. Between the 8- and 16-cell stages the cells compact as the pre-embryo becomes a "morula" and passes from the oviduct (fallopian tube) into the uterus. Throughout this period the pre-embryo remains within the zona pellucida, and at each cleavage division the cytoplasmic volume per cell is halved, so the pre-embryo does not increase in size but remains approximately 100 μm in diameter. The cleavage divisions are not exactly synchronous and, for example, three- and five-cell stages are also seen. Also, human pre-embryos commonly have anuclear fragments so that cleavage stage pre-embryos often appear more irregular than those of mouse pre-embryos. Some stages of human pre-implantation development are illustrated in Figure 1. At around the 32-cell stage (typically on day 5 of culture) the human pre-embryo forms a blastocele cavity and thereby becomes a blastocyst, comprising an outer layer of trophectoderm cells that surrounds the blastocele and the inner cell mass (ICM). (Hardy et al. [1989a] have suggested that blastulation of human pre-embryos may begin before the fifth cleavage division is completed, because some blastocysts have fewer than 32 cells.) The ICM subsequently divides into a primitive ectoderm layer (or epiblast) and a primitive endoderm layer (or hypoblast). The blastocyst hatches from the surrounding zona pellucida and implants into the uterus, thereby ending the preimplantation period of development at 6–7 days. Further details of preimplantation development are given in the chapters on oogenesis and embryogenesis in this book and some landmarks of human preimplantation development are listed in Table I.

Most of what we know about preimplantation development comes from studies of pre-embryos of other mammalian species, especially the mouse. Significant differences exist between the species, so care must be taken when extrapolating to human pre-embryos, and this highlights the need for more research on human pre-embryos. The developmental lineages are now well established for the mouse, thanks to painstaking experimental embryology [Gardner and Papaioannou, 1975; Gardner 1988], but the situation in the human is less clear. There are some obvious differences between the two species. For example, in the mouse the abembryonic pole of the blastocyst makes the first contact with the uterus at implantation but the human blas-

tocyst implants in the opposite orientation, with the embryonic pole making the first contact. The scheme of developmental lineages outlined by Crane and Cheung [1988], shown in Figure 2, is based on studies of human and mouse conceptuses and provides a guide to the developmental lineages of the human conceptus but must still be considered tentative. From this it can be seen that the entire fetus, as well as certain extraembryonic membranes, is derived from the primitive ectoderm layer.

The pre-embryos of rabbits, sheep, and cattle do not implant until they have several thousand cells, whereas implantation of mouse and human embryos occurs when the blastocyst has only about 100 cells. In both human and mouse pre-embryos, only about 40% of these are ICM cells [Hardy et al., 1989a], and in mice only about 50% of the ICM cells (20% of the total) are in the primitive ectoderm layer of the ICM, which eventually forms the fetus. Cells are allocated to trophectoderm or ICM developmental lineages earlier in mouse and human pre-embryos than in sheep or cattle. The later commitment of cells to one or the other of these developmental lineages in sheep is thought to explain why twins or quads can be produced by "pre-embryo splitting" more easily with sheep (and possibly cow and horse) pre-embryos than with mouse pre-embryos [e.g., Willadsen, 1979, 1981; Lehn-Jensen and Willadsen, 1983; Ozil, 1983; Allen and Pashen, 1984]. The poorer viability of single blastomeres from four- or eight-cell mouse pre-embryos is thought to be a consequence of reducing the total cell number below a critical number such that when cells are allocated to ICM or trophectoderm there are too few cells to have an "inside" population of cells, so an ICM does not form. (The human pre-embryo is likely to be similar to the mouse in this respect.) The production of lambs from single cells of eight-cell sheep pre-embryos provides good evidence that at least some cells remain totipotent up to

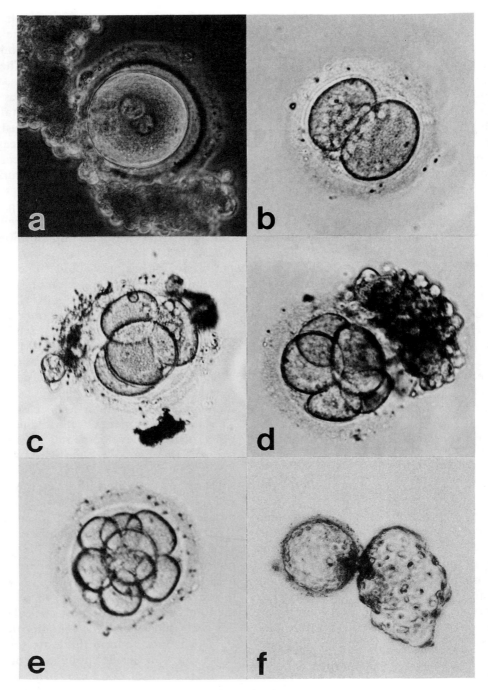

Fig. 1. *Photographs of human pre-embryos at various stages.* **a:** *Fertilized egg with two pronuclei and attached cumulus cells.* **b:** *Two-cell stage,* **c:** *Four-cell stage,* **d:** *Approximately six-cell stage with attached cumulus cells.* **e:** *approximately eight-cell stage.* **f:** *Hatching blastocyst, part of which has herniated out of the zona pellucida. (Courtesy of Dr. R.R. Angell.)*

TABLE I. Some Landmarks in Human Preimplantation Development

Embryonic stage	Days after ovulation		Biological events	Possible manipulations
	In vivo	In vitro*		
Secondary oocyte	0	0		Oocyte collection, first polar body biopsy, insemination for IVF
		.25		
Fertilization	0.5–1.5	.25–1	Meiosis completed, pronuclei visible	Second polar body biopsy, cryopreservation
2-cell stage	1.5–2	2		Transfer to uterus, cryopreservation
4-cell stage	2	2–3	Gene expression begins	Transfer to uterus, cryopreservation
8-cell stage	2.5	3–4	Cells still totipotent in mice	Blastomere biopsy, transfer to uterus, cryopreservation
Morula stage (~16 cells)	4	4–5	Compaction of cells	
Blastocyst (32 or more cells)	5	5–6	X-inactivation begins in mice	Collection by flushing uterus, trophectoderm biopsy, transfer to uterus, cryopreservation
Implantation	6–7			

*Times are approximate. In vivo times are taken from Moore [1977], McLaren [1986], and Penketh and McLaren [1987]. In vitro times are from Sundstrom et al. [1981], Bolton et al. [1989], Hardy et al. [1989a], and West et al. [1988, 1989a].

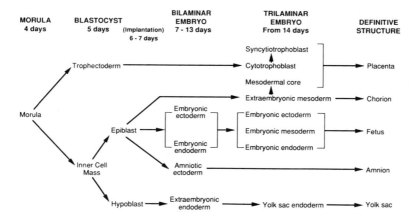

Fig. 2. *Tentative developmental lineages in the human conceptus.*

the eight-cell stage [Willadsen, 1981]. Less direct evidence suggests that cells of mouse pre-embryos are also totipotent up to the eight-cell stage [Kelly, 1977].

There is general agreement that in mouse pre-embryos gene expression begins at the two-cell stage [Johnson, 1981; Magnuson and Epstein, 1981; Flach et al., 1982; Bolton et al., 1984] but this appears to be delayed until the four- to eight-cell stage in human pre-embryos [Braude et al., 1988] and to the 8- to 16-cell stage in sheep pre-embryos [Crosby et al., 1988]. X-chromosome inactivation begins at the blastocyst stage in mouse pre-embryos [e.g., Epstein et al., 1978], but it is not known whether this is also true of human embryos. The timing of these events may be important when considering the feasibility of various biopsy procedures (section IV) and genetic tests that are based on embryonic gene products (section V).

B. How Could Preimplantation Pre-Embryos With Genetic Disorders Be Identified?

Penketh and McLaren [1987] considered three approaches to obtaining genetic information about the pre-embryo: 1) non-invasive testing of the intact pre-embryo, 2) predicting the genotype of the pre-embryo from a test of the genotype of the second polar body, and 3) testing the genotype of the pre-embryo by removing one or more cells from the pre-embryo. Recently, Monk and Holding [1990], Strom et al. [1990] and Verlinsky et al. [1990] suggested a fourth possibility: that the first polar body could be used to predict whether an unfertilized secondary oocyte would be likely to produce an affected zygote ("preconception genetic diagnosis"). The third option (pre-embryo biopsy) has received the most attention and, in conjunction with sensitive genetic tests, has been used for preimplantation diagnosis of both mouse and human pre-embryos [Monk et al., 1988; Holding and Monk,

1989; Gomez et al., 1990; Handyside et al., 1990; Lindeman et al., 1990]. These options are discussed in more detail in the following sections.

C. When Would Preimplantation Diagnosis Be Used?

Preimplantation diagnosis is likely to benefit couples who are at high risk for sex-linked disorders or autosomal single-gene defects and anxious to avoid the need for a therapeutic abortion. Preimplantation diagnosis may be used less frequently for couples at lower risk for genetic disorders, such as many of the spontaneous chromosomal aberrations, where the risk of therapeutic termination is correspondingly lower. For couples at risk, the risk of conceiving an affected conceptus is typically 25% for an autosomal recessive, 50% for an autosomal or X-linked dominant and 25% (50% of the males) for an X-linked recessive trait. Individuals who carry certain chromosome translocations in the balanced state also run a high risk of producing chromosomally unbalanced offspring, although many of these do not survive to term. In contrast, the frequency of the most common de novo chromosomal disorder (trisomy 21) is only 2.42% at term for mothers over age 40 years and is lower still for younger mothers [e.g. Brock, 1982]. Although this represents a high population frequency, the risk to an individual couple is lower than for a known carrier of a mendelian disorder or chromosome translocation.

There would be an additional advantage of using preimplantation diagnosis for couples at risk for a recessive disorder if both heterozygous (carriers) and homozygous (affected) pre-embryos could be identified. It would then be possible to avoid the conception of any conceptus carrying the deleterious gene and thus eliminate it from a family in a single generation without the need for induced abortion.

III. OBTAINING PRE-EMBRYOS

Pre-embryos for genetic tests may be obtained either by IVF of secondary oocytes aspirated from preovulatory follicles or by transcervical uterine flushing (uterine lavage).

A. Obtaining Pre-Embryos by IVF

Techniques for aspiration of secondary oocytes, fertilization by washed spermatozoa, and culture of the resultant pre-embryos are now commonplace in numerous IVF clinics worldwide. Superovulation is achieved by the administration of hormones designed to prolong the period of elevated levels of follicle-stimulating hormone (FSH). This allows more than one follicle to reach the stage of maturation, when it can respond to FSH and develop into a preovulatory follicle. The period of high FSH levels may be prolonged, either by direct administration of preparations containing FSH, such human menopausal gonadotrophin (hMG), or by the use of clomiphene citrate, an antiestrogen compound that prevents the negative feedback of estrogen to the pituitary and thus prevents the normal inhibition of FSH secretion from the pituitary. The maturation of the follicles is monitored by assessing growth by ultrasound scans and by measuring plasma estradiol (E2) levels. Follicles are usually considered mature if they are more than 16 mm in diameter and there is a peripheral estradiol concentration of more than 500 pmol/liter per follicle. The levels of LH are also monitored to ensure that an endogenous LH surge does not occur undetected, as this would cause the oocytes to be ovulated. When the follicles are judged to be sufficiently mature (typically, once there are at least three follicles with a diameter greater than 16 mm) human chorionic gonadotrophin (hCG) is given to stimulate ovulation approximately 36 hours later. The oocytes are aspirated from the preovulatory follicles between 32 and 34 hours after hCG administration to ensure that ovulation has not already occurred.

In recent years there has been a trend away from the original technique of laparoscopy toward ultrasound-guided techniques for follicle aspiration, which can be done without a general anaesthetic [Booker and Parsons, 1990]. New protocols for hormone-induced ovulation have been introduced, such as the use of luteining hormone-releasing hormone (LHRH) agonist, often in the form of Buserelin nasal spray (Suprefact, Hoechst Ltd.), in the pretreatment cycle [Porter et al., 1984; Fleming and Coutts, 1986; Neveu et al., 1987; Rutherford et al., 1988]. This initially stimulates and then downregulates the pituitary gland, thus depleting the endogenous supply of FSH and LH and thereby improving the response to subsequent ovarian stimulation with hMG by avoiding any interference from endogenous hormones. The main advantages of using Buserelin are 1) to prevent premature LH surges, which otherwise occur in 20%–30% of superovulated cycles; 2) to increase the yield of mature secondary oocytes and thereby increase the number of pre-embryos that are likely to be available for preimplantation diagnosis in a given treatment cycle; 3) to program the treatment cycles to a chosen timetable and thus, for example, minimize the weekend workload by choosing the appropriate time to start the hMG treatment. (Buserelin can also be used in a "short protocol," which makes use of its initial agonist effect by starting treatment with hMG much sooner after starting Buserelin treatment.) Although Buserelin does not seem to reduce the risk of hyperstimulation during superovulation and adds to the cost of treatment, it is now being widely used in IVF units for the advantages listed above.

Procedures used for IVF and handling human pre-embryos have been reviewed elsewhere [e.g., Braude, 1987; Edwards, 1990]. Techniques vary in different medical centers, but, in general, follicular fluid is aspirated into sterile tubes and the oocytes, each surrounded by cumulus cells, are located under a dissecting microscope and incubated for about 6 hours in a suitable culture medium at 37°C in a CO_2 incubator. Motile spermatozoa are prepared from the semen sample, and the oocyte is inseminated, typically with approximately 10^5 spermatozoa in a volume of 1 ml. After overnight culture, overlying cumulus cells are often removed from the eggs so that they can be examined, under the dissecting microscope, for the presence of two pronuclei. After another 24 hours in culture, each fertilized zygote will have cleaved once or twice and be at the two-, three-, or four-cell stage. Further development occurs in culture (Table I), but human pre-embryos often arrest at the eight-cell stage. Oocyte retrieval and fertilization rates are generally high (around 80%–90%) and most fertilized eggs develop successfully to the two- to four-cell stage, but only about 25%–30% form blastocysts [Fehilly et al., 1985; Braude et al., 1986; Bolton and Braude, 1987; Bolton et al., 1989]. However, several recent publications have reported improved success in culturing human pre-embryos beyond the eight-cell stage by either using Ham's F12 medium containing hypoxanthine [Muggleton-Harris et al., 1990] or coculturing the pre-embryos with oviduct cells [Bongso et al., 1989].

Although IVF is the most obvious means of obtaining pre-embryos for preimplantation diagnosis, the pregnancy rate following IVF remains fairly low: Approximately 10% of the pre-embryos transferred implant. (Figures for implantations per pre-embryo or egg transferred, calculated for transfers of up to four pre-embryos or eggs from the British 1988 data in Table VI of the Fifth Report of the Interim Licensing Authority [ILA, 1990], showed similar success rates for IVF [1,721/17,645; 9.8%] and gamete intrafallopian transfer [GIFT; 846/9,779; 8.7%].)

It has been suggested that the procedures used for IVF may increase the incidence of chromosome abnormalities in human pre-embryos [e.g., Bolton and Braude, 1987]. If so, there would be a risk of causing a new genetic disorder while using preimplantation diagnosis to avoid a specific genetic disease. However, the MRC Working Party on Children Conceived by In Vitro Fertilisation [1990] concluded that the incidence of congenital malformations (35 among 1,581 babies; 2.2%) was comparable with that in the population at large, although there is some evidence to suggest that the frequency of neural tube defects could be elevated, possibly as a result of ovulation induction [Cornel et al., 1989a, b].

B. Obtaining Pre-Embryos by Uterine Flushing

An alternative source of pre-embryos for preimplantation diagnosis is the collection of pre-embryos from the uterus by transcervical uterine flushing (uterine lavage) after normal in vivo fertilization [Buster et al., 1983; Formigli et al., 1990; Brambati and Tului, 1990]. Buster and Carson [1989] discussed the relative merits of IVF and uterine flushing as a source of pre-embryos. They maintained that uterine flushing had the advantages of a higher pregnancy rate (per pre-embryo transferred), not requiring ultrasound guidance or anaesthesia, and a lower cost, but that there was a risk of retained pregnancy (pre-embryo not flushed out).

One attraction of this approach is that the pre-embryos could be collected at the appropriate stage for preimplantation diagnosis without first being subjected to in vitro culture and thus would not be compromised by any inadequacies in the culture conditions. Although the transfer of morulae and blastocysts recovered by this procedure has resulted in pregnancy rates of about 40%, the yield of pre-embryos was low, even after ovarian stimulation [Sauer et al., 1989; Formigli et al., 1990]. The pre-embryos also varied from the two-cell to blastocyst stage, even though the uterus was flushed at a predetermined time after the endogenous LH surge (or administration of hCG in stimulated cycles). The new ovarian stimulation regimes used during IVF for oocyte collection, such as those combining pretreatment with an LHRH agonist followed by hMG (discussed above), produce a much greater yield of oocytes, and, with current fertilization rates, this also results in a higher yield of pre-embryos. Consequently, Formigli et al. [1990] concluded that pre-embryos for preimplantation diagnosis may be more successfully obtained by IVF procedures than by uterine flushing. Uterine flushing has the added drawback that some pre-embryos may not be recovered, and thus pregnancies may occur with pre-embryos that were not screened for the genetic disorder. This risk may be greater in stimulated cycles [Sauer et al., 1989]. Bolton [1991] has also pointed out that the procedure may involve a risk of ectopic pregnancy or salpingitis (inflammation of the fallopian tubes) caused by the passage of bacteria from the cervix to the upper genital tract. From these considerations, it seems that any advantages of obtaining pre-embryos by uterine flushing are outweighed by the disadvantages. Since IVF is now widespread, it seems likely that IVF will be the most common source of pre-embryos for preimplantation diagnosis.

IV. BIOPSY AND PRE-EMBRYO MANIPULATIONS

There are three approaches to obtaining genetic information about the pre-embryo 1) noninvasive testing of the intact pre-embryo, 2) predicting the genotype of the pre-embryo from a test of the genotype of the

first or second polar body, and 3) testing the genotype of the pre-embryo by removing one or more cells from the pre-embryo.

The latter two approaches are invasive. The prime consideration with any invasive test is the possibility of damaging the oocyte or pre-embryo. The loss of a polar body would not be expected to be detrimental but the removal of blastomeres could have adverse effects and this is discussed below (section IV.C,) Another consideration is the effect of puncturing the zona pellucida. Nichols and Gardner [1989] have shown that one-cell and two-cell stage mouse pre-embryos survive poorly after transfer to the oviduct if the zona pellucida is punctured. They argue that an intact zona protects the cleavage stage pre-embryo from mechanical damage in the oviduct. If the damage is dependent on the size of the reproductive tract rather than on the stage of the pre-embryo, this is unlikely to be a problem, because, after IVF, human pre-embryos are usually transferred to the uterus, not the oviduct. If necessary, however, transfers could be delayed to the morula or blastocyst stage. Others have suggested that there may be an advantage in puncturing the human zona pellucida, in the course of biopsy, because it may assist the human embryo to hatch from the zona pellucida (by overcoming the effects of culture-induced hardening of the zona). This may aid implantation and thereby improve the pregnancy rate. Other potential problems are attack by leukocytes or bacteria [Moore et al., 1968] and reduced survival after cryopreservation (see section V). However, as pregnancies have already been obtained following the transfer of biopsied human pre-embryos [Handyside et al., 1990; discussed in section IV.C], some of these concerns may be academic.

A. Noninvasive Genetic Tests

Although noninvasive procedures can be used for studying a variety of metabolic parameters [Leese et al., 1986; Leese,

1987a,b; Wales et al., 1987; Hardy et al., 1989b; Gott et al., 1990], we know of no work that offers the prospect of being able to test the genotype of a pre-embryo by a noninvasive genetic test. This possibility has been discussed by Adinolfi and Polani [1989].

B. Removing the First or Second Polar Body

The polar bodies have no known function, so could be removed without compromising the developmental potential of the pre-embryo. For this reason, the possibility of gaining genetic information about the zygote from the polar bodies is attractive. Obviously, the polar bodies would not be useful for diagnosing conditions in which the deleterious gene was inherited only from the father (e.g., when the father but not the mother carried an autosomal dominant disorder such as Huntington chorea). Similarly, the polar bodies would be of no use for determining the sex of the pre-embryo. However, information from the polar bodies could be useful in any cases in which the defective gene was inherited from the mother, including sex-linked transmission of deleterious genes from female carriers to their sons, as well as autosomal dominant and recessive genes inherited from the mother. In the case of autosomal recessive disorders, the conceptus would only be affected if it inherited an abnormal gene from both parents, but the polar body would only give information about the gene inherited from the mother. Thus, to avoid the conception of homozygous pre-embryos, it would be necessary also to prevent the conception of some heterozygous pre-embryos (which received the normal allele from the father). The polar bodies could be used to provide information about the maternal contribution to the genotype of the pre-embryo but, as shown in Figure 3, the genotypes of the polar bodies are not identical to the maternal pronucleus in the fertilized egg or zygote.

1. First polar body. Two groups have suggested using the first polar body for "preconception genetic diagnosis" [Monk and Holding, 1990; Strom *et al.*, 1990; Verlinsky *et al.*, 1990]. This could be done by removing the first polar body from unfertilized secondary oocytes, testing them, and only fertilizing oocytes that could be guaranteed to be free from the genetic disorder under test. (Although the secondary oocyte and first polar body are haploid, they have a 2c DNA content and each chromosome has two chromatids, so the chromosome may be heterozygous if genetic crossing-over has occurred. Heterozygosity in this instance means that the two chromatids carry different alleles and not that there are two homologous chromosomes with different alleles.) Figure 3 illustrates the segregation of alleles at two loci (A and B), the polar bodies, secondary oocyte, and female pronucleus of the egg. For the purpose of illustration, we assume that alleles a and b are normal but that A and B alleles are deleterious and are to be avoided in the egg. If the genetic locus of interest did not undergo recombination at meiosis (locus A with alleles A and a in Fig. 3), and a test showed that the first polar body was homozygous a/a, then the secondary oocyte must be homozygous A/A and the maternal pronucleus in the egg must always inherit the deleterious A allele (as in eggs i and ii). Conversely, if the first polar body is shown to be A/A, then the egg must inherit the normal maternal a allele (iii and iv). If the locus is very close to the centromere, the chance of recombination is very low. Otherwise there may be no crossover, a single crossover event, or multiple crossover events between the centromere and the locus of interest.

Locus B (Fig. 3) illustrates the effect of a single crossover between the centromere and the locus. Both the first polar body and the secondary oocyte are heterozygous B/b,

Homologous pairing (Meiotic arrest in primary oocyte)

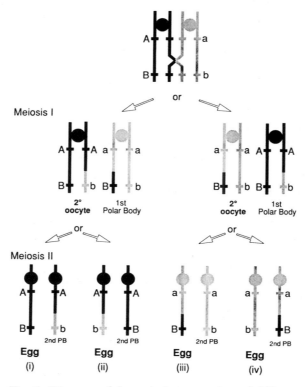

Fig. 3. *Diagram of the meiotic segregation of different alleles at two genetic loci (A and B) in a heterozygous mother. The genotype of the first and second polar bodies, the secondary oocyte ("2° oocyte"), and the female pronucleus of the fertilized egg ("egg") are all affected by genetic recombination between the centromere and the genetic locus, as shown by the different segregation patterns for locus A (no recombination) and locus B (with recombination). Unless it is possible to determine whether recombination has occurred, the polar bodies cannot be used to predict the maternal contribution to the genotype of the pre-embryo (see text for details).*

and the egg has an equal chance of inheriting the deleterious *B* allele (eggs i and iii in Fig. 3) or the normal *b* allele (examples ii and iv). Thus the first polar body can only be useful if it is shown to be homozygous for the deleterious allele of the locus under test (*A* in our example). Unless the locus is so close to the centromere that the chance of crossing over is negligible, the genetic test must be able to distinguish the three possible genotypes (heterozygote and the two types of homozygotes) in the polar body.

Monk and Holding [1990], Strom et al. [1990], and Verlinsky et al. [1990] have shown that specific DNA sequences can be detected in the first polar body after amplification with the PCR (see below), and, crucially, Strom et al. [1990] and Verlinsky et al. [1990] claimed that the three genotypes could be distinguished in first polar bodies of oocytes from women who were heterozygous (for cystic fibrosis and α-1-antitrypsin, respectively). Strom et al. reported that one oocyte, with a homozygous normal first polar body, was fertilized, and subsequent analysis of a blastomere, removed at the six-cell stage, showed that the pre-embryo was homozygous for the mutant gene. This is equivalent to hypothetical examples i and ii for the A locus in Figure 3, where the first polar body is homozygous normal (a/a); the egg contributes the abnormal allele (A). In this case the oocyte was fertilized by a sperm also carrying the abnormal allele, to produce a pre-embryo that was homozygous for the mutant cystic fibrosis allele. Despite the removal of both the first polar body and, subsequently, a blastomere at the six-cell stage, the pre-embryo developed to the blastocyst stage [Strom et al., 1990].

Nevertheless, there may be biological difficulties in using the first polar body. Monk and Holding [1990] removed the zona pellucida completely before dislodging the polar body, but Verlinsky et al. [1990] removed the polar body with a micromanipulator, via a hole made in the zona (see discussion of zona drilling in section IV.C.2). The first defence against polyspermy (changes in the zona induced by the cortical granule reaction) will have been impaired because additional sperm will be able to pass through the hole unless it is sealed. If this first defence against polyspermy is breached, polyspermic fertilization of eggs, judged to be free of the genetic disorder, may occur and produce polyploid pre-embryos. It should be possible to reduce this risk by fertilizing the oocytes in lower concentrations of sperm, and it is unlikely that removal of the first polar body would cause any local changes to the vitelline membrane that would impair the second defence against polyspermy. Although

clearly the preimplantation procedure should not be allowed to *increase* significantly the probability of inducing genetic abnormalities, if this risk was low and confined to additional cases of polyploidy it should be possible to screen out polyploid pre-embryos by examining the number of pronuclei after fertilization. An alternative approach might be to remove the first polar body after fertilization, but this is likely to be difficult both because the first polar body often degenerates and because it would be easily confused with the second polar body, which is not genetically equivalent (Fig. 3). It is also possible that puncture of the zona could allow leukcocytes, present in the sperm preparation, to attack the oocyte or pre-embryo [Moore et al., 1968], but the successful development of a blastocyst following removal of the first polar body [Strom et al., 1990] suggests that the risk of this may be low. Also, Verlinsky et al. [1990] reported that fertilization occurred in six of the seven oocytes after polar body biopsy and that two pronuclei were visible. This encourages the hope that the risk of polyspermy may also be low.

Despite the encouraging technical advances discussed above, the use of the first polar body for genetic tests seems unlikely to become the preferred biopsy method because the first polar body would only be informative if it was shown to be homozygous for the deleterious allele. It would therefore be necessary to exclude many oocytes that could develop into normal pre-embryos. Genetic recombination would result in a heterozygous first polar body and a heterozygous secondary oocyte. The deleterious allele (B in Fig. 3) could pass either to the egg (Fig. 3i, iii) or to the second polar body (Fig. 3ii, iv). Although in principle it would be possible to identify the normal eggs as those with the abnormal allele in the *second* polar body, in practice this would be demanding because it would require sequential biopsy of both polar bodies and accurate diagnosis on each sample.

2. Second polar body. Penketh and McLaren [1987] considered the prospect of using the second polar body as the sole means of diagnosis. If no recombination occurs between the centromere and the genetic locus (locus A in Fig. 3), the second polar body will carry the same allele as the female pronucleus of the egg (whereas the first polar body was homozygous for the other allele). If recombination occurs (locus B) the female pronucleus and the second polar body will have different alleles. Diagnosis is only possible if it is known whether genetic recombination has occurred. In most cases this will be unknown, although loci very close to the centromere will recombine rarely, and thus a diagnosis might, for example, be correct in 95% of the cases, assuming 100% technical accuracy. The genetic distance (in centimorgans [cM]) between the centromere and the locus is proportional to the probability of recombination, but unfortunately this cannot be used to argue that recombination is almost bound to occur between the centromere and distant genetic loci because multiple crossovers will be more common as the genetic distance increases. Genetic recombination occurs if there is an odd number of crossovers but not if there is an even number of crossovers. It therefore seems unlikely that the second polar body would provide reliable information about the genotype of the pre-embryo unless used in conjunction with other biopsy procedures (see section VIII.B).

3. Nondisjunction and diagnosis of aneuploidy with the polar bodies. The above discussion relates to normal segregation of genetic variants inherited from the mother but genetic anomalies may also arise de novo at meiotic division to produce aneuploidy (such as trisomy 21). The frequency of trisomic conceptuses is positively correlated with maternal age, and this implies that older mothers produce a higher proportion of disomic eggs as a result of a meiotic error. The error usually occurs at

the first meiotic division [Hassold et al., 1984] and is generally believed to involve nondisjunction of whole bivalents. Figure 4 illustrates the expected chromosome complements of the first and second polar bodies of an egg that is disomic for one chromosome as a result of nondisjunction at the first meiotic division. From this it can be seen that if the number of copies of a given chromosome could be accurately determined, by either cytogenetic or molecular methods, it should be possible to predict whether a zygote would become trisomic or monosomic (the chromosome segregation pattern generating a monosomic zygote is not shown). The absence of a particular chromosome (nullisomy) from the first polar body would imply that the secondary oocyte would be disomic, and after fertilization the zygote would be trisomic for that chromosome. If the second polar body was used for preimplantation diagnosis, disomy in the second polar body would imply disomy of the maternal contribution to the egg, and, after the maternal and paternal chromosomes had merged, the zygote would be trisomic. Although polar body chromosomes have on occasions been visualized (RR Angell, personal communication), the chromosome morphology is almost invariably very poor. Usually it is not even possible to see the chromosomes well enough to count them, so classical cytogenetics is unlikely to be applicable. It may be possible to determine the number of copies of a given chromosome by another method (such as in situ hybridization or DNA amplification; see later). In principle, therefore, this could provide a means of avoiding the transfer of aneuploid pre-embryos when the aneuploidy is a result of maternal nondisjunction. However, this assumes that the aneuploidy arises as a result of nondisjunction.

If aneuploidy did arise in this way, cytogenetic studies of the secondary oocyte should reveal a small proportion displaying the products of nondisjunction (i.e., extra

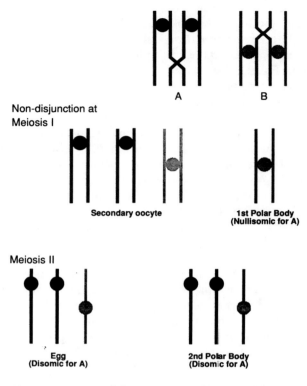

Homologous pairing (two pairs of chromosomes shown)

A   B

Non-disjunction at
Meiosis I

Secondary oocyte   1st Polar Body
                                    (Nullisomic for A)

Meiosis II

Egg   2nd Polar Body
(Disomic for A)   (Disomic for A)

Fig. 4. *Diagram of the meiotic segregation of different alleles of two pairs of homologous chromosomes (A and B), illustrating nondisjunction of chromosome A at the first meiotic division and normal disjunction of chromosome B. The maternal contribution shown for the egg is disomic for A and monosomic for B, so the zygote is trisomic for A and disomic for B. The abnormal segregation of chromosome A could be predicted from the first or second polar bodies if the numbers of chromosomes could be determined (see text). Nondisjunction at the first meiotic division could also produce the reciprocal segregation pattern (not shown). This would result in a disomy in the first polar body, nullisomy in the second polar body, and secondary oocyte and consequently a haploid zygote after fertilization.*

whole chromosomes), but thus far there has been no convincing data to confirm this prediction. On the other hand, Angell [1991] showed that about 10% of oocytes were hyperhaploid, with extra single chromatids but not extra whole chromosomes.

From this she has proposed that trisomic conceptuses arise by a mechanism other than nondisjunction [Angell, 1991], in which precocious division of centromeres takes place at the first meiotic division. This kind of chromosome behavior arises when univalents, rather than bivalents, are present at meiotic metaphase I (i.e., the homologous chromosomes are not paired in the usual way) and has been described in plants [Darlington, 1957, cited by Rieger, et al., 1976]. Univalents that lie far from the equator of the spindle segregate randomly (without division) to one or the other pole, but univalents that are close to the equator divide into the two chromatids, which segregate to opposite poles (as they normally would at meiosis II). Thus, after the first meiotic division, when the univalents have divided, the secondary oocyte and first polar body would each contain chromatids. Consequently these would not divide at all at the second meiotic division, but would be passively distributed either to the egg or second polar body and hence give rise to trisomy after fertilization [Angell, 1991]. Since this mechanism predicts a different range of chromosome complements in the polar bodies of trisomic zygotes from that expected on the basis of nondisjunction, it is important to test which mechanism is correct before attempting to diagnose aneuploidy by polar body biopsy.

Thus our present uncertainty about the mechanisms of formation of trisomy in humans presents a problem for the use of polar bodies to detect aneuploidy. Also, as was pointed out above (section II.C), the frequency of even the most common *de novo* chromosomal disorder (trisomy 21) is only a few percent. Thus, in our view, preimplantation diagnosis would be less likely to be used for those at risk for these relatively low-risk disorders than for those at high risk for a genetic disorder inherited as a mendelian trait or chromosome translocation from the mother.

C. Biopsy of the Pre-Embryo

At present, the most likely means of performing a preimplantation genetic test is to use a cell or cells biopsied from a pre-embryo. This could be done either at the blastocyst stage or at an earlier cleavage stage, as shown in Figure 5. This topic has been reviewed in depth by Bolton [1991]. One concern is that some human pre-embryos contain anucleate or binucleate cells [Lopata et al., 1983; Sathananthan et al., 1990] so that the biopsied cell or cells may not always be representative of the whole pre-embryo.

1. Trophectoderm biopsy. The first biopsy of mammalian pre-embryos for diagnostic purposes was performed by Gardner and Edwards [1968] by removing 200–300 trophectoderm cells from rabbit blastocysts in order to determine whether they were male or female by sex chromatin analysis. Live young were subsequently born. A similar approach has been used for the biopsy and cytogenetic sexing of cattle blastocysts [reviewed by Betteridge et al., 1981], although it is not clear whether only trophectoderm cells were removed. Biopsies were done either after hatching from the zona pellucida (12–15 days) or before hatching (6–7 days) when groups of 7–17 cells were aspirated. One attraction of trophectoderm biopsy is that trophectoderm cells are genetically identical to ICM cells but only the ICM cells contribute to the fetus, so removal of trophectoderm cells is unlikely to impair the pre-embryo's capacity to form a normal fetus. In this respect, biopsy of extra-embryonic trophectoderm cells is equivalent to a very early chorionic villus biopsy. The trophectoderm forms the outer layer of the blastocyst and encloses the ICM and blastocelic cavity; the trophectoderm that overlies the ICM is known as *polar* trophectoderm and that overlying the cavity is known as *mural* trophectoderm. The mural trophectoderm is the most suitable for biopsy both because it can be removed without damaging the ICM and because it contributes relatively little to the postimplantation conceptus; thus removal of a few cells is likely to be tolerated. In the mouse, mural trophoblast cells stop dividing early and pro-

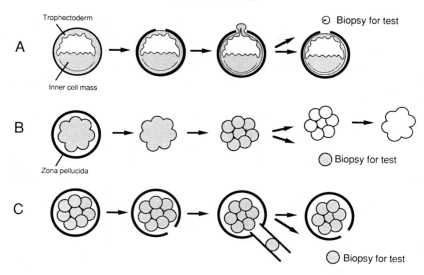

Fig. 5. *Diagram showing three possible approaches to pre-embryo biopsy.* **A:** *A portion of trophectoderm may be removed after causing it to herniate through a hole made in the zona pellucida. This approach is based on the micromanipulation methods described by Gardner and Edwards [1968] and Papaioannou [1982] and has been used to biopsy marmoset, mouse and human pre-embryos [Summers et al., 1988; Monk et al., 1988; Dokras et al., 1990].* **B:** *A single blastomere from a cleavage stage (about the eight-cell stage) pre-embryo could be removed after dissolving the surrounding zona pellucida and if necessary decompacting the blastomeres in a medium lacking calcium and magnesium [described for mouse pre-embryos by Monk and Handyside, 1988].* **C:** *A single blastomere from a cleavage stage pre-embryo could be removed with a micropipette, through a hole made in the surrounding zona pellucida, as described for human pre-embryos by Handyside et al. [1989] (Reproduced from West, 1990, with permission of the publisher.)*

duce only a small number of trophoblast giant cells. In the human, the mural trophectoderm is on the side closest to the uterine lumen and is covered by a thin layer of maternal decidua capsularis. The mural trophectoderm becomes cytotrophoblast and is then replaced by trophoblastic villi that subsequently degenerate, leaving the thin layer of chorion laeve (smooth chorion), which does not participate in the formation of the placenta [Harrison, 1978; Moore, 1977].

Although, as noted above (section II.A), the rabbit blastocyst is much larger than either the mouse or human blastocyst, it is possible to biopsy the smaller blastocysts with the aid of a micromanipulator. Papaioannou [1982] developed a microsurgical technique to produce trophectoderm vesicles by making a cut in the zona pellucida overlying the ICM, allowing both the ICM and polar trophectoderm to herniate through the hole and then removing the extruded tissue. Summers et al. [1988] adapted this technique for trophectoderm biopsy by positioning the hole in the zona to overly the mural trophectoderm rather than the polar trophectoderm. In this way they were able to remove a portion of mural trophectoderm from marmoset blastocysts and thus achieve the first primate pre-embryo biopsy. Normal offspring were subsequently born. The method of Summers et al. [1988] has also been tried on mouse pre-embryos [Monk et al., 1988; Gomez et al., 1990] and human pre-embryos [Edwards and Hollands, 1988; Dokras et al., 1990]. While Edwards and Hollands [1988] only reported trophectoderm biopsy of one human pre-embryo, which subsequently hatched from the zona pellucida and produced hCG, the more recent study of Dokras et al. [1990] involved 47 human blastocysts from which 10–30 cells were removed. As yet no human pre-embryos that have been biopsied in this way have been transferred to the uterus for further development.

The implantation rate of biopsied mouse pre-embryos, after surgical transfer to pseudopregnant recipients, was lower than nonoperated controls in both studies, but the numbers were small. Monk et al. [1988] reported an implantation rate of 7/33 (21%), or 7/21 (33%) for females that became pregnant versus 16/29 (55%) for control transfers. Gomez et al. [1990] reported that 93/133 (70%) of the blastocysts survived biopsy and 96% cavitated after 12 hours in culture. After surgical transfer, 22/37 (59%) resulted in implantations at 6.5 days compared with 30/34 (88%) of the controls; at 12–15 days the implantation rates were, respectively, 10/42 (24%) for biopsied blastocysts and 15/42 (36%) for controls.

The feasibility of human blastocyst biopsy for preimplantation diagnosis has only recently been investigated [Dokras et al., 1990], and most attention has been focused on biopsy of cleavage stage pre-embryos. Nevertheless, biopsy of blastocyst stages is attractive because it yields more cells than biopsy of cleavage stages and, as noted above, trophectoderm cells are extraembryonic.

2. Biopsy of cleavage stages. Unlike the mosaic development of some organisms, in which every cell has a predetermined fate, mammals have a more flexible, regulative system of morphogenesis, and the cells of mouse and sheep pre-embryos are totipotent up to the eight-cell stage (see section II.A). It is therefore feasible to consider removing a blastomere of a cleavage stage pre-embryo for preimplantation diagnosis. The simplest method is to remove the zona pellucida completely (in acidic Tyrode's solution), disaggregate the cells (in medium lacking calcium and magnesium or in the presence of a chelating agent, such as EDTA), remove one, and reaggregate the remaining cells (in medium containing calcium and magnesium). This has been done with eight-cell stage mouse embryos by Monk et al. [1987] and by Monk and Handyside [1988], and live fetuses were produced when pre-embryos were transferred 1 or 2 days after the biopsy. These transfers resulted in 16 fetuses from 51 transferred (31%) in one experiment [Monk et al., 1987] and in 15/21 (71%) in the other [Monk and Handyside 1988], with 19/26 (73%) in the controls.

The technique could be used at even earlier stages, but, with human and mouse pre-embryos, removal of a cell at the two-cell stage would risk reducing the cell number below that needed to form an ICM (see section II.A). The proportion of live mouse fetuses developing from such half embryos was low [Tsunoda and McLaren, 1983]. According to Rands [1985], halving the cell number by destruction of one cell at the two-cell stage reduces the number of cells in the ICM by more than 50%. She also reported that "half pre-embryos," produced by mechanical lysis of one blastomere, were smaller than controls in the early postimplantation period but usually achieved size regulation by 10.5 days, although they tended to fall behind again [Rands, 1986]. Tsunoda and McLaren [1983] also found that fetuses from half pre-embryos tended to be somewhat smaller than controls near term (1.04 ± 0.23 compared with 1.17 ± 0.20 g at 18.5 days), but no morphological abnormalities were reported.

The zona pellucida is necessary for the transit of cleaving eggs through the oviduct [Bronson and McLaren, 1970; Modlinski, 1970; Nichols and Gardner, 1989]. One potential problem with the simple disaggregation–reaggregation method, even at the eight-cell stage, is that the zona pellucida is completely destroyed. Willadsen [1979] recognized that this was often detrimental to further development of cleavage stage embryos and overcame this problem by transferring denuded sheep pre-embryos to host zonae pellucidae. However, evidence suggests that the problem only arises if denuded cleavage stage pre-embryos are transferred to the oviduct rather than to the uterus [Nichols and Gardner, 1989] (discussed in section IV). Thus, if biopsied pre-embryos were cultured to the morula or blastocyst stage and returned to the uterus, the lack of a zona may not have any significant adverse effect.

An alternative means of biopsying cleavage stage pre-embryos, illustrated in Figure 5C, requires a micromanipulator. A hole is made in the zona pellucida, by local application of acidic Tyrode's solution (the "zona drilling" technique), and a blastomere is dislodged from the pre-embryo and aspirated into a micropipette inserted through the hole. Experiments with mouse pre-embryos [Wilton and Trounson, 1989] showed that removal of one blastomere at the four-cell stage caused no significant reduction in the proportion of pre-embryos that developed to the blastocyst stage after 48 hours in culture (94% [135/143] compared with 98% [136/139] for controls), but, on transfer to pseudopregnant recipients, fewer biopsied pre-embryos implanted (53% of 76 transferred compared with 82% of 80). However, those pre-embryos that did implant showed normal fetal development, and in a second series of biopsied four-cell pre-embryos [Wilton et al., 1989] there was no significant reduction in fetal survival.

In another study, Krzyminska et al. [1990] compared the developmental potential of mouse pre-embryos biopsied at the four-cell, eight-cell, and morula stages. In contrast to the study of Wilton and Trounson [1989], removal of one cell from four-cell embryos reduced the proportion that formed blastocysts. After removal of two cells, eight-cell stage pre-embryos fared better than four-cell pre-embryos with one cell removed. Biopsy of eight-cell stage pre-embryos had no significant effect on the proportion that formed blastocysts, the implantation rate after transfers, or the fetal and placental weights (at 17 days). There was a trend toward a lower implantation rate in the biopsied group, which, coupled with a higher resorption rate, resulted in a statistically significant reduction in the proportion of transferred blastocysts that developed into viable day 17 fetuses (48% [88/182] compared with 63% [74/118] in the unoperated controls). Biopsy of the morulae produced lower success than with either the four-cell or eight-cell pre-embryos, but at this stage cells may no longer be totipotent and the biopsy procedure was different (the zona pellucida was completely removed, and approximately five cells were sliced from the zona-free pre-embryo without knowing how many were presumptive ICM cells).

These results, showing that biopsy of mouse eight-cell pre-embryos had minimal detrimental effects, are reassuring for the prospects of human preimplantation diagnosis. They support the widely held belief that damage caused to the pre-embryo during the preimplantation period of development is of an "all-or-none" nature. Damage may sometimes cause death but is unlikely to cause the type of fetal abnormality associated with teratogenic insults during the period of organogenesis. This belief is also supported by the absence of malformations in a fetus that resulted from the transfer of a cryopreserved, human pre-embryo in which only five of the eight blastomeres survived thawing [Trounson and Mohr, 1983].

This biopsy approach has also been tried

on human pre-embryos. Handyside et al. [1989] removed a single cell from each of thirty 6- to 10-cell stage human pre-embryos via a hole in the zona pellucida. Of 27 biopsied pre-embryos returned to culture, 10 (37%) developed to the blastocyst stage within 3 days (6 days after fertilization), which was comparable to the control rate. In a later series of experiments from the same group, Hardy et al. [1990] reported that removal of one or two cells at the eight-cell stage had no adverse effect on further preimplantation development, as judged by the number of pre-embryos that developed to blastocysts (79% and 71% [22/28 and 10/14], respectively, compared with 59% [19/32] of controls) and subsequently hatched (56%). The ratio of inner cell mass to trophectoderm cells was normal, and the total number of cells and the uptake of glucose and pyruvate was only reduced by the proportion expected following the biopsy. It has already been pointed out (section II.A) that human pre-embryos commonly have anuclear fragments. For this reason it is important to ensure that the biopsied "cell" has a nucleus. This can be checked by interference microscopy or labeling of DNA with a polynucleotide-specific fluorochrome and fluorescence microscopy [Handyside et al., 1990].

The ultimate test of pre-embryo biopsy has now been performed. Handyside et al. [1990] biopsied fifty 6- to 10-cell stage human pre-embryos produced by IVF, and, after diagnosis of sex (see below), returned 17 of these to the patients (one or two returned in each of 10 treatment cycles). Two of these 10 cases had been confirmed as clinical pregnancies (two normal sets of twins seen by ultrasonography at 20 and 22 weeks). According to Dr A. Handyside (personal communication), five healthy girls have now been delivered from a total of 22 transferred. Further pregnancies will be required to evaluate fully the biopsy procedure for human pre-embryos. Nevertheless, these preliminary results are ex-

tremely encouraging and indicate that the biopsy of human pre-embryos can be safely performed without compromising the ability of the pre-embryo to develop into a normal baby. Exciting times lie ahead!

D. Proliferation of Cells Biopsied From Pre-embryos

Having removed a cell from a pre-embryo, it may be necessary to culture the cell in order to increase the material available for the genetic test. Wilton and Trounson [1989] showed that a single blastomere from a four-cell mouse pre-embryo, cultured in human amniotic fluid in the presence of extracellular matrix components (fibronectin, laminin, or a complex of laminin and nidogen), can replicate to form a monolayer of about 20 cells. Recent advances in the culture of cells from human pre-embryos [e.g., Muggleton-Harris et al., 1990] are also encouraging. Muggleton-Harris [1990] has reported that when two blastomeres from a human four-cell pre-embryo were cultured in Ham's F-12 medium with L-glutamine or in amniotic fluid, using microdrops of medium on fibronectin or gelatin coated dishes, the blastomeres replicate at least two or three times. Similarly, abembryonic trophectoderm biopsied from human blastocysts attached to gelatin and fibronectin substrates and replicated in conditioned media (AL Muggleton-Harris, I Findlay, DG Whittingham, and VN Bolton, personal communication). A number of research groups are trying to improve the culture conditions for human pre-embryos, so it may soon be possible to produce sufficient cells from a single human blastomere for a variety of genetic tests.

V. CRYOPRESERVATION AND STORAGE OF BIOPSIED PRE-EMBRYOS

If the genetic test is performed on cells removed from the pre-embryo, it will be necessary to store the pre-embryo until the test result is known. If the genetic test can be completed quickly, the pre-embryo may be safely left in culture, since pregnancies can be achieved after transfer of human blastocysts to the uterus [Bolton et al 1990]. However, it is unlikely that the pre-embryo could be maintained in culture beyond 5 days after oocyte collection in an IVF cycle. So, if the genetic test cannot be completed within this time, it would be necessary to preserve the pre-embryos frozen and then return those that are free of the genetic disorder at an appropriate time in a subsequent cycle.

The methods used for cryopreservation of human pre-embryos stem from the initial successes with mouse pre-embryos [Whittingham et al., 1972; Wilmut, 1972], which involved slow cooling in the presence of a cryoprotectant (dimethylsulfoxide [DMS]), storage in liquid nitrogen (−196°C), and slow thawing. The frozen pre-embryos are usually stored in plastic straws. Successful freezing of pre-embryos depends on the prevention of excessive intracellular ice during freezing or thawing. Ice formation is reduced by dehydration of the pre-embryo: Water flows out to restore the equilibrium, across the membrane, when the concentration of extracellular solutes is increased by freezing, or by the addition of a nonpermeating solute, such as sucrose. This in turn raises the intracellular concentration of solutes. Since this could cause cell damage, cryoprotectants such as glycerol, DMSO, and 1,2-propanediol (PROH) are used to replace the intracellular water and to protect the pre-embryo from the high intracellular concentrations of solutes. Current methods include 1) slow cooling [Whittingham et al., 1972; Wilmut, 1972]; 2) rapid, two-step cooling, involving a slow cool to −30°C to −40°C followed by a plunge into liquid nitrogen at −196°C [Wood and Farrant, 1980]; 3) ultra-rapid cooling [Trounson et al., 1987]; and 4) vitrification [Rall and Fahy, 1985]. Thawing rates are also critical; slow thawing is used in conjunction with slow cooling (method

1) and rapid thawing is used with the other techniques. Most IVF clinics use either of the first two methods to freeze cleavage stage pre-embryos or pronucleate stage oocytes in PROH or DMSO, but for blastocysts glycerol may be a better cryoprotectant [Fehilly et al., 1985]. The last two methods have not yet been evaluated properly with human pre-embryos.

Vitrification is a relatively new and potentially simpler method that uses high concentrations of cryoprotectants to avoid crystallization when cooled to low temperatures. The cryoprotectant solutions are supercooled and become so viscous that they "vitrify" to form a glassy solid. Pre-embryos are equilibrated in a mixture of cryoprotectants, and, when the cytoplasm is dehydrated, the straw containing the pre-embryo and vitrification solution is immersed into liquid nitrogen to vitrify. Thawing must be sufficiently rapid to avoid crystallization. Vitrification requires high concentrations of a mixture of cryoprotectants that is toxic at room temperature, so pre-embryos can only be exposed to them at low temperatures (around O°C). Although it has been used for mouse pre-embryos, as far as we are aware only Quinn and Kerin [1986] have tried this technique with human pre-embryos. Only 1 of the 22 pre-embryos survived vitrification, and this did not produce a pregnancy. Trounson [1986] and Feichtinger et al. [1987] have also tried vitrification with human oocytes and have shown that they can survive, be fertilized, and develop to at least the eight-cell stage. Feichtinger et al. [1987] transferred some, but no pregnancies resulted.

The ultra-rapid freezing method is a simple approach that has the advantage of avoiding the toxic concentrations of cryoprotectants used in vitrification. The pre-embryos are exposed to cryoprotectants at room temperature and plunged directly into liquid nitrogen; they are subsequently thawed rapidly at 37°C. This method has resulted in good survival and pregnancy rates for cryopreserved mouse pre-embryos [Wilton et al. 1989], and Trounson et al. [1988] reported that 11/12 human pre-embryos survived ultra-rapid freezing and thawing with more than 50% of their blastomeres intact, but no pregnancies resulted when these were transferred to six patients. The updated figures for 1988 and 1989 [Trounson, 1990] revealed that of 75 human pre-embryos frozen by the ultra-rapid technique, 55 were successfully thawed and transferred to 37 patients. This resulted in three pregnancies, but none came to term (one ectopic pregnancy and two spontaneous abortions of developmentally and genetically normal fetuses at 17 and 22 weeks, respectively). Although this is clearly a promising approach for studies with mouse pre-embryos and the preliminary studies with human pre-embryos produced pregnancies, there were no live births, and overall the results are less encouraging than with mouse pre-embryos.

Current methods of cryopreservation of mouse pre-embryos have been reviewed by Wood et al. [1987], and application to human pre-embryos has been extensively reviewed [e.g., Ashwood-Smith, 1986; Trounson, 1986, 1990; Friedler, et al., 1988; Siebzehnrubl, 1989]. The first report of a pregnancy following cryopreservation of a human pre-embryo was published by Trounson and Mohr [1983]. This pregnancy did not continue to term, and the first live birth following cryopreservation was achieved in 1983 by Zeilmaker et al. [1984]. Cryopreservation has been used with a large number of cleavage stage pre-embryos and has also been used successfully with pronucleate eggs [e.g., van Steirteghem et al., 1987], blastocysts [e.g., Fehilly et al., 1985], and, so far with less success, unfertilized oocytes [reviewed by Friedler et al., 1988; Fugger, 1989]. Although it is possible that cryopreservation of oocytes and pronucleate eggs may be useful if polar body biopsy was used for diagnosis, it seems more likely that cryopreservation of cleavage stages and blastocysts will be more appropriate for preimplantation diagnosis. There is also a concern that cryopreservation of unfertilized oocytes could result in an increased risk of aneuploidy, because the procedures that are used affect the meiotic spindle [Johnson and Pickering, 1987; Pickering and Johnson, 1987]. One careful cytogenetic study revealed no increase in aneuploidy after cryopreservation of mouse oocytes [Glenister et al., 1987], but another study showed an increase in aneuploidy among vitrified mouse oocytes [Kola et al., 1988]. Until this risk is better understood, it is unlikely that oocyte freezing will be adopted.

Cryopreservation will only be useful in conjunction with preimplantation diagnosis if a high proportion of human pre-embryos survive the freeze–thaw procedures to produce normal pregnancies. At present the survival rate is around 60% (see below). It is often difficult to compare the results from different medical centers because data are presented in different ways and the same data set is sometimes included in consecutive publications that update the results of a particular center. The most relevant statistic is the number of babies produced per pre-embryo frozen. Since many pregnancies are ongoing when the data are written up for publication, results are often given as the number of fetuses (or implanted conceptuses seen as amniotic sacs by ultrasound) per frozen pre-embryo. This is an acceptable alternative, but it must be remembered that even after IVF (without cryopreservation) the number of live births is reduced by abortion, ectopic pregnancy, and perinatal death and this is only about 70%–80% of the number of pregnancies [ILA, 1990]. Some data relating to survival of human pre-embryos after cryopreservation are summarized in Tables II and III. These data are for illustrative purposes only and do not represent a systematic search of the literature. A more extensive list is given in Table 3 of Friedler et al. [1988]. It is difficult to compare the results with frozen and fresh pre-embryos. Bias arises because some published studies are no more than case reports of the successful examples and thus the true success rate is lower. On the other hand, in most centers the "best" three pre-embryos are transferred "fresh" and the remainder are cryopreserved. Since the "best" pre-embryos are less often cryopreserved, and "poor quality" embryos are more likely to be damaged by freezing and thawing [Van Steirteghem et al., 1987; Camus et al., 1989], this will probably reduce the success rate compared with "fresh" pre-embryos.

The results given in Tables II and III show that pregnancies can be achieved by transferring either fresh or cryopreserved pre-embryos at cleavage or blastocyst stages. The number of implanted conceptuses per cleavage pre-embryo transferred is similar for fresh and frozen pre-embryos, although about 40% of the frozen pre-embryos failed to survive the cryopreservation. Consequently, the number of implantations per pre-embryo frozen ($\geq 4.1\%$) is lower than the number per fresh pre-embryo transferred (9.8%). The results with frozen pronucleate eggs and blastocysts are based on smaller numbers of pre-embryos but appear to be at least as good. However, this obscures the fact that many human cleavage stage pre-embryos do not survive to the blastocyst stage in culture (see section III), and thus the additional loss of pre-embryos needs to be taken into account. Nevertheless, these results are encouraging because biopsies for pre-implantation diagnoses that require cryopreservation could be done at either the cleavage or blastocyst stage without compromising the success of the cryopreservation. Moreover, since the time that many of these studies were undertaken cryopreservation methods have continued to be refined and tailored to different developmental stages. Some groups have re-

ported higher success rates. For example Testart [1988] reported 38 conceptuses from a group of 400 pre-embryos that were cryopreserved (9.5%), which was similar to the success rate in the same study (52/456 or 11.4%) for transferring fresh pre-embryos.

For cryopreservation to be of use for pre-implantation diagnosis, it must be applicable to human pre-embryos after the biopsy procedure, which would involve either the removal or puncture of the zona pellucida. The first successfully cryopreserved bovine embryo was frozen at the hatched blastocyst stage (comprising several hundred cells), without a zona pellucida [Wilmut and Rowson, 1973], and, more recently, Blakewood et al. [1986] found that there were no differences in survival of bovine embryos frozen with intact zonae or no zonae at the late morula or early blastocyst stage. There is some evidence to suggest that the presence of a damaged zona may be detrimental. Although Kanagawa et al. [1979] reported that bovine pre-embryos did not survive slow cooling cryopreservation well if their zonae pellucidae had been punctured, Niemann et al. [1986] claimed that the success could be improved by placing the pre-embryo within a second zona pellucida in order to cover the puncture hole and, encouragingly, Wilton et al. [1989] have recently reported good results with ultrarapid freezing for biopsied mouse pre-embryos. A high proportion of fetuses were produced from cryopreserved four-cell stage mouse pre-embryos frozen within 1 hour after the zona had been punctured and a blastomere removed. After thawing, the pre-embryos were cultured for a further 40 hours before being transferred to a pseudopregnant female recipient. Biopsied (frozen–thawed), nonbiopsied (frozen–thawed), and nonfrozen mouse pre-embryos had implantation rates of 81% (30/37), 74% (26/35), and 74% (20/27) and formed fetuses in 62%, 63% and 67% cases, respectively.

TABLE II. Some Results After Cryopreservation of Human Pre-Embryos

| | No. of pre-embryos frozen and thawed | Condition of pre-embryos on thawing | | | | Pregnancies per transfer | Implantation[b] per pre-embryo (excluding preclinical pregnancies) | |
| | | Cell survival | | | Suitable for transfer[a] | | | |
Study	thawed	None	Some	All			Transferred	Frozen
Pronucleate eggs								
1	778	—	—	—	540	47/270	≥47/540	≥47/778
2	56	16	0	40	40	7/26	7/26?	7/56
3	16	2	0	14	14	Included in cleavage stage data for ref. 3		
Total					594/850 (69.9%)	54/296 (18.2%)	≥54/566 (≥9.5%)	≥54/834 (≥6.5%)
Cleavage stages								
1	1,567	—	—	—	1,089	68/544	≥68/1089	≥68/1567
2	232	?	45–170?	62	107?	13/84	12/100?	12/232
3	37	6	7	11	28	7/32[c]	7/42[c]	7/53[c]
4	396	—	—	—	229	16/144	(13+3[d])/229	(13+3[d])/396
5	21	4	12	5	16	1/15	1/16	1/21
6	15	0–7?	6–13?	2	8	2/8	3[e]/8	3[e]/15
7	18	—	—	—	10	1/8	1/10	1/18
8	10	4	2	4	6	3/6?	(1+1[d])/6	(1+1[d])/10
9	?	—	—	—	—	4/13	(4+1[d])/22	?
10	88[f]+11[g]	—	—	—	10+1[g]	3/7	(2+1[h])/10	(2+1[h])/88
11a	347	—	—	—	226	21/123	(16+2[d]+2[h])/226	(16+2[d]+2[h])/347
11b	75	—	—	—	55	3/37	(2+1[h])/75	(2+1[h])/75
12	74	38?	19	17	26	4/17	4/26?	4/74?
Total					1,811/2,891 (62.6%)	146/1,038 (14.1%)	≥145/1,859 (≥7.8%)	≥119/2,896 (≥4.1%)
Blastocysts								
1	186	—	—	—	138	3/69	≥3/138	≥3/186
12	38	2?	9	17	25	8/15	8/25?	8/38?
Total					163/224 (72.8%)	11/84 (13.1%)	≥11/163 (≥6.7%)	≥11/224 (≥4.9%)

Studies are 1) Compilation of U.S.A. data reported by Fugger [1989]; 2) Van Steirteghem et al. [1987]; 3) Testart et al.[1986]; 4) Freemann et al. [1986]; 5) Trounson and Mohr [1983]; 6) Zeilmaker et al. [1984]; 7) Siebzehrubl et al [1986], cited by Siebzehrubl, [1989]; 8) Lucena et al. [1986]; 9) Salat-Baroux et al. [1988]; 10) Quinn and Kerin [1986]; 11a, 11b) Trounson [1990] (results shown as study 11b are for ultra-rapid freezing); 12) Fehilly et al. [1985]. Other studies are listed in Table III of Friedler et al. [1988].
[a]Pre-embryos with ≥50% cell survival were usually judged to be suitable for transfer.
[b]Data from study 1 were given as numbers of pregnancies, so the numbers of implantations were probably greater than those shown.
[c]Pregnancy data for study 3 include pre-embryos frozen as pronucleate eggs. Testart[1988] later recorded 38 amniotic sacs from a group of 400 cryopreserved pre-embryos (9.5%).
[d]First trimester abortions.
[e]Includes one set of monozygotic twins.
[f]Includes 8 blastocysts.
[g]Vitrified pre-embryos.
[h]Ectopic pregnancy.

VI. EMBRYO TRANSFER

In IVF treatment cycles, the pre-embryo is usually transferred at the two- or four-cell stage two days after oocyte recovery. The pre-embryos are loaded into an appropriate catheter, which is inserted into the uterus, via the cervix, where the pre-embryos are

TABLE III. Pregnancy Rates After Transfer of "Fresh" and Cryopreserved Human Pre-embryos

	"Fresh" pre-embryos		Cryopreserved pre-embryos	
Cleavage stage pre-embryos				
Pregnancies/transfer	1,354/6,553	(20.7)	146/1,038	(14.1)
Implantations/pre-embryos transferred	1,720/17,645	(9.8)	≥145/1,859	(≥7.8)
Implantations/pre-embryos frozen	Not applicable		≥119/2,896	(≥4.1)
Blastocysts				
Pregnancies/transfer	3/29	(10.3)	11/84	(13.1)
Implantations/pre-embryos transferred	4/58	(6.9)	≥11/163	(≥6.7)
Implantations/pre-embryos frozen	Not applicable		≥11/224	(≥4.9)

Values in parentheses are percentages. Data for "fresh" cleavage stage pre-embryos are for 1,988 IVF pregnancies in the United Kingdom taken from Table II (pregnancies/transfer) and calculated from Table VI (implantations/pre-embryo) of the Fifth Report of the ILA [1990]. Data for "fresh" blastocysts are from Bolton et al. [1991]. Data for cryopreserved pre-embryos are taken from Table II in this chapter.

expelled. Preimplantation diagnosis on eight-cell or blastocyst stage pre-embryos will require embryo transfers to be delayed for 12 hours to 3 days. If the diagnosis takes several days cryopreservation will be needed (discussed above), and pre-embryos will be transferred in a subsequent cycle. It has been suggested that it may be beneficial to delay the transfer until a later menstrual cycle in order to improve the synchrony between the pre-embryo and the uterus and to avoid exposing the pre-embryo to the abnormal endocrine environment of a stimulated IVF cycle [e.g. Fehilly et al., 1985; Frydman et al., 1988]. In the context of preimplantation diagnosis, this would have the added advantage of removing the pressure on laboratory staff to produce a reliable diagnosis by a set deadline. It would, of course, be important to ensure that the patient avoided unprotected sexual intercourse that might result in a pregnancy with a pre-embryo that was produced in the later cycle and thus had not been monitored by preimplantation diagnosis.

The two twin pregnancies, following preimplantation diagnosis, reported by Handyside et al. [1990] clearly show that transfer of the pre-embryo can safely be left to day 3 of the treatment cycle. Furthermore, Bolton et al. [1991] showed that human pre-

embryos could develop to term if they were transferred at the blastocyst stage on day 5 of an IVF treatment cycle. This offers two options for preimplantation diagnosis. First, it means that biopsy of mural trophoectoderm at the blastocyst stage would be possible. (The possible advantages of this are discussed above.) Second, if biopsy were performed at the eight-cell stage, there is a period of 2 days in which to perform the diagnosis. Not only would this be welcomed by those responsible for the diagnosis but also it might help to select the "best" pre-embryos (those that were both free of the genetic disorder and were capable of developing to the blastocyst stage) for transfer.

VII. SENSITIVE GENETIC TESTS

Several types of genetic tests are now being evaluated for use in preimplantation diagnosis. These include chromosome analysis [e.g., Angell et al. 1988], biochemical assays [e.g., Monk, 1988; Braude et al., 1989], and molecular tests involving either in situ hybridization [Jones et al. 1987; West et al., 1987, 1988; Penketh et al., 1989], or amplification of target DNA by PCR [Handyside et al., 1989].

A. Cytogenetic Tests

Because preimplantation diagnosis is performed on one or a few cells biopsied from an early cleavage pre-embryo, classical chromosome analysis will only be possible after culture to produce metaphase figures. This creates delays in diagnosis and also runs the risk of introducing artifacts resulting from the culture procedure.

Cattle pre-embryos have been sexed by sex chromosome analysis of groups of 7–17 cells aspirated from 6–7 day blastocysts. In the two studies reviewed by Betteridge et al. [1981], 45/62 (73%) of preparations had more than one metaphase spread and 32/62 (52%) were sexed. Of 353 cattle pre-embryos biopsied after hatching from the zona pellucida (12–15 days) in five laboratories, 241 (68%) were sexed, and 110 were transferred and resulted in 36 pregnancies (33% of those transferred).

Three cytogenetic analyses of biopsied preimplantation mouse embryos have been reported. Bacchus and Buselmaier [1988] and Nijs et al. [1988] biopsied four-cell and two-cell mouse embryos, respectively. Roberts et al. [1990] compared biopsy of four-cell, eight-cell, and morula stages and had most success with two cells removed from eight-cell stage pre-embryos. After these pairs of cells were exposed to colchicine and cultured for a total of 18–26 hours, 56% of 149 samples contained at least one metaphase spread (26% had only interphase nuclei, 7% died, and 11% were lost during the processing). The present success rate is not acceptable for use in preimplantation diagnosis. It is also possible to sex human cells by the presence of sex chromatin (Barr body) in cells with more than one X chromosome, but it has been frequently shown [West et al., 1988] to be subject to error and consequently not sufficiently reliable for sexing pre-embryos.

B. In Situ Hybridization

The technique of hybridizing chromosome-specific or gene-specific probes, labeled with a radioactive or immunogenic marker, to cells in order to identify the presence or absence of specific sequences or the risk of aneuploidy of specific chromosomes has recently undergone considerable improvement in sensitivity and resolution. While "classic" in situ hybridization (if we may use the term for a technique that is barely 21 years old) depended on labeling DNA probes with tritium, with consequent lack of resolution, most investigations today use immunogenic moieties (in particular biotin, but also digoxigenin and alkylated nucleotides) to label the probes.

These nonisotopic methods have been shown to have wide applications in the investigation of aneuploidy and chromosome rearrangement (translocation) [Lichter et al., 1988a,b]. However, the primary limitation to their use in preimplantation diagnosis is in the number of cells available for such investigation. Chromosome-specific probes are now available [e.g., Pinkel et al., 1988], and these are capable of identifying monosomy or trisomy, or some unbalanced translocations, in interphase nuclei ("interphase cytogenetics"). But to do this, the investigator must be aware of specific risks. It is not possible to carry out 24 separate hybridizations (one for each chromosome type), each of which requires a number of cells to be analyzed, to eliminate the risk of aneuploidy. Even less feasible is the analysis of potential unbalanced translocations, unless one of the parents carries a balanced translocation, the components of which are susceptible to analysis (see below). For these, as for less gross cytological abnormalities (deletions, duplications, or inversions), molecular analysis, using the power of PCR is the only practical approach (see section VII.D).

The feasibility of using in situ hybridization for preimplantation diagnosis has been tested by sexing whole human pre-embryos by hybridization to a DNA probe specific for repeated DNA sequences on the Y chromosome [Jones et al., 1987; West et al. 1987, 1988; Handyside et al., 1989; Penketh et al., 1989; Pieters et al., 1990]. In our own studies [West et al., 1987, 1988], a few pre-embryos were sexed by hybridization to biotinylated Y probe, but most were analyzed with tritiated Y probe. In all, excluding five pre-embryos that either failed to cleave or had been previously mounted in DPX or hybridized to biotinylated Y probe, 27 pre-embryos were analyzed by in situ hybridization to tritiated Y probe, pHY2.1 (Fig. 6). Of these, 7 (26%) were technical failures (4 were lost during processing and 3 gave equivocal results). The remaining 20 pre-embryos (74%) were sexed (8 females and 12 males), and in 4 cases the sex was confirmed either by karyotype analysis or by identification of a fluorescent Y chromosome after staining with spermidine *bis*-acridine or quinacrine dihydrochloride. Four of the 27 pre-embryos had been pretreated with either spermidine *bis*-acridine or quinacrine dihydrochloride and UV light (two males, one equivocal, and one lost during processing), and another three were lost during processing. Eight of the remaining 20 pre-embryos were classified as morphologically abnormal (4 male, 2 female and 2 that gave equivocal results), and 12 were classified as morphologically normal. All 12 morphologically normal pre-embryos (ranging from the two-cell stage to a blastocyst with 107 cells) were sexed with confidence (6 females and 6 males). The incidence of nuclei with Y hybridization sites among the 12 male pre-embryos (2 pre-treated, 4 with abnormal morphology, and 6 normal) is shown in Table IV. Although 97% of interphase nuclei from a large blastocyst had a Y hybridization site, only 54% of the interphase nuclei were positive from morphologically normal cleavage stage pre-embryos that were classified as male.

Some of the morphologically abnormal pre-embryos had small (probably fragmenting) nuclei without hybridization bodies, and two of these pre-embryos were not sexed. All four male morphologically abnormal pre-embryos (4–8-cell stages) had one or more nucleus with multiple hybridization bodies, and about 10% of nuclei in the morphologically normal blastocyst had two hybridization bodies. It seems likely that polyploidy occurs both during the normal development of human blastocysts and in early cleavage stages of abnormally developing pre-embryos [Angell et al., 1987; West et al., 1987, 1988]. The polyploidy in the latter group may be a culture-induced phenomenon. In addition, some triploid pre-embryos were produced when two sperm fertilized a single egg [Angell et al., 1987], and two Y hybridization sites were detectable by in situ hybridization in 69 XYY triploids [West et al., 1987, 1988].

Three other laboratories have also used DNA–DNA in situ hybridization to sex human pre-embryos [Jones et al., 1987; Penketh et al., 1989; Pieters et al., 1990]. Jones et al [1987] reported three pre-embryos that were sexed with a different tritiated Y probe (p102d). Penketh et al. [1989] sexed 51 pre-embryos (33 cleavage stage, 8 morulae, and 10 blastocysts) by in situ hybridization with biotinylated pHY2.1 and subsequent detection with streptavidin-linked alkaline phosphatase. The in situ hybridization confirmed the sex for 25 of 27 pre-embryos that had already been sexed either cytogenetically or by Y-body fluorescence (16 females and 9 males) and provided ambiguous results for the remaining two that had been sexed as females. Of 21 pre-embryos that had not previously been sexed, 8 were scored as males and 13 as females, although there was some doubt about three of the females. Overall, 66% of male interphase nuclei hybridized to the Y probe. In addition, two pre-embryos were biopsied to allow the biopsies to be analyzed by in situ hybridization while the re-

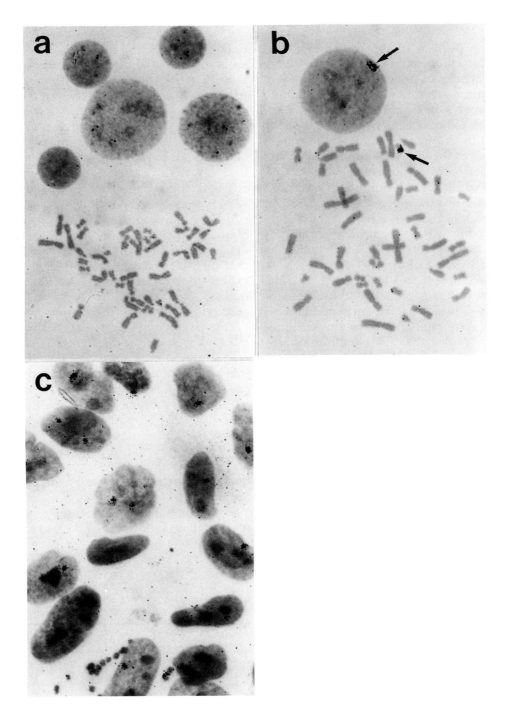

Fig. 6. *Autoradiographic detection of tritiated Y probe hybridised to spreads of* **a:** *control female cells,* **b:** *control male cells, and* **c:** *a human blastocyst. Clusters of silver grains (arrowed in b) denoting the presence of the Y chromosome are seen in b and c, indicating that the blastocyst was male. Some cells in c have two such clusters of silver grains and are probably tetraploid cells with two Y chromosomes.*

maining portion of the pre-embryo was used for cytogenetic sexing. In one case both techniques indicated a male pre-embryo and in the other case the pre-embryo was shown to be chromosomally female but one of the three nuclei hybridized to the Y probe. In situ hybridization was also used to confirm the sex of 19 pre-embryos that had been biopsied for sexing by PCR DNA amplification [Handyside et al., 1989]. In 17 cases the two techniques agreed (4 males, 11 females, and two parthenogenetic females), but two polyspermic pre-embryos were sexed as males by PCR and female by in situ hybridization. This discrepancy is discussed in section VII.D.

Finally, Pieters et al. [1990] used a fluorescent immunocytochemical method to detect sulphonated probes for chromosome 1 (probe pUC 1.77) and the Y chromosome (probe pHY2.1) on human oocytes and pre-embryos. Twenty-three oocytes were tested with the chromosome 1 probe, but only seven showed a positive hybridization signal, and in four of these the signal was weak. Of the eight pre-embryos tested with the chromosome 1 probe, four gave positive hybridization signals, two were lost during processing, and two gave no signal. In two diploid pre-embryos, two strong positive signals were detected in about half of the nuclei (2/4 in one case; no

details given for the other), and one pre-embryo that had three pronuclei produced three hybridization signals with the chromosome 1 probe. The only pre-embryo (one-cell zygote) tested with the Y probe had four pronuclei and produced four essentially haploid metaphase spreads, three of which were positive after in situ hybridization with the Y probe.

These studies show that it would be feasible to sex pre-embryos by DNA–DNA in situ hybridization, but it is not clear how many cells would be required for a reliable sex determination. In our studies we found that early cleavage embryos had large nuclei with relatively diffuse hybridization bodies, but approximately 50% were readily scored with the tritiated probe (Table IV) [West et al., 1987, 1988]. Calculations based on the incidence of labeled nuclei in male and female lymphocytes suggest that it should normally be possible to sex morphologically normal pre-embryos with samples of three or more nuclei [West et al. 1988]. The frequency of false-negative interphase nuclei may be higher for early cleavage nuclei than for the control lymphocytes in which case more than three nuclei may be required for reliable sex determination.

With tritiated DNA probe, sexing would take 4–8 days (allowing 3–7 days to expose

TABLE IV. Frequency of Nuclei With Y Bodies in 12 Male Pre-Embryos After *In Situ* Hybridization With Tritiated Y Probe*

	Pre-embryos	Metaphases	Interphase nuclei
Morphologically normal			
Cleavage stages	5	10/10	6/13 (46)
Cleavage stages (pretreated**)	2	3/3	2/2 (100)
Total cleavage stages			(54)
Blastocysts	1	20/21	83/86 (97)
Morphologically abnormal			
Cleavage stages	4	1/1	12/30 (40)

*Data are from West et al. [1988]. Values in parentheses are percentages.
**Pre-treated spermidine *bis*-acridine or quinacrine dihydrochoride and UV light.

the autoradiographs), but with sulphonated DNA [Pieters et al., 1990] or DNA labeled with biotin [e.g., Burns et al., 1985; West et al., 1988; Handyside et al. 1989; Penketh et al., 1989] or digoxygenin, this could be reduced to 1–2 days. The sensitive fluorescent methods [Lichter et al, 1988a,b; Pinkel et al., 1988; Pieters et al., 1990] are becoming increasingly popular and are suitable for diagnostic work that requires a rapid result but does not demand a permanent preparation. Such methods should be fast enough to allow the identification of female pre-embryos in time for replacement in the same menstrual cycle, especially if transfer can be delayed until the blastocyst stage.

If a Y probe is used clinically to sex pre-embryos by in situ hybridization to interphase nuclei, it would be important to test blood samples from both parents to screen for any abnormalities. For example, interphase nuclei of male Y/autosome translocation carriers have two Y hybridization signals [Ellis et al., 1990], and so interphase nuclei from a female conceptus with a Y/autosome translocation would have a single hybridization body and, on this basis, be scored as a male unless the parents are screened. The identification of inherited chromosome translocations in interphase nuclei by in situ hybridization [Pinkel et al., 1988; Ellis et al., 1990] raises the possibility of whether this would be a useful approach for preimplantation diagnosis.

An individual who carries a balanced translocation is clinically normal because this simply involves exchange of chromosome segments between two chromosomes without loss or gain of genetic material. However, a balanced chromosome carrier is at risk for producing chromosomally unbalanced offspring who will have either gained or lost a chromosomal segment and thus be genetically abnormal. Because the ends of chromosomes (telomeres) do not readily stick to other chromosomes, translocations invariably occur by exchanging

chromosome material between two or more breakpoints. Translocations are classified as 1) reciprocal translocations (exchange of two chromosome segments that include the telomeres), 2) insertional translocations (an interstitial segment is deleted from one chromosome and inserted into another), and 3) robertsonian translocations or centric fusions (fusion of two chromosomes at breakpoints close to the centromere). A robertsonian translocation between two acrocentric chromosomes generates a dicentric chromosome and an acentric fragment that comprises only the genetically inert heterochromatin that was close to the centromeres. Such a fragment is usually lost. Centric fusion of chromosomes 13 and 14 and of chromosomes 14 and 21 are the two most common human translocations. Translocations involving chromosome 21 (e.g., 13/21, 14/21, and 21/22) are of particular importance because they are involved in the inherited form of trisomy 21 (Down syndrome).

Fluorescence in situ hybridization can be done with chromosome-specific probes [e.g., Pinkel et al., 1988] labeled with up to three different colored fluorochromes [Nederlof et al., 1989] to identify chromatin from the different chromosomes and this has been termed *chromosome painting* [Pinkel et al., 1988]. It may be possible to use this approach on interphase nuclei to detect pre-embryos with inherited unbalanced translocations, as illustrated in Figure 7 for a 14/21 robertsonian translocation. Of the six classes of pre-embryos, four would be unbalanced translocation carriers (three would normally be lethal later in pregnancy, and the other would be trisomy 21), one would be a balanced carrier, and one would be normal. Only the latter two classes of pre-embryos would be suitable for transferring to the uterus for further development. If a chromosome 21 probe was used alone then there would be a risk of transferring pre-embryos that had a normal complement of chromosome 21 but

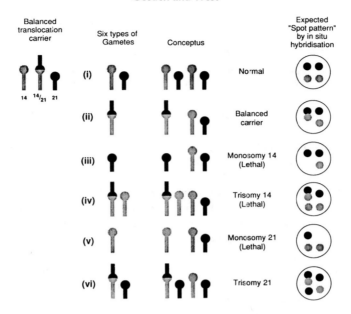

Fig. 7. *Diagram showing the six chromosomally distinct types of pre-embryos expected for a balanced 14/21 robertsonian translocation carrier and the theoretically expected pattern of spots in interphase nuclei after in situ hybridization with probes for chromosomes 14 and 21 (lighter shading represents chromosome 14 and darker shading represents chromosome 21). Four classes of pre-embryos would be unbalanced translocation carriers (three would normally be lethal later in pregnancy and the other would be trisomy 21), one would be a balanced carrier, and one would be normal.*

were either monosomic or trisomic for chromosome 14. If chromosome 21 and chromosome 14 probes were labeled with different fluorochromes, each chromosome would be identified as a different colored spot in interphase nuclei, and in theory the six types of pre-embryo could be distinguished by the number and color of spots (translocation chromosomes would produce two-colored spots), but this would be technically demanding. Lichter et al. [1988] found that the frequency of trisomy 21, interphase nuclei with three spots after hybridization with chromosome 21–specific probes was only 55%–65% (<5% had

no signal, 5%–15% had one spot, and 25%–35% had two spots). The use of several probes would compound these difficulties, and clearly more than one cell would be needed for such an analysis.

Although there have been technical improvements in in situ hybridization since the early studies of pre-embryos, this approach is only feasible for preimplantation diagnosis that involves the detection of highly repeated sequences. It may therefore be used to detect numerical chromosome anomalies, including inherited translocations, but is not suitable for the diagnosis of single-gene defects. For these types of anomalies PCR provides a more promising approach.

C. Biochemical Assays as Genetic Tests for Preimplantation Diagnosis

In principle, biochemical assays of specific gene products (usually enzymes) could be used to diagnose some specific genetic disorders or to determine the sex of a pre-embryo. Monk [1988] listed three requirements for this approach: 1) the gene is expressed (protein product) or can be induced in pre-embryos of the appropriate stage; 2) there is little residual oocyte-coded gene product that could confuse the diagnosis; 3) a sensitive and accurate assay, with an appropriate control (e.g., assay of another enzyme) is available. To this we could add that the developmental and genetic variation among pre-embryos sampled at a particular stage should not be large enough to confuse the diagnosis. Assays for a second gene product may reduce the variation attributable to developmental asynchrony but this will only help if the developmental profiles of the two gene products are identical.

Polar bodies would be unsuitable for biochemical diagnosis because they would only have residual oocyte-coded activity, but cells from cleavage stage pre-embryos or blastocysts could be used if the proportion of oocyte-coded gene product was negligible compared with the level of that produced by the genome of the pre-embryo. Residual oocyte-coded products are more likely to be a problem in early cleavage stages than in the blastocyst, but some gene products will persist for longer than others. For X-linked enzymes, expectations will depend on whether X chromosome inactivation has occurred in the cells removed for the assay. In mouse pre-embryos, biochemical evidence of X-inactivation is first seen at the blastocyst stage [reviewed by West, 1990] and suggests that X-inactivation occurs earlier in the trophectoderm and primitive endoderm lineages than in the primitive ectoderm [Kratzer and Gartler, 1978; Monk and Harper, 1979]

In the mouse, there is a large store of maternally derived RNA and protein in the one-cell zygote. Most maternal mRNA is degraded at the two- or four-cell stage [Johnson, 1981; Magnuson and Epstein 1981; Piko and Clegg, 1982], but some persists until the blastocyst stage [Bachvorova and DeLeon, 1980], and at least one oocyte-coded protein (glucose phosphate isomerase) has been detected after implantation [West and Green, 1983]. The genome of the pre-embryo is activated at the two-cell stage in the mouse [Johnson, 1981; Magnuson and Epstein, 1981; Flach et al., 1982; Bolton et al., 1984] and probably between the four- and eight-cell stage in the human [Braude et al., 1988]. However, different genes are activated at different times during preimplantation development, and many tissue-specific genes are not activated until later in development.

Epstein [1975] compared the reported activities of 15 different enzymes in preimplantation mouse pre-embryos (Fig. 8). From this comparison it can be seen that some enzymes are initially present at low levels but the activity rises rapidly, whereas other enzymes are initially present at much higher levels that may remain stable or even decline during preimplantation development. Although the rapid rise in the activity of some enzymes probably represents new enzyme synthesis after the activation of the embryonic genome, it is not safe to assume that this is so because, in some cases, it can represent synthesis of new protein from maternally inherited mRNA. Genetic polymorphisms have been used to show that the paternally derived genes encoding β-glucuronidase and β-galactosidase (which shows a similar rapid rise in activity) are expressed at early cleavage stages concomitant with this rapid rise [Wudl and Chapman, 1976; Esworthy and Chapman, 1981]. Genetic variants of glucose phosphate isomerase (GPI)

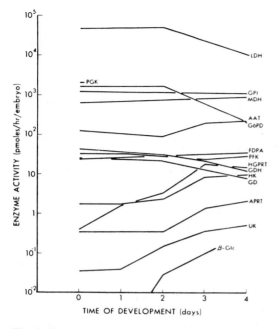

human preimplantation development [West et al., 1989a]. A larger study of the X chromosome–coded enzyme hypoxanthine guanine phosphoribosyltransferase (HGPRT or HPRT) and the autosomal enzyme adenine phosphoribosyltransferase (APRT), involving 56 unfertilized human oocytes and 46 human pre-embryos (between one-cell and blastocyst stages) revealed no major increase in activity during the preimplantation period [Braude et al., 1989]. The HPRT and APRT activity levels in human pre-embryos showed a downward trend between the four-cell and blastocyst stages rather than the dramatic rise in activity seen in early mouse pre-embryos (Fig. 8). This difference in activity profile between mouse and human pre-embryos means that it will not be safe to extrapolate from mouse data to decide whether oocyte-coded enzyme persists in human pre-embryos. This again highlights the need for more research on

Fig. 8. *Enzyme activities in preimplantation mouse pre-embryos, plotted on a logarithmic scale. The enzymes (from top to bottom) are LDH, lactate dehydrogenase; PGK, phosphoglycerate kinase; GPI, glucose phosphate isomerase; MDH, malate dehydrogenase; AAT, aspartate aminotransferase; G6PD, glucose-6-phosphate dehydrogenase; FDPA, fructose-1,6-diphosphate aldolase; PFK, phosphofructokinase; HGPRT, hypoxanthine guanine phosphoribosyltransferase; GDH, glutamate dehydrogenase; HK, hexokinase; GD, guanine deaminase; APRT, adenine phosphoribosyltransferase; UK uridine kinase; β-Glc, β-glucuronidase. (Reproduced from Epstein, 1975, with permission of the publisher.)*

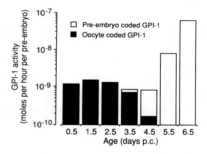

Fig. 9. *Changes in oocyte-coded and embryo-coded activity of the autosomal-coded enzyme glucose phosphate isomerase during the preimplantation (0.5–4.5 days) and early postimplantation (5.5 and 6.5 days) stages of mouse development. Data are from West et al. [1986]. Some 5.5 day pre-embryos have low levels of oocyte-coded enzyme [West and Green, 1983], but this is not shown in the figure because it is difficult to quantify. (Reproduced, from West, 1991, with permission of the publisher.)*

have been used in a similar way. This enzyme is initially present at quite high levels (Fig. 9), and the total activity declines slightly around the blastoctyst stage. Genetic analysis shows that relatively high levels of oocyte-coded enzyme persist for several days after the genes of the pre-embryo are activated.

A small study of GPI in human oocytes and pre-embryos suggests that high levels of this enzyme are also maintained during

human pre-embryos. Moreover, neither human HPRT nor human APRT activity was inhibited by the transcriptional inhibitor α-amanitin, implying that the activity was likely to be oocyte coded. In contrast, the corresponding mouse enzymes are largely the product of the genome of the pre-embryo after the eight-cell stage [Harper and Monk, 1983].

1. Sexing by quantitative biochemical assays of X-linked gene products. For biochemical sexing to be feasible, the gene product must be assayed at a time when the residual oocyte-coded activity is insignificant compared with the embryo-coded activity (as noted above) but before X-chromosome inactivation. This is not a trivial requirement, and, to judge whether this approach is feasible, it is relevant to consider the theoretical problems.

Figure 10 shows two hypothetical situations, both of which assume that X chromosome inactivation begins at the blastocyst stage as it does in the mouse (although the timing of X chromosome inactivation in human pre-embryos is unknown). In the first case (Fig. 10A) the oocyte-coded activity is low, and the activity encoded by the pre-embryo rises rapidly. By the morula and early blastocyst stage the expected activity in females (two X chromosomes active) is twice that in males (one X active), but by the late blastocyst stage this difference is reduced because X-inactivation in many cells of the female reduces the effective dose to a single X chromosome. (In reality, the situation is likely to be more complex because, for assays at the protein level, there will be a lag between transcription and a measurable increase in activity and also a lag between X-inactivation and a measurable decrease in activity per cell.

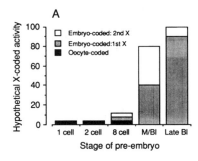 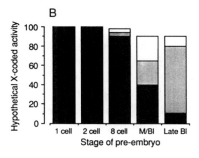

Fig. 10. *Hypothetical changes in activity of an X-linked gene product during preimplantation development (M, morula; Bl, blastocyst). The shading distinguishes oocyte-coded activity and pre-embryo–coded activity resulting from transcription of genes on both X chromosomes. The total activity expected in male pre-embryos (one X chromosome) is shown by the black plus grey shading, and the expected activity in females (two X chromosomes) is shown by the total height of the bar (black plus grey plus white). In **A** the oocyte-coded activity is low and the pre-embryo-coded activity rises rapidly. By the morula and early blastocyst stage the expected activity in females (two X chromosomes active) is twice that in males (one X active), but by the late blastocyst this difference is reduced because X-inactivation in many cells of the female reduces the effective dose to a single X chromosome. In **B** the oocyte-coded activity is high and persists for longer. In this case there is no time at which measurements of X-linked activity would provide a reliable prediction of the sex of a pre-embryo. (Reproduced from West, 1991, with permission of the publisher.)*

Assay variation and variation in the rate of embryonic development may also cause problems.) In the second case (Fig. 10B) the oocyte-coded activity is high and persists for longer. The activity encoded by the pre-embryo increases during preimplantation development, but the expected activity in females is never twice that in males because significant levels of oocyte-coded activity remain until X-inactivation reduces the 2:1 (female:male) activity ratio below a measurable difference. In this case there is no time at which measurements of X-coded activity would provide a reliable prediction of the sex of a pre-embryo.

Experimental studies of X-coded enzymes in mouse pre-embryos have laid the foundation for the development of methods that may be applicable to sexing the human pre-embryo. These experiments with mouse pre-embryos have been discussed in detail elsewhere [West, 1991]. As a result of the work with mouse pre-embryos, two enzymes (α-galactosidase and HPRT) seem the most likely to provide a "developmental window" when the enzyme activity in female pre-embryos is close to double that in males (as in the hypothetical example shown in Fig. 10A). Plots of enzyme activity per mouse pre-embryo during this "developmental window" (eight-cell to early blastocyst stage) gave characteristic bimodal distributions [reviewed by West, 1991]. The two modes of activity were subsequently shown to reflect activity in XX (high activity) and XY (low activity) embryos, respectively [Epstein et al., 1978].

Monk and Handyside [1988] used HPRT assays to develop a mouse model for sexing pre-embryos by preimplantation diagnosis. They isolated single blastomeres from zona-free eight-cell pre-embryos and cultured them for 12 hours. In this time the sample would probably have divided to form two cells, and the whole pre-embryo would have been equivalent to the 16-cell stage. (This is within the "developmental window," when all the HPRT activity should be coded by the pre-embryo and X chromosome inactivation has not yet begun.) The biopsied cells were assayed for HPRT and APRT. The sex of each pre-embryo (now at the blastocyst stage) was decided from the distribution of HPRT:APRT activities in the biopsied samples for all of the pre-embryos in a litter of 40 (shown in Fig. 11a). Putative male and female pre-embryos were surgically transferred to separate recipient females, and 15 fetuses were recovered at 14.5 days gestation and sexed by gonad morphology. The correct gonadal sex was predicted in 14 of the 15 fetuses examined [cross a in Table V].

If the developmental profile of HPRT was similar in mouse and human pre-embryos, a similar approach would be possible with human pre-embryos. The biopsy could be performed at the eight-cell stage, the assay performed the following day, and, if transfers could be done at the blastocyst stage there would be no need for cryopreservation of the pre-embryos. However, it seems that this approach will be more difficult for human pre-embryos and may not be feasible for several reasons.

First, the availability of inbred mice means that the mouse experiments can be designed to avoid genetic variability in enzyme activity that could confound the results. The human population could be more heterogeneous in this respect. Second, in a clinical situation a patient will produce a relatively small number of pre-embryos so that the results for each pre-embryo cannot be viewed against an overall distribution of a large number of contemporaneous samples. Several patients could be tested at the same time, particularly if the pre-embryos were stored frozen, but this might introduce another variable. If assay variability or developmental variation is significant it will be difficult to compare results obtained on different occasions and it will be difficult to sex a human pre-embryo that is one of a small group. Even for mouse pre-embryos,

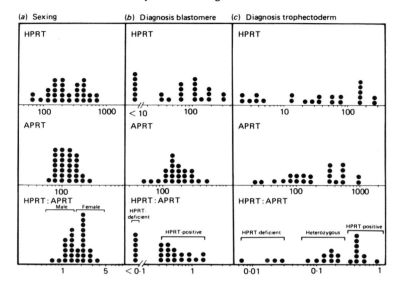

Fig. 11. *The use of sensitive HPRT assays to* **a:** *sex normal mouse pre-embryos and* **b, c:** *diagnose those that are hemizygous or heterozygous for an X chromosome-linked deficiency (hprt⁻) of HPRT (a model for the human Lesch-Nyhan syndrome) produced by crossing heterozygous hprt⁺/hprt⁻ carrier females (X/X*) with normal hprt+/Y males (X/Y). Distributions of HPRT and APRT activities and HPRT:APRT ratios: a, single blastomeres taken from eight-cell stage embryos from a normal cross and cultured for 12 hours; b, single blastomeres from eight-cell embryos from an hprt+/hprt− carrier female (X/X*) and cultured for 24 hours; and c, 5–10 trophectoderm cells taken from blastocysts from an hprt+/hprt− carrier female (X/X*). HPRT and APRT activities are given in femtomoles per hour per sample. Sexing is only possible with cells assayed before X chromosome inactivation (a); hemizygous affected males were identified after biopsy of eight-cell or blastocyst stages from the appropriate genetic cross (b, c), but heterozygous hprt+/hprt− carrier females were only distinguished after biopsy of blastocysts (c). (Reproduced from Monk, 1988, with permission of the Company of Biologists,Ltd..)*

where large numbers can be obtained at a given stage of development to produce a bimodal distribution, there is an area of overlap, and it is uncertain whether the pre-embryos that fall in the middle of the distribution are males or females. Bimodal distributions were not always demonstrated convincingly for mouse pre-embryos with either α-galactosidase or HPRT [reviewed by West, 1991], and the difficulties in sexing human pre-embryos would be greater.

Braude et al. [1989] studied HPRT activity in human oocytes, pre-embryos, and single blastomeres removed from four- to eight-cell stage pre-embryos. As predicted from the above considerations, there was a wide variation in activity per pre-embryo, which would have made sexing difficult. However, as noted above, this study revealed a second, more fundamental reason why assays of HPRT activity in human pre-embryos are unlikely to provide a means of preimplantation diagnosis of sex. (Unlike

TABLE V. Sexing of Mouse Pre-Embryos and Preimplantation Diagnosis of HPRT Deficiency by Assays of HPRT*

Experiment and biopsy method	Age at assay (days)	HPRT result	Fetus classification (14.5 days)			
			X*Y	X*X	XY	XX
a. Sexing (cross: XX × XY)						
1 blastomere from 8-cell (2.5 days)	3	Low	NA	NA	3	0
		High	NA	NA	1	11
b. Diagnosis of HPRT-deficiency (Cross: X*X × XY)						
1 blastomere from 8-cell (2.5 days)	3.5	Very low	4	0	0	0
		Positive	0	4	2	6
5–10 trophectoderm cells (4.5 days)	4.5	Very low	2	0	0	0
		Intermediate	0	3	0	0
		High	0	0	1	1

*The X chromosome carrying the mutant *hprt⁻* gene causing HPRT deficiency is designated X* and the normal X chromosome as X. NA, not applicable. Data are from Monk [1988].

the developmental profile of HPRT in mouse pre-embryos, human HPRT showed a downward trend rather than a rapid increase in activity. Also, experiments with the transcriptional inhibitor α-amanitin implied that the activity in human pre-embryos was likely to be oocyte coded.) Thus, despite the encouraging results with attempts to sex mouse pre-embryos [Monk and Handyside, 1988], the present evidence forces us to conclude that there is little prospect of reliably sexing human pre-embryos biochemically with products of X-linked genes, such as α-galactosidase or HPRT, in the near future.

2. Biochemical preimplantation diagnosis of specific genetic defects. About 200 of approximately 1,500 defined monogenic, genetic disorders result in enzyme deficiencies, and about 60 of these can be diagnosed by enzyme assays on cultured amniotic fluid cells or chorionic villus samples [Galjaard, 1980, 1984; Patrick, 1983, 1984; Benson and Fensom, 1985]. One advantage of this approach is that the genetic defect present in the family does not have to be defined as rigorously as is necessary for DNA methods. Some assays can be used with very small samples and may be suit-

able for preimplantation diagnosis. For example, fluorescent endpoints, notably those using 4-methylumbelliferyl substrates, have provided sensitive assays for α-galactosidase, β-galactosidase, and β-glucuronidase in individual mouse pre-embryos and in small samples of human material for prenatal diagnosis [Galjaard et al., 1974a,b; Wudl and Paigen, 1974; Wudl and Chapman, 1976; Kleijer et al., 1976; Adler et al., 1977; Galjaard, 1980; Esworthy and Chapman, 1981]. NADPH fluorescence has also been used as an endpoint for measuring glucose-6-phosphate dehydrogenase activity in mouse pre-embryos [Leese, 1987b] and fluorescence of NADH or NADPH provides the endpoint for a sensitive assay system, described by Lowry and Passoneau [1972], that could be used for a variety of enzymes. This system involves enzymatic cycling of co-enzymes, thereby chemically amplifying the amount of product that can then be assayed by fluorescence of NADH or NADPH. Radiolabeled end-points have been used for microassays of HPRT and adenine deaminase [Hosli et al., 1974; Monk, 1978, 1987; Kratzer and Gartler, 1978; Epstein et al., 1978; Benson and Monk, 1988], and mouse models have been used to explore whether such assays would be feasible for preimplantation diagnosis of enzyme deficiencies.

Monk and colleagues [1987, 1988] have measured the activity of the X chromosome–coded enzyme HPRT in samples biopsied from mouse pre-embryos as a model for diagnosis of the human X-linked condition, Lesch-Nyhan syndrome (HPRT deficiency), using a similar approach to that used for sexing normal pre-embryos (see above). This work has also been reviewed by Monk [1988] and Monk et al. [1990]. The model for Lesch-Nyhan syndrome used embryos produced by mating female mice, heterozygous ($hprt^+/hprt^-$) for a mutant that causes a deficiency of HPRT [Hooper et al., 1987], with normal males ($hprt^+/Y$). Two experiments have been described, involving biopsy of cleavage stages and blastocysts, respectively [cross b in Table V].

In the first experiment, Monk et al. [1987] removed a single cell from eight-cell embryos after lysing the zona pellucida and assayed HPRT levels after culturing the cell for 24 hours. This is similar to the experiment [Monk and Handyside, 1988] described above for sexing mouse pre-embryos except that the biopsied cell was allowed to divide for 24 rather than 12 hours. By the time the cell had divided to produce two or three cells, almost all of the HPRT would be encoded by the genome of the pre-embryo and X-inactivation may have begun in some pre-embryos. Three groups of pre-embryos were expected (high, medium, and zero HPRT activity) in the ratio of 1:2:1, representing 1) normal XX females, 2) normal XY males plus heterozygous X*X females, and 3) affected X*Y males (where X* is the X chromosome carrying the $hprt^-$ mutation). In practice the embryos with medium and high levels of HPRT could not be clearly distinguished, and the expected trimodal distribution was not seen. Although heterozygous females could not be identified, those that were HPRT deficient (hemizygous af-

fected males) could be distinguished from those that were HPRT positive (all others) even though there was some residual oocyte-coded enzyme (Fig. 11b). Pre-embryos were transferred to pseudopregnant female mice at the blastocyst stage, and fetuses were subsequently recovered and typed. Four pre-embryos that were diagnosed as HPRT deficient grew into fetuses and were confirmed as HPRT-deficient males (Table V).

In a second experiment [Monk et al., 1988] the biopsy was performed later in order to try to identify heterozygous females as well as hemizygous males. Mouse blastocysts were collected at 3.5 days and cultured overnight before trophectoderm biopsy and assay. Five to 10 trophectoderm cells were removed from the blastocysts and assayed immediately for HPRT and APRT (to control for variable numbers of cells in the biopsy and developmental asynchrony between embryos). Consequently, at the time of assay the cells were 1 day more advanced than in the previous experiment (biopsy at 2.5 days, at the eight-cell stage, and another 24 hours in culture before the assay) and 1.5 days more advanced than those used for sexing. Different results are predicted for assays of trophectoderm cells of 4.5 day blastocysts than for assays of earlier pre-embryos, because X chromosome inactivation would have already occurred in the trophectoderm cells. The paternally derived X chromosome is always inactivated in the trophectoderm developmental lineage [Frels et al., 1979; Frels and Chapman, 1980; Papaioannou et al., 1981], so four activity groups would be expected, as follows: 1) highest activity from normal XX females with one normal $hprt^+$ gene expressed per cell (plus some residual activity from the normal $hprt^+$ gene on the recently inactivated, paternally derived X chromosome); 2) normal XY males with one X chromosome and therefore one normal $hprt^+$ gene active per cell; 3) heterozygous X*X females with only the maternally derived X* chromosome, carrying the deficient *hprt*⁻ gene, expressed (all of the HPRT activity will be residual activity from the normal *hprt*⁺ gene on the recently inactivated, paternally derived X chromosome); 4) affected X*Y males with only the deficient *hprt*⁻ gene. In practice the normal females and normal males (groups 1 and 2) were indistinguishable, so only three activity groups (HPRT:APRT ratio) were identified (Fig. 11c). This is unimportant provided that the affected X*Y males and heterozygous (carrier) X*X females can be distinguished from one another and from the normal (XX and XY) pre-embryos. After overnight culture of the embryo the blastocysts were divided into three groups, according to HPRT:APRT activity ratio in the corresponding biopsy, and transferred to pseudopregnant females as before. Seven fetuses were subsequently recovered, and all had been correctly diagnosed with respect to these three groups (Table V).

HPRT assays of trophectoderm samples (Fig. 11c) could not be used for sexing because the cells were equivalent to 4.5 days and X-inactivation would have begun to equalize the HPRT activities (dosage compensation), as shown in Figure 10. (The pre-embryos had developed beyond the appropriate "developmental window" for sexing.) In the first eight-cell biopsy experiment (Fig. 11a), discussed in the previous section, the assayed cells were equivalent to 3 days and normal XX embryos were distinguished from normal XY embryos, but in the second eight-cell biopsy experiment (Fig. 11b) the cells were equivalent to 3.5 days at the time of assay and normal XX embryos were indistinguishable from normal XY embryos. This difference may be explained if X-inactivation was beginning in the 3.5 day cells. The results from the various experiments with HPRT in mouse pre-embryos (Fig. 11 and Table V) show that HPRT-deficient male mice and hetero-

zygous female carriers can be identified by enzyme assays of trophectoderm biopsies. However, as noted above, the results of Braude et al. [1989] make it unlikely that HPRT assays could be used either to sex human pre-embryos or to identify those with Lesch-Nyhan syndrome.

Benson and Monk [1988] measured adenine deaminase (ADA) in mouse pre-embryos as a model for the preimplantation diagnosis of adenine deaminase deficiency, which is responsible for one form of severe combined immunodeficiency disease (SCID) in humans. They also cited experiments in which they had successfully assayed individual cells removed from eight-cell mouse pre-embryos. ADA activity was inhibited by α-amanitin in mouse pre-embryos beyond the two-cell stage, implying that the activity was not likely to be oocyte coded. However, the differences in HPRT expression in mouse and human pre-embryos [Braude et al., 1989] prevent us from predicting whether this mouse model for preimplantation diagnosis of ADA deficiency will be applicable to human pre-embryos.

More research on human pre-embryos is needed to determine whether biochemical assays offer any feasible genetic tests that are suitable for preimplantation diagnosis. Even gene products that are synthesized by the early human pre-embryo may be difficult to use in this way because of the likelihood of developmental and technical variability.

D. The Polymerase Chain Reaction

The genes for an increasing number of inherited diseases have been identified, and in many cases the specific lesion (e.g., mutation, deletion) is known. When the disease is homogeneous or when family studies have isolated the genetic lesion and its neighboring DNA sequence, DNA analysis is often possible, using gene-specific probes to identify the restriction endonuclease fragments altered by the mutation;

in diseases with a high rate of mutation, closely linked markers can be used to identify restriction fragment length polymorphisms (RFLPs) linked to the disease condition in a particular family. Both of these approaches require a detectable amount of DNA, which may be at minimum several nanograms, but more often are in microgram quantities. This represents fetal genomic DNA from 1 million or more cells [Old, 1986]. This can be obtained from fetal trophoblast by chorionic villus sampling in the eighth week of gestation [Gosden et al., 1982]. Closely linked RFLPs are only informative in some families, and DNA from family members has to be tested to check whether the RFLP is informative for any given family. Also, genetic recombination may occur between the gene responsible for the disorder and the marker RFLP and result in incorrect diagnosis. Thus, ideally either the DNA probe should detect the disease gene itself or probes should be available for a closely linked marker RFLP on either side of the disease gene.

In preimplantation embryos, the small number of cells and the limitations on the number that can be biopsied without compromising the development of the embryo mean that conventional approaches are not possible. DNA analysis on such small numbers of cells has only recently become feasible with the introduction of a technique that has revolutionized many aspects of DNA analysis, not only in this field but also in the wider area of molecular genetics. This technique, the polymerase chain reaction, is capable of amplifying short segments of DNA from very small initial amounts up to 10^7-fold in a few hours [White et al., 1989].

PCR uses a process involving cycles of target DNA denaturation, followed by oligonucleotide primer annealing, then enzymatic primer extension, and a second cycle begins with denaturation. The oligonucleotide primers are chosen so that they flank the target sequence, with their 3' ends directed toward each other. Thus, the primer extension step copies the region that the primers enclose. With each cycle the amount of the target sequence is doubled so that about 40 cycles, taking a total of 6–7 hours, can produce sufficient amount of the target DNA for analysis [Handyside et al., 1990]. Many variants of the basic method have been developed, in which multiple pairs of primers are used [Mullis and Faloona, 1987].

The sensitivity of the method and the speed with which results can be obtained make this the procedure of choice whenever the criteria can be fulfilled. These criteria include knowledge of the DNA sequence of the target and neighboring sequences; knowledge of the precise nature of the mutation, and a suitable labeled probe. One of the earliest applications of PCR to human genetics was in the detection of sickle cell anemia on DNA from 75 cells [Saiki et al., 1985a, b, 1986]. In this case the probe was an allele-specific oligonucleotide (ASO) capable of distinguishing between the normal and mutant allele. The results were obtained within 6 hours. The use of ASOs eliminates any risk of error arising from the accidental amplification of sequences other than the specific target.

Although the application of this technique to pre-implantation diagnosis is limited by the number of genetic conditions whose molecular biology is understood, this limitation is expected to be rapidly overcome with current fury of investigation into human molecular genetics, and the exponential increase in the range of diseases that have been successfully analyzed. It is already feasible to use PCR for the diagnosis of a significant number of human genetic diseases, including sickle cell anemia [Saiki et al., 1985a, b, 1986], α-thalassemia [Saiki et al., 1985b; Chehab et al., 1987], β-thalassemia [Saiki et al., 1988], α_1-antitrypsin deficiency [Dermer and Johnson, 1988], Tay-Sachs disease [Myerowitz, 1988; Myerowitz and Costigan, 1988], cystic fibrosis [Kerem et al., 1989], and Duchenne muscular dystrophy [Speer et al., 1989].

The extreme sensitivity of PCR has a down side in that it also increases the risk of error due to the inadvertent amplification of sequences other than the intended target. Errors can arise from the presence of minute contaminations in the reaction, for example, a scrap of dandruff from the investigator. For this reason meticulous care to ensure absolute cleanliness and sterility is essential in any laboratory attempting to use this approach [Kwok and Higuchi, 1989]. With this caveat, however, the PCR technique is the most powerful currently available for DNA analysis, and it has the potential to solve most of the analytical problems arising from the limitations on material available from pre-implantation embryos. PCR has already been used to amplify specific DNA sequences from single spermatozoa [Li et al., 1988], oocytes [Coutelle et al., 1989], polar bodies [Monk and Holding, 1990; Strom et al., 1990], and cells biopsied from mouse or human pre-embryos [Handyside et al., 1989, 1990; Holding and Monk, 1989; Gomez et al., 1990; Handyside, 1991; Lindeman et al., 1990]. False-negative results are more likely for single-cell samples, and in approximately 10% of such samples the DNA sequence fails to amplify (A. Handyside, personal communication). It may be possible to identify samples that fail to amplify by coamplifying a second DNA sequence (e.g., an X-chromosome sequence and a Y-chromosome sequence). Also, by analyzing two samples per pre-embryo it should be possible to reduce significantly the risk of false-negative results (e.g., from 10% to 1%).

Handyside et al. [1989] showed that it is feasible both to biopsy human pre-embryos and to sex them by amplification of target DNA from a single blastomere with PCR. A single blastomere was removed from each of 38 human pre-embryos at the 6–10 cell stage via a micropipette inserted through a hole made in the zona pellucida, and PCR was used to sex the single blastomere. The whole procedure, from biopsy to sexing, took only about 8 hours. The remaining portion of the pre-embryo was cultured to test its viability and to confirm the diagnosis of sex by in situ hybridization and/or fluorescent F-body staining with spermidine bis-acridine. Ten of the 27 pre-embryos (37%) that were cultured developed into blastocysts, which is similar to the proportion expected from cultured pre-embryos that are not biopsied. PCR was used to amplify a 149 bp DNA sequence that is repeated 800–1500 times on the Y chromosome. Sixty cycles of amplification were used and then the amplified DNA was stained with ethidium bromide and visualized under ultraviolet light. A cell was judged to be male if visualization under ultraviolet light revealed a strong band corresponding to the 149 bp Y chromosome-specific sequence. If the band was absent or very faint the cell was judged female.

Stringent precautions were taken to avoid contamination of the single-cell sample with foreign DNA that could have produced a false-positive result. These precautions included washing the blastomere three times, the use of a class II containment cabinet throughout, filtration of buffers through a 0.2 μm filter, and digestion of any contaminating target sequences present in the oligonucleotides and salmon sperm carrier DNA with restriction enzyme EcoR1. Negative controls were also run as a check on the effectiveness of the anticontamination procedures. The authors commented that without these procedures they frequently produced false-positive results from female samples or blank controls.

Of the single-cell biopsies from the 38 pre-embryos, 13 were lost because of technical problems and 25 were sexed by PCR. Nineteen of these 25 pre-embryos were also sexed by an alternative method (in situ hybridization and/or F-body fluorescence on the remainder), and the sex was confirmed in 17 (90%) and contradicted in 2 (10%) cases. The two contradictory cases were both pre-embryos that had had three or more pronuclei after fertilization and were sexed as male by PCR but female by in situ hybridization (see section VII.B). The authors suggested that this contradiction could be explained if the pre-embryo was a polyploid mosaic in which the Y chromosome segregated to only a proportion of the cells. This explanation requires that the single blastomere biopsied from both pre-embryos was the only one at the 6–10-cell stage that had a Y chromosome. Despite these two contradictory cases, this preliminary study was very encouraging and this approach has now been put into clinical practice [Handyside et al., 1990], as described in section VIII.A.

Amplification of single-copy DNA sequences for clinical preimplantation diagnosis will pose further problems, but PCR has already been used to detect single-copy sequences in individual human spermatozoa, oocytes, or polar bodies [Li et al. 1988; Coutelle et al., 1989; Monk and Holding, 1990; Strom et al., 1990, Verlinsky, 1990], and Strom et al. [1990] and Verlinsky et al. [1990] have claimed that the heterozygous and two homozygous genotypes could be distinguished among first polar bodies of oocytes from women who were heterozygous for cystic fibrosis and α_1-antitrypsin (see section IV. B).

In addition to amplification of these human DNA sequences, mouse models have been developed that incorporate some of the various steps required for pre-implantation diagnosis. Ideally these would include 1) biopsy of cells from the pre-embryo, 2) identification of affected pre-embryos by amplification of single-copy sequences, 3) transfer of unaffected pre-embryos, and 4) verification that subsequent conceptuses are free of the genetic disorder. Holding and Monk [1989] addressed the first two of these four steps.

They amplified sequences from the mouse β-major hemoglobin gene in individual blastomeres from 4–8-cell stage pre-embryos from matings between a) normal × normal and (b) homozygous thalassemic × homozygous thalassemic (Hbb^{th-1}/Hbb^{th-1} × Hbb^{th-1}/Hbb^{th-1}) mice. They found positive results for 25/30 normal pre-embryos (30/30 expected), 1/26 thalassemic pre-embryos (0/26 expected), 1/14 wash medium control samples (0/14 expected), and 1/25 water blanks (0/25 expected). Overall 50/56 (89%) pre-embryos were correctly identified, although heterozygous pre-embryos were not included in the study. The identification of homozygous thalassemic pre-embryos was inferred from the absence of the normal allele after amplification. This could be misleading since this result would also occur if the PCR reaction failed. A similar experiment was reported by Lindeman et al. [1990]. They found positive results for 11/20 normal pre-embryos (20/20 expected) and 2/16 thalassemic pre-embryos (0/16 expected).

Gomez et al. [1990] tested another model system. They biopsied one to five trophectoderm samples from mouse blastocysts and amplified DNA to test for normal and abnormal sequences of the myelin basic protein gene that is abnormal in the shiverer (*shi*) mutant. Two sets of primers were used in order to distinguish between *shi/shi*, *shi/+* and +/+ genotypes. One set was used to amplify a sequence that is deleted in homozygous *shi/shi* mice, and the other set was used to amplify a sequence unique to the *shi* gene. The test took only 7 hours, so pre-embryos could be transferred to pseudopregnant recipients for further development after the diagnosis had been made. Both the normal (+) and mutant (*shi*) alleles were amplified in four of the seven samples from *shi/+* heterozygotes. In the remaining three cases no amplification occurred. The *shi* allele was detected in 6/10 trophectoderm biopsies from homozygous *shi/shi* pre-embryos (10/10 expected), but in 3 of these 6 cases contamination with normal cells led to the false conclusion that the pre-embryos were *shi/+* heterozygotes. No amplification occurred in 4/10 cases, so no diagnosis was possible. The + allele was detected in 15/19 samples from homozygous normal (+/+) pre-embryos, and in 4 cases no amplification was detected. Two false-positive samples were found among 10 medium or water controls. Thus 22/36 (4/7 *shi/shi*, 3/10 *shi/+*, and 15/19 +/+) pre-embryos were correctly identified (61%). The authors do not state the genotypes of the parent mice or how the genotypes were verified. If *shi/+* × *shi/+* matings were used verification could only have been done by testing the progeny after transfer. However, *shi/shi* homozygotes are fertile so it is possible that the three genotypes could have been generated by three separate matings (*shi/shi* × *shi/shi*, *shi/shi* × +/+ and +/+ × +/+). Although results from these model systems are encouraging and show that diagnosis of polymorphisms of single-copy DNA sequences may be feasible, the diagnostic accuracy is not yet adequate for clinical use.

E. Feasibility of Different Genetic Tests

The above discussion leads us to the conclusion that, for the foreseeable future, molecular rather than biochemical or cytogenetic tests for genetic disorders are likely to be most useful for preimplantation diagnosis. Sexing is feasible with Y-chromosome-specific probes in conjunction with in situ hybridization or PCR. If sexing was done to allow female pre-embryos to be selected for transfer to the uterus (to avoid conceiving a son affected by an X-linked genetic disease), false-negative test results (males scored as females) would pose a more serious clinical problem than false positives. More cells are likely to be needed for sexing by in situ hybridization than for PCR in order to avoid false-negative results, and PCR can also produce a more rapid result. However, even PCR can produce false negatives (by failure to amplify) so that it would be preferable to have at least two samples of each pre-embryo for DNA amplification, even if each sample comprised only one cell. If false positives can be reduced to a minimum by taking rigorous precautions against the possibility of contaminating the tiny sample with foreign DNA, PCR seems to be a more efficient means of sexing pre-embryos than in situ hybridization, but further experimental comparisons are needed.

It is already clear that other types of preimplantation diagnosis are likely to be feasible in the near future. In situ hybridization could be used for the identification of pre-embryos with specific numerical chromosome anomalies, caused by inherited translocations. However, we have already argued that preimplantation diagnosis may be less appropriate for chromosome anomalies that arise de novo because the risk to a particular couple is much lower. The use of PCR to diagnose specific single-gene defects is likely to have the biggest impact, and this has already been used to detect single-copy DNA sequences in single cells, as discussed above.

VIII. FUTURE TRENDS IN PREIMPLANTATION DIAGNOSIS

A. Current Experience of Preimplantation Diagnosis

At the time of writing (1990–1991), only the group at the Hammersmith Hospital in London has published an account of clinical preimplantation diagnosis [Handyside et al., 1990]. This therefore must be considered as the current "state of the art." Handyside and colleagues used preimplantation diagnosis to help couples at risk for sex-linked (X-chromosome-linked) diseases (X-linked mental retardation, adrenoleukodystrophy, Lesch-Nyhan syndrome, and Duchenne muscular dystrophy) start a pregnancy with female conceptuses. (In these cases only boys are affected by the sex-linked disease.)

They obtained the pre-embryos by standard IVF procedures, cultured them to the 6–10-cell stage, and removed one or two blastomeres with a micropipette via a hole in the zona pellucida. The biopsied cell was checked to ensure that it had a nucleus, and DNA amplification was carried out by PCR, using conditions that were specific for the amplification of certain reiterated Y-chromosome-specific DNA sequences. Thus, if the single blastomere was from an XY male pre-embryo, the Y chromosomal sequences were amplified and could be detected subsequently by polyacrylamide gel electrophoresis and ethidium bromide staining. The whole procedure (biopsy and diagnosis) took about 8 hours [Handyside et al., 1989, 1990], so the pre-embryo could be safely left in culture and the female pre-embryos transferred to the uterus on the day of diagnosis.

Because it was planned to check the diagnosis of sex later in gestation by chorionic villus sampling (CVS), no more than two female pre-embryos were returned to the mother. (Chorionic villus samples can be obtained from each twin with only a small risk of miscarriage, but sampling more than two conceptuses would be more difficult.) Levels of hCG were monitored from 12 days after transfer to check for successful implantation (this occurred in 5 of the 10 cases, but in 2 of these the levels declined and the pregnancy was not maintained), and ultrasonography was performed at intervals from 3 weeks to monitor the conceptus. CVS was performed on two ongoing twin pregnancies at 10 weeks and cytogenetic analysis confirmed the preimplantation diagnosis that each conceptus was female. Detailed monitoring by ultrasonography, at 20 and 22 weeks respectively, revealed all four fetuses were morphologically normal and that both sets of twins were probably dizygotic, as expected. Handyside et al. [1990] reported

two or possibly three pregnancies (one early pregnancy with elevated hCG) following preimplantation diagnosis in 10 treatment cycles (5 couples).

The updated clinical results from the team at the Hammersmith Hospital (A. Handyside, personal communication) [Handyside and Delhanty, 1991] are as follows. Of a total of 50 pre-embryos that were biopsied for sex determination by PCR, 22 (female by PCR) were transferred, and 7/22 were implanted and continued to develop. Six of these seven were subsequently confirmed as females by chorionic villus biopsy, but one singleton was found to be a male and was terminated. All of the remaining 6 females have now progressed to term (including two sets of twins); five have been successfully delivered as healthy girls but one twin was stillborn (obstetric complication). This pioneering work paints a remarkably optimistic picture for clinical preimplantation diagnosis, and it is appropriate to consider what may lie ahead.

The work by the Hammersmith team is undoubtedly a major achievement, but the methodology is not simple. Although this means that it is unlikely that many clinics will be able to offer preimplantation diagnosis as a routine clinical procedure for some time, no doubt the required expertise will become more widespread just as the skills required for IVF itself have become widespread. The biopsy, diagnosis, and transfer of pre-embryos were done within 1 working day. This is likely to cause difficulties and put laboratory staff under pressure when for technical reasons the results of the diagnosis are not clear cut. When this occurs several courses of action are open: 1) Repeat the biopsy and diagnosis the next day. This may be feasible if only one cell is removed each time and transfers can be safely postponed to the morula or blastocyst stage. 2) Accept that diagnosis is not possible and transfer two pre-embryos. If a pregnancy results, prenatal diagnosis can be undertaken by CVS at the appropriate time. 3) Abandon the treatment cycle. It may also be possible to plan for a second attempt at diagnosis by dividing the sample into two, using one for diagnosis and one to culture and produce a group of cells for further attempts. This would be most feasible if the biopsy comprised more than two cells (or was cultured to produce more than two cells), since we have already argued that, ideally, each diagnosis should be done on duplicate samples to avoid false negatives. If such attempts are time consuming it would also be possible to cryopreserve the pre-embryo until a sure diagnosis was achieved.

B. Technical Progress for Preimplantation Diagnosis

Progress in developing techniques that could be applied to preimplantation diagnosis has been extraordinarily rapid and, for such a fast moving field, it is dangerous to predict future trends. We started to evaluate in situ hybridization as a method to sex preimplantation pre-embryos at the beginning of 1985 [West et al., 1987], and at that time it seemed the most realistic means of performing a genetic diagnosis with a small number of cells. Only a few years later PCR arrived on the scene and rapidly eclipsed this approach. The following comments are made on the basis of how the field seems to us in 1991 but, as we have found from our own experience, they will probably seem dated and inappropriate by the time this chapter is published. It seems likely that most pre-embryos for preimplantation diagnosis will be produced by IVF rather than by uterine flushing, but improvements are likely to be made in many areas, including biopsy, culture of cells removed from the pre-embryo, cryopreservation, and the genetic tests.

For preimplantation diagnosis of autosomal traits it may be possible to combine polar body and pre-embryo biopsy. One possible three-stage scheme is outlined in Table VI for couples at risk for an autosomal recessive disease. At stage 1, identification of a first polar body that was homozygous for the disease allele would imply that the secondary oocyte was homozygous for the normal allele and so could be fertilized and transferred (either via IVF or gamete intrafallopian transfer [GIFT] procedures) without requiring further tests. Even if the sperm carried the disease allele the zygote would only be a heterozygous carrier. A secondary oocyte that was deduced to be heterozygous for the disease allele (because the first polar body was heterozygous) could be tested in a second stage by biopsy of the second polar body after fertilization. If the second polar body contained the disease allele then the female pronucleus must contain the normal allele and such zygotes could be safely transferred. Alternatively, if the second polar body contained the normal allele then the female pronucleus must contain the disease allele and the zygote could only be safely transferred if the sperm contribution was the normal allele. This could be tested by pre-embryo biopsy (stage 3). A secondary oocyte that was deduced to be homozygous for the disease allele (because the first polar body was homozygous for the normal allele) would only produce an unaffected zygote if it was fertilized by a sperm bearing the normal allele. Consequently there would be no purpose in testing the genotype of the second polar body, and the next test would be done by pre-embryo biopsy at stage three.

Although this scheme is attractive because the diagnosis is made as early as possible, in some cases before fertilization, it is technically very demanding. For this reason it seems unlikely that multistage schemes involving sequential sampling of polar bodies and blastomeres will prove popular in the short term. The use of polar bodies alone for genetic tests seems unlikely to become widespread because only some are informative (see above and section IV. B), so

TABLE VI. Possible Scheme for Three-Stage Preimplantation Diagnosis for Autosomal Conditions: Hypothetical Example of a Couple Who Are Both Heterozygous (d/+)* for an Autosomal Recessive Genetic Disease

Stage	Genotypes			Comments
Maternal genotype	d/+			Aim: Avoid d/d conceptions
Stage 1: First polar body biopsy (after ovulation)				Three possible genotypes:
	(i)	(ii)	(iii)	(i) or (ii) if there is no crossing-over between disease locus and centromere; (iii) if crossing-over occurs**
First polar body	+/+	d/d	d/+	Genotype determined from first polar body biopsy
Secondary oocyte	d/d	+/+	d/+	Deduced from biopsy of first polar body
Possible conceptus	d/d or d/+	d/+ or +/+	d/d, d/+ or +/+	Secondary oocyte (ii) cannot produce a d/d conceptus
(sperm may be d or +)	To stage 3	Normal†	To stage 2	Secondary oocytes (i) and (iii) require further tests
				Secondary oocyte (iii) produces genotypes (iii) and (iii')
Stage 2: Second polar body biopsy (after fertilization)				
	(iii)	(iii')		
Second polar body	d	+		Genotype determined from second polar body biopsy
Female pronucleus	d	+		Deduced from biopsy of second polar body
Possible conceptus	d/+ or +/+	d/d or d/+		Zygote (iii) cannot be d/d
(sperm may be d or +)	Normal†	To stage 3		Zygotes (i) and (iii') require a further test
Stage 3: Pre-embryo biopsy (after cleavage to the 8-cell stage):				
	(i)	(iii')		
Pre-embryo	d/d or d/+	d/d or d/+		Genotype determined from pre-embryo biopsy

*d is the disease allele and + is the normal allele at the disease locus.

**See also Figure 3.

†Secondary oocytes (stage 1) that cannot produce d/d zygotes could be used in IVF (without pre-embryo biopsy) or GIFT procedures. Zygotes (stage 2) that cannot be d/d zygotes could be used in IVF without pre-embryo biopsy.

most centers are likely to opt for biopsy of cleavage stage pre-embryos or blastocysts.

While the initial experience of transferring fresh human blastocysts was discouraging [Dawson et al., 1988], other more recent results [Bolton et al., 1991] and the experience with cryopreserved blastocysts (Table III) are encouraging. The option of trophectoderm biopsy at the blastocyst stage therefore remains open. Only 10% of fresh pre-embryos transferred in IVF cycles implant, and about 40% of frozen pre-embryos fail to survive cryopreservation procedures (Table III), so, after pre-implantation diagnosis, most of the unaffected pre-embryos would fail to develop to term. Improvements in the pregnancy rate following IVF would greatly improve the usefulness of preimplantation diagnosis. The use of cryopreservation in conjunction with preimplantation diagnosis is attractive for several reasons. Cryopreservation would remove the pressure on laboratory staff to produce a reliable diagnosis by a set deadline, and when there are sufficient unaffected pre-embryos they could be stored and transferred on several occasions. Hopefully, further research will increase the proportion of pre-embryos that survive cryopreservation. For all types of genetic tests discussed above, it would be preferable to have more than one cell for the diagnosis, to avoid the risk of false negatives. This could be achieved by removing more than one cell or by culturing the cell(s). The culture of cells removed from the pre-embryo is one area where improvements seem likely in the near future.

Thus even without polar body biopsy, several options may soon be available These include 1) biopsy at the cleavage stage, rapid diagnosis on one or two cells, and transfer to the uterus within one working day [as currently employed by Handyside et al., 1990]; 2) biopsy at the cleavage stage, culture of the cells for 1 or 2 days, rapid diagnosis on part of the material, allowing time for a second test if required,

and transfer to the uterus at the blastocyst stage; 3) biopsy at the cleavage stage, cryopreservation of the pre-embryos at cleavage or blastocyst stage, culture of the cells for as long as is necessary, unhurried diagnosis, transfer at cleavage or blastocyst stage; 4) biopsy of trophectoderm cells at the blastocyst stage, rapid diagnosis on several cells, and transfer to the uterus within one working day; and 5) biopsy at the blastocyst stage, cryopreservation, culture of the cells for as long as is necessary, unhurried diagnosis, transfer at blastocyst stage.

There will undoubtedly be improvements in the range of sensitive genetic tests that could be used for preimplantation diagnosis. As noted earlier, preimplantation diagnosis may be less appropriate for couples at relatively low risk for genetic disorders, such as many of the spontaneous chromosomal aberrations where the risk is associated with maternal age. However, preimplantation diagnosis is likely to benefit couples who are at high risk for sex-linked disorders, autosomal single-gene defects and possibly inherited chromosome translocations. Sexing by PCR is the first step and we do not yet know how much more difficult it will be to use PCR to diagnose single-gene defects in one or two cells in a clinical setting. The amplification of Y chromosome DNA in only one or two blastomeres is undoubtedly an impressive achievement, but between 500 and 8,000 copies of the sequence were present on the Y chromosome. Amplification of single-copy sequences will undoubtedly be even more demanding, particularly for clinical diagnosis. However, only a few years ago it would have seemed impossible to expect to be able to diagnose sex by DNA amplification, and already PCR has been used to detect single copy DNA sequences in one spermatozoa [Li et al. 1988], oocyte [Coutelle et al., 1989] or polar body [Monk and Holding, 1990; Strom et al., 1990]. Techniques are bound to improve even further. The availability of suitable DNA polymorphisms may be a limiting factor with some genetic diseases, but this problem is likely to diminish as human molecular genetics continues to progress at a breathtaking pace.

Although PCR has largely eclipsed the use of in situ hybridization as a sensitive genetic test, Griffin et al. [1991] have recently sought to improve its applicability to sexing pre-embryos by using a fluorescent endpoint with probes for both sex chromosomes. The two probes were used on separate samples, but if the X and Y chromosomes could be routinely detected simultaneously, with different colored fluorescent endpoints, this could herald a comeback for in situ hybridization. However, as discussed in section VII. B, its use is likely to be limited to sexing and for detecting numerical chromosome anomalies.

C. Is There a Role for Preimplantation Gene Therapy?

The prospect of preimplantation gene therapy has been raised by some authors [eg., Buster and Carson, 1989]. Couples at risk for the familiar mendelian genetic disorders will always produce some pre-embryos without the abnormal gene. The goal of preimplantation diagnosis is to select those pre-embryos and return them to the uterus for further development. In our opinion, there can be no justification for attempting gene therapy on pre-embryos in order to correct a genetic defect caused by a defective mendelian nuclear gene, since unaffected pre-embryos are also produced.

There are less straightforward cases, however, where a form of gene therapy would provide the only means for a couple to be the biological parents of a child who was free of the genetic disorder. A number of neuromuscular diseases are caused by mutations, in the mitochondrial DNA (mtDNA). In some cases the mutations arise "spontaneously" during development, probably as a result of action of a nuclear gene product [e.g., Zeviani et al.,

1989]. In other cases, however, the disease is *maternally inherited* [Ozawa et al., 1988; Wallace, 1989], like mitochondrial DNA itself. (Although Gresson, [1940] suggested that the mitochondria in the sperm midpiece may get incorporated into the egg cytoplasm at fertilization, Szollosi [1965] concluded that they disintegrated. Molecular studies confirm that paternal mtDNA does not contribute to the genotype of the progeny of various mammals, including humans [Hutchison et al., 1974; Hayashi et al., 1978; Kroon et al., 1978; Avise et al., 1979; Giles et al., 1980].) Maternally inherited diseases that are known to be caused by mtDNA mutations include Lebers hereditary optic neuropathy (LHON), infantile bilateral striatal necrosis (IBSN), which may also be accompanied by symptoms of LHON, and myoclonic epilepsy associated with ragged red muscle fibres (MERRF) [Wallace et al., 1988a,b; Wallace, 1989]).

Women affected by these conditions will pass the defective mitochondria to *all* of their children. Prenatal or preimplantation diagnosis is of no help because all of the conceptuses will be genetically affected, even though the severity of the disease is variable and, in the case of LHON, tends to affect males more severely than females. In principle, these conditions could be corrected microsurgically at the pronucleus stage by transplanting the male and female pronuclei from the patient's fertilized egg to an enucleated, donated egg (cytoplast) with normal mitochondria. Alternatively, it might be possible to replace some of the cytoplasm of the patient's fertilized egg with donor cytoplasm or mitochondria, but this would be unlikely to cure the disorder unless all the mutant mitochondria were replaced because a mixed population of normal and mutant mitochondria within the cell (heteroplasmy) is thought to occur naturally in MERRF and IBSN patients. Such procedures would be unlikely to gain ethical approval, so women afflicted by these conditions could only have children who were free of the disorder and fathered by their partner if they opted for IVF or GIFT with donated oocytes and their partner's spermatozoa.

D. Will There Be a Demand for Preimplantation Diagnosis?

At present the impetus to develop methods for preimplantation diagnosis is coming from clinicians and scientists who view this prospect as an exciting challenge. Developments have been welcomed by couples who run a high risk of conceiving a child affected by a genetic disease. If preimplantation diagnosis reaches the point where it can be offered as a diagnostic service, couples will have the option to choose between preimplantation diagnosis and chorionic villus sampling (CVS) or amniocentesis. At present, couples may have to start several pregnancies before a conceptus is diagnosed unaffected. Since each pregnancy has to progress to the stage when prenatal diagnosis can be performed there may be a considerable delay before an unaffected conceptus is confirmed. Although not all cycles of preimplantation diagnosis will result in pregnancy, if the pregnancy rate is reasonably high after preimplantation diagnosis, this approach may allow a couple to have an unaffected child more rapidly than if they had to undergo several rounds of prenatal diagnosis and therapeutic abortions. Despite this advantage, it seems to us that most couples are likely to opt for either CVS or amniocentesis followed, if necessary, by a therapeutic abortion rather than subject themselves to the rigours of IVF and preimplantation diagnosis. However, there are many couples for whom an abortion would be unacceptable and so would benefit from preimplantation diagnosis. If preimplantation diagnosis is to be provided for these couples the financial cost has to be considered. In the United Kingdom, prenatal diagnosis is offered, where appropriate, free of charge on the National Health Service (NHS), but preimplantation diagnosis would almost certainly require IVF. There are only two NHS clinics that offer IVF, and the great majority of patients pay approximately £1,000 to £2,000 per cycle for treatment [ILA, 1990] although the costs could be reduced somewhat if oocytes were collected in natural cycles. The situation in other countries is different and in some cases public funds are available to subsidise the cost of IVF [ILA, 1990]. Given that IVF still has a relatively low success rate (about 20%–30%; see Table III), it is clear that many patients will opt for alternative forms of prenatal diagnosis, purely on financial grounds, unless the charge for preimplantation diagnosis is brought into line with charges for later prenatal diagnosis. In the United Kingdom this would involve bringing preimplantation diagnosis into the NHS sector, but in the present financial climate this seems unrealistic.

In conclusion, scientific advances during the past 5 years have already made preimplantation diagnosis a reality for a handfull of patients at risk for sex-linked disorders [Handyside et al., 1990]. Although preimplantation diagnosis is initially only likely to be feasible in a small number of specialized centers, from a purely technical point of view the rapid rate of progress is bound to improve the prospects for preimplantation diagnosis. However, financial constraints are looming on the horizon and may curtail the introduction of these scientific and medical advances into the clinic. Thus, despite enormous technical advances, the prospect of widespread use of preimplantation diagnosis is still uncertain.

IX. CONSTRAINTS ON PREIMPLANTATION DIAGNOSIS RESEARCH

Research on human pre-embryos has been restricted to different extents in different countries, and as a consequence much of the research has been concentrated in a few countries. In some (e.g., West

Germany, Norway, and Spain), research on pre-embryos is not permitted. In France, the National Ethics Committee imposed a 3 year moratorium on developing pre-implantation diagnosis (from 1986). There are no legal restrictions on research on pre-implantation diagnosis in the United States but the major funding agencies are unwilling to support this type of research. In the United Kingdom, the government convened a Committee of Inquiry, chaired by Dame Mary Warnock, in 1982 to consider issues related to human IVF and embryology. The report was published in 1984 [Warnock, 1984], but legislation was delayed until 1990. In the interim period the Medical Research Council and the Royal College of Obstetricians and Gynaecologists set up the "Voluntary Licensing Authority for Human In Vitro Fertilisation and Embryology" (VLA). The VLA was established in 1985 and became the "Interim Licensing Authority for Human In Vitro Fertilisation and Embryology" (ILA) in 1989. Its role was taken over by a statutory licensing authority, the "Human Fertilisation and Embryology Authority" (HFEA) in 1991. Further details of the legal position in the United Kingdom and other countries can be found in the Fifth Report by the Interim Licensing Authority [1990].

The following types of work were not approved by the VLA/ILA and are prohibited by the British Government's legislation that was introduced in 1991:

1. The modification of the genetic constitution of a pre-embryo
2. The placing of a human pre-embryo in the uterus of a member of another species for gestation
3. Replacing a nucleus of a cell of a pre-embryo with a nucleus from a cell of another individual (pre-embryo, tissue, or person)
4. Growing a pre-embryo beyond 14 days after fertilization

A combination of political, legal, and financial pressure has resulted in human pre-embryo research being concentrated in a few countries. Despite these restrictions rapid progress toward preimplantation diagnosis has been made over the last few years, and it is to be hoped that progress will continue, on a broader front, in the future.

ACKNOWLEDGMENTS

We thank Dr R.R. Angell for the photographs shown in Figure 1; Professor C.J. Epstein and Dr M. Monk for allowing us to reproduce figures; Drs. R.R. Angell, V.N. Bolton, A.H. Handyside, W. Ledger, M. Monk, A.L. Muggleton-Harris, and M.J. Wood for helpful discussions and, in some cases, providing information prior to publication; Professor H.J. Evans for reading the manuscript; and T. McFetters and E. Pinner for preparing the illustrations. J.D.W. is grateful to the Wellcome Trust for financial support.

BIBLIOGRAPHY

It will be obvious to anyone who has even glanced through this chapter that preimplantation diagnosis involves an unusually wide range of clinical and laboratory techniques. Some are described in the following bibliography of selected texts and laboratory manuals.

In Vitro Fertilization

Braude PR (1987): Fertilization of human oocytes and culture of human pre-implantation embryos in vitro. In Monk M (ed): Mammalian Development. A Practical Approach. Oxford: IRL Press. pp 281–306.
Edwards RG (1990): Assisted human conception. Bri Med Bull 46(3):1–864.

Mammalian Pre-Embryo Culture, Manipulations, and Biopsy

Hogan B, Costantini F, Lacy E (eds) (1986): Manipulating the Mouse Embryo. A Laboratory Manual.

Cold Spring Harbor, NY: Cold Spring Harbor
Laboratory Press.

Monk M (ed) (1987) Mammalian Development. A
Practical Approach. Oxford: IRL Press.

Cryobiology

Wood MJ, Whittingham DG, Rall WF (1987): The low
temperature preservation of mouse oocytes and
embryos. Monk M (ed): In Mammalian Development. A Practical Approach. Oxford: IRL Press,
pp 255–280.

Cytogenetics

Evans EP (1987): Karyotyping and sexing of gametes,
embryos and fetuses and in situ hybridisation to
chromosomes. In Monk M (ed): Mammalian Development. A Practical Approach. Oxford: IRL
Press, pp 93–114.

Rooney DE Czepulkowski BH (1986): Human Cytogenetics: A Practical Approach. Oxford: IRL
Press.

In Situ Hybridization

Lichter P, Cremer T, Borden J, Manuelides L. Ward
DC (1988a): Delineation of individual human
chromosomes in metaphase and interphase cells
by in situ suppression hybridisation using recombinant DNA libraries. Hum Genet 80: 224–234.

Lichter P, Cremer T, Tang C-JC, Watkins PC,
Manuelides L, Ward DC (1988b): Rapid detection
of human chromosome 21 aberrations by in situ
hybridisation. Proc Natl Acad Sci USA 85: 9664–
9668.

Pinkel D, Landegent J, Collins C, Fuscoe J, Segraves
R, Lucas J, Gray J (1988): Fluorescence in situ
hybridization with human chromosome-specific
libraries: Detection of trisomy 21 and translocations of chromosome 4. Proc Natl Acad Sci USA
85: 9138–9142.

Biochemical Assays

Galjaard H (1980): Genetic Metabolic Disease: Early
Diagnosis and Prenatal Analysis. Amsterdam:
Elsevier-North Holland. See chapter III, part 7 (pp
554–596) and the appendices (pp 791–858) for
methods.

Monk M (1987): Biochemical microassays for X-chromosome-linked enzymes HPRT and PGK. In
Monk M (ed): Mammalian Development. A Practical Approach. Oxford: IRL Press, pp 139–161.

Shapira E, Blitzer MG, Miller JB, Africk DK (1989):
Biochemical Genetics: A Laboratory Manual.
New York: Oxford University Press, 145 pp. (This
is a compilation of standard biochemical assays

used in prenatal diagnosis and does not give protocols for microassays.)

Polymerase Chain Reaction and Other Molecular Techniques

Innis MA, Gelfand DH, Sninsky JJ (eds) (1990): PCR
Protocols—A Guide to Methods and Applications. New York: Academic Press, 482 pp.

Erlich HA, Gibbs R, Kazazian HH Jr (eds) (1990):
Polymerase Chain Reaction. Current Communi-
cations in Molecular Biology. Cold Spring Harbor,
NY: Cold Spring Harbor Laboratory Press.

REFERENCES

Adinolfi M, Polani PE (1989): Prenatal diagnosis of
genetic disorders in preimplantation embryos: Invasive and non-invasive approaches. Hum Genet
83:16–19.

Adler DA, West JD, Chapman VM (1977): Expression of alpha-galactosidase in preimplantation
mouse embryos. Nature 267:838–839.

Allen WR, Pashen RL (1984): Production of monozygotic (identical) horse twins by embryo
micromanipulation. J Reprod Fertil 71:607–613.

Angell RR (1991): Predivision in human oocytes at
meiosis I: A mechanism for trisomy formation in
man. Hum Genet. 86:383–387.

Angell RR, Summer AT, West JD, Thatcher SS, Glasier AF, Baird DT (1987): Post-fertilization polyploidy in human preimplantation embryos fertilized in vitro. Hum Reprod 2:721–727.

Angell RR, Hillier SG, West JD, Glasier AF, Rodger
MW, Baird DT (1988): Chromosome anomalies in
early human embryos. J Reprod Fertil (Suppl)
36:73–81.

Ashwood-Smith MJ (1986): The cryopreservation of
human embryos. Hum Reprod 1:319–332.

Avise JC, Lansman RA, Shade RO (1979): The use of
restriction endonucleases to measure mitochondrial DNA sequence relatedness in natural populations. I. Population structure and evolution in the
genus *Peromyscus*. Genetics 92:279–295.

Bacchus C, Buselmaier W (1988): Blastomere
karyotyping and transfer of chromosomally selected embryos. Implications for the production of
specific animal models and human preimplantation diagnosis. Hum Genet 80:333–336.

Bachvarova R, DeLeon V (1980): Polyadenylated
RNA of mouse ova and loss of maternal RNA in
early development. Dev Biol 74:1–8.

Benson C, Monk M (1988): Microassay for adenosine
deaminase, the enzyme lacking in some forms of
immunodeficiency, in mouse preimplantation embryos. Hum Reprod 3:1004–1009.

Benson PF, Fensom AH (1985): Genetic Biochemical Disorders. Oxford Monographs on Medical Genetics No. 12. Oxford: Oxford University Press.

Betteridge KJ, Hare WCD, Singh, EL (1981): Approaches to sex selection in farm animals. In Brackett BG, Seidel SM (eds): New Technology in Animal Breeding. New York: Academic Press, pp 109–125.

Blakewood EG, Rorie RW, Pool SH, Godke RA (1986): Freezing bovine embryos with and without a zona pellucida. Theriogenology 25:141.

Bolton VN (1991): Embryo biopsy. In Chapman M, Grudzinskas JG, Chard T (eds): The Embryo: Normal and Abnormal Development and Growth. Berlin: Springer-Verlag, pp 63–79.

Bolton VN, Braude PR. (1987): Development of the human preimplantation embryo in vitro. Curr Top Dev Biol 23: 93–114.

Bolton VN, Hawes SM, Taylor CT, Parsons JH (1989): Development of spare human preimplantation embryos in vitro: An analysis of the correlations among gross morphology, cleavage rates and development to the blastocyst. J. In Vitro Fertil Embryo Transfer 6:30–35.

Bolton VN, Oades PJ, Johnson MH (1984): The relationship between cleavage, DNA replication and gene expression in the mouse 2-cell embryo. J Embryol Exp Morphol 79:139–163.

Bolton VN, Wren ME, Parsons JH (1991): Pregnancies following in vitro fertilization and transfer of human blastocysts. Fertil Steril 55:830–832.

Bongso A, Ng S-C, Sathananthan H, Poh Lian N, Rauff M, Ratnam S (1989): Improved quality of human embryos when co-cultured with human ampullary cells. Hum Reprod 4:706–713.

Booker M, Parsons J (1990): The perurethral technique for ultrasound directed follicle aspiration in an in-vitro fertilization and embryo transfer programme: A report of 636 patient cycles. Br J Obstet Gynaecol 97:499–505.

Brambati B, Tului L (1990): Preimplantation genetic diagnosis: A new simple uterine washing system. Hum Reprod 5:448–450.

Braude PR (1986): The use of human embryos for infertility research. In Human Embryo Research: Yes or No? Ciba Foundation. London: Tavistock Publications, pp 63–82.

Braude PR (1987): Fertilization of human oocytes and culture of human pre-implantation embryos: in vitro. In Monk M (ed): Mammalian Development. A Practical Approach. Oxford: IRL Press, pp 281–306.

Braude P, Bolton V, Moore S (1988): Human gene expression first occurs between the four cell and eight cell stages of preimplantation development. Nature 332:459–461.

Braude PR, Monk M, Pickering SJ, Cant A, Johnson MH (1989): Measurement of HPRT activity in the human unfertilised oocyte and pre-embryo. Prenat Diagn 9:839–850.

Brock DJH (1982): Early Diagnosis of Fetal Defects. Edinburgh: Churchill Livingstone.

Bronson RA, McLaren A (1970): Transfer to the mouse oviduct of eggs with and without the zona pellucida. J Reprod Fert 22:129–137.

Burns J, Chan VTW, Jonasson JA, Flemming KA, Taylor S, McGee JOD (1985): Sensitive system for visualising biotinylated DNA probes hybridised in situ: Rapid sex determination of intact cells. J Clin Pathol 38:1085–1092.

Buster JE, Bustillo M, Thorneycroft IH, Simon JA, Boyers SP, Marshall JR, Louw JA, Seed RW, Seed RG (1983): Non-surgical transfer of in vivo fertilized donated ova to five infertile women: report of two pregnancies. Lancet 2:223–224.

Buster JE, Carson SA (1989): Genetic diagnosis of the preimplantation embryo. Am J Med Genet 34:211–216.

Camus M, Van den Abbeel E, Van Waesberghe L, Wisanto A, Devroey P, Steirteghem A. (1989): Human embryo viability after freezing with dimethysulfoxide as a cryoprotectant. Fertil Steril 51:460–465.

Chehab FF, Doherty, M, Cai S, Kan YW, Cooper S, Rubin EM (1987): Detection of sickle cell anaemia and thalassaemias. Nature 329:293–294.

Cornel MC, Ten Kate LP, Dukes MNG (1989a): Ovulation induction and neural tube defects. Lancet 2:1386.

Cornel M, Ten Kate LP, Te Meernan GJ (1989b): Ovulation induction, in vitro fertilisation and neural tube defects. Lancet 2:1530.

Coutelle C, Williams C, Handyside A, Hardy K, Winston R (1989): Genetic analysis of DNA from single human oocytes: A model for preimplantation diagnosis of cystic fibrosis. Bri Med J 299:22–24.

Crane JP, Cheung SW (1988): An embryogenic model to explain cytogeneic inconsistencies observed in chronic villus versus fetal tissue. Prenat Diagn 8:119–129.

Crosby IM, Gandolfi F, Moor RM (1988): Control of protein synthesis during early cleavage sheep embryos. J Reprod Fertil 82:769–775.

Dawson KJ, Rutherford AJ, Winston NJ, Subak-Sharpe R, Winston RML (1988): Human blastocyst transfer. Is it a feasible proposition? In Human Reproduction: Abstracts From the Fourth Meeting of the European Society of Human Reproduction and Embryology. Oxford: IRL Press, abstract 145, pp 44–45.

Darlington CD (1957): Messages and Movements in the Cell. Conference on Chromosomes, Wageningen: Willink Zwolle.

Dokras A, Sargent IL, Ross C, Gardner RL, Barlow DH (1990): Trophectoderm biopsy in human blastocysts. Hum Reprod 5:821–825.

Dermer SJ, Johnson EM (1988): Rapid DNA analysis of alpha-1-antitrypsin deficiency: Application of an improved method for amplifying mutated gene sequences. Lab Invest 59:403–408.

Edwards JH (1956): Antenatal detection of hereditary disorders. Lancet 1:579.

Edwards RG (1990): Assisted human conception. Br Med Bull 46 3:1–864.

Edwards RG, Hollands P (1988): New advances in human embryology: Implications for the preimplantation diagnosis of genetic disease. Hum Reprod 3:549–556.

Ellis PM, West JD, West KM, Murray RS, Coyle MC (1990): Relevance to prenatal diagnosis of the identification of a human Y/autosome translocation by Y-chromosome-specific in situ hybridisation. Mol Rep Dev 25:37–41.

Epstein CJ (1975): Gene expression and macromolecular synthesis during preimplantation embryonic development. Biol Reprod 12:82–105.

Epstein CJ, Smith S, Travis B, Tucker G (1978): Both X chromosomes function before visible X chromosome inactivation in female mouse embryos. Nature 274:500–503.

Esworthy S, Chapman VM (1981): The expression of β-galactosidase during preimplantation mouse embryogenesis. Dev Genet 2:1–12.

Fehilly CB, Cohen J, Simons RF, Fishel SB, Edwards RG (1985): Cryopreservation of cleaving embryos and expanded blastocysts in the human: A comparative study. Fertil Steril 44:638–644.

Feichtinger W, Banko I, Kemeter P (1987): Freezing human oocytes using rapid techniques. In Feichtinger WE, Kemeter P (eds). Future Aspects in Human In Vitro Fertilization. Heidelberg: Springer-Verlag, pp 101–110.

Flach G, Johnson MH, Braude PR, Taylor AS, Bolton VN (1982): The transition from maternal to embryonic control in the 2-cell mouse embryo. EMBO J 1:681–686.

Flemming R, Coutts JRT (1986): Induction of multiple follicular growth in normally menstruating women with endogenous gonadotropin-releasing hormone agonist and gonadotropins for in vitro fertilization. Fertil Steril 45:226–230.

Formigli L, Roccio C, Belotti G, Stangalini A, Coglitore MT, Formigli G (1990): Nonsurgical flushing of the uterus for pre-embryo recovery: possible clinical applications. Hum Reprod 5:329–335.

Freemann L, Trounson A, Kirby C (1986): Cryopreservation of human embryos: Progress on the clinical use of the technique in human in vitro fertilization. In Vitro Fertil Embryo Transfer 3:53–61.

Frels WI, Chapman VM (1980): Expression of the maternally derived X Chromosome in the mural trophoblast of the mouse. J Embryol Exp Morphol 56:179–180.

Frels WI, Rossant J, Chapman VM (1979): Maternal X-chromosome inactivation in mouse chorionic ectoderm. Dev Genet 1:123–132.

Friedler S, Guidice LC, Lamb EJ (1988): Cryopreservation of embryos and ova. Fertil Steril 49:743–764.

Frydman R, Forman RG, Belaisch-Allart J, Hazout A, Testart J (1988): An assessment of alternative policies for embryo transfer in an in vitro fertilization-embryo transfer program. Fertil Steril 50:466–470.

Fuchs F, Philip J (1963): Mulighed for antenatal undersogelse af fosterets kromosomer. Nord Med 9(V):572–573.

Fuchs F, Riis P (1956): Antenatal sex determination. Nature 177:330.

Fugger EF (1989): Clinical status of human cryopreservation in the United States of America. Fertil Steril 52:986–990.

Galjaard H (1980): Genetic Metabolic Disease: Early Diagnosis and Prenatal Analysis. Amsterdam: Elsevier-North Holland.

Galjaard H (1984): Early diagnosis and prevention of genetic disease. In Aspects of Human Genetics, With Special Reference to X-Linked Disorders. Basel: Karger, pp 1–15.

Galjaard H, Niermeijer MF, Hahnemann N, Mohr J, Sorensen SA (1974a): An example of rapid prenatal diagnosis of Fabry's disease using microtechniques. Clin Genet 5:368–377.

Galjaard H, Van Hoogstraten JJ, De Josselin, De Jong JE, Mulder MP (1974b): Methodology of quantitative cytochemical analysis of single or small numbers of cultured cells. Histochem J 6:409–429.

Gardner RL (1988): Cell fate in the developing embryo. Jones CT (ed): In Fetal and Neonatal Development. Perinatology Press, pp 10–23.

Gardner RL, Papaioannou VE (1975): Differentation in the trophectoderm and inner cell mass. In Balls M, Wild AE (eds): The Early Development of Mammals. Cambridge: Cambridge University Press, pp 107–132.

Gardner RL, Edwards RG (1968): Control of the sex ratio at full term in the rabbit by transferring sexed blastocysts. Nature 218:346–348.

Giles RE, Blanc H, Cann HM, Wallace DC (1980): Maternal inheritance of human mitochondrial DNA. Proc Natl Acad Sci USA 77:6715–6719.

Glenister PH, Wood MJ, Kirby C, Whittingham DG (1987): The incidence of chromosome anomalies in first-cleavage mouse embryos obtained from frozen-thawed oocytes fertilized in vitro. Gamete Res 16:205–216.

Gomez CM, Muggleton-Harris AL, Whittingham DG, Hood LE, Readhead C (1990): Rapid preimplantation detection of mutant (*shiverer*) and normal alleles of the mouse myelin basic protein gene allowing selective implantation and birth of live young. Proc Natl Acad Sci USA 87:4481–4484.

Gosden JR, Mitchell AR, Gosden CM, Rodeck CH, Morsman JM (1982): Direct vision chorion biopsy and the use of chromosome-specific DNA probes for first trimester fetal sex determination in prenatal diagnosis. Lancet 2:1416–1419.

Gott AL, Hardy K, Winston RML, Leese HJ (1990): Non-invasive measurement of pyruvate and glucose uptake and lactate production by single human preimplantation embryos. Hum Reprod 5:104–108.

Gresson RAR (1940): Presence of the sperm middlepiece in the fertilized egg of the mouse (*Mus musculus*). Nature 145:425.

Griffin DK, Handyside AH, Penketh RJH, Winston RML, Delhanty JDA (1991): Fluorescence in situ hybridization to interphase nuclei of human preimplantation embryos with X and Y chromosome specific probes. Hum Reprod 6:101–105.

Hahnemann N (1974): Early prenatal diagnosis; a study of biopsy techniques and cell culturing from extraembryonic membrane. Clin Genet 6:294–306.

Handyside AH (1991): Preimplantation diagnosis by DNA amplification. In Chapman M, Grudzinskas JG, Chard T (eds): The Embryo: Normal and Abnormal Development and Growth. Berlin: Springer-Verlag, pp 81–90.

Handyside A, Delhanty J (1991): Cleavage stage biopsy of human embryos and diagnosis of X-linked recessive disease. In Edwards RG (ed): Preimplantation Diagnosis of Genetic Disease (in press).

Handyside AH, Kontogianni EH, Hardy K, Winston RML (1990): Pregnancies from biopsied human preimplantation embryos sexed by Y-specific DNA amplification. Nature 344:768–770.

Handyside AH, Penketh RJA, Winston RML, Pattinson JK, Delhanty JDA, Tuddenham EGD (1989): Biopsy of human preimplantation embryos and sexing by DNA amplification. Lancet 1:347–349.

Hardy K, Handyside A, Winston RML (1989a): The human blastocyst: Cell number, death and allocation during late preimplantation development in vitro. Development 107:597–604.

Hardy K, Hooper MAK, Handyside AH, Rutherford AJ, Winston RML, Leese HJ (1989b): Non-invasive measurement of glucose and pyruvate uptake by individual human oocytes and preimplantation embryos. Hum Reprod 4:188–191.

Hardy K, Martin KL, Leese HJ, Winston RML, Hand-

yside AH (1990): Human preimplantation development in vitro is not adversely affected by biopsy at the 8-cell stage. Hum Reprod 5:708–714.

Harper M, Monk M (1983): Evidence for translation of HPRT enzyme on maternal mRNA in early mouse embryos. J Embryol Exp Morphol 74:15–28.

Harrison RG (1978): Clinical Embryology. London: Academic Press.

Hassold T, Chiu D, Yamane JA (1984): Parental origin of autosomal trisomies. Ann Hum Genet 48:129–144.

Hayashi JI, Yonekawa H, Gotoh O, Watanabe J, Tagashira Y (1978): Strictly maternal inheritance of rat mitochondrial DNA. Biochem Biophys Res Commun 83:1032–1038.

Holding C, Monk M (1989): Diagnosis of beta-thalassaemia by DNA amplification in single blastomeres from mouse preimplantation embryos. Lancet 2:532–535.

Hooper M, Hardy K, Handyside A, Hunter S, Monk M (1987): HPRT-deficient (Lesch-Nyhan) mouse embryos derived from germ-line colonization by cultured cells. Nature 274:503–504.

Hosli P, de Bruyn CHMM, Oei TL (1974): Development of a micro HG-PRT activity assay: preliminary complementation studies with Lesch-Nyhan cell strains. In Sperling O, DeVries A, Wyngaarden JB (eds): Purine Metabolism in Man: Advances in Experimental Medicine and Biology, New York: Plenum Press, pp 811–815.

Hutchison CA, Newbold JE, Potter SS, Edgell MH (1974): Maternal inheritance of mammalian mitochondrial DNA. Nature 251:536–538.

Interim Licensing Authority (1990): The Fifth Report of the Interim Licensing Authority for Human In Vitro Fertilisation and Embryology. Sponsored by the Medical Research Council and the Royal College of Obstetricians and Gynaecologists. 76 pp.

Jacobson CB, Barter RH (1967): Intrauterine diagnosis and management of genetic defects. Am J Obstet Gynecol 99:796–807.

Johnson MH (1981): The molecular and cellular basis of preimplantation mouse development. Biol Rev 56:463–498.

Johnson MH, Pickering SJ (1987): The effect of dimethylsulphoxide on the microtubular system of the mouse oocyte. Development 100:313–324.

Jones KW, Singh L, Edwards RG (1987): The use of probes for the Y chromosome in preimplantation embryo cells. Hum Reprod 2:439–445.

Kanagawa H, Frim J, Kruuv J (1979): The effect of puncturing the zona pellucida on freeze–thaw survival of bovine embryos. Can J Anim Sci 59:623–626.

Kelly S (1977): Studies of the developmental potential

of 4- and 8-cell stage mouse blastomeres. J Exp Zool 20:365–376.

Kerem B-S, Rommens JM, Buchanan JA, Markiewicz D, Cox TK, Chakravarti A, Buchwald M, Tsiu L-C (1989): Identification of the cystic fibrosis gene: genetic analysis. Science 245:1073–1080.

Kleijer WJ, Van der Veer E, Niermeijer MF (1976): Rapid prenatal diagnosis of GM₁-gangliosidosis using microchemical methods. Hum Genet 33:299–305.

Kola I, Kirby C, Shaw J, Davey A, Trounson A (1988): Vitrification of mouse oocytes results in aneuploid zygotes and malformed fetuses. Teratology 38:467–474.

Kratzer PG, Gartler SM (1978): Hypoxanthine guanine phosphoribosyl transferase expression in early mouse development. In Russell LB (ed): Genetic Mosaics and Chimaeras in Mammals. New York: Plenum Press, pp 247–260.

Kroon AM, DeVos WM Bakker H (1978): The heterogeneity of rat liver mitochondrial DNA. Biochim Biophys Acta 519:269–273.

Krzyminska UB, Lutjen J, O'Neill C (1990): Assessment of the viability and pregnancy potential of mouse embryos biopsied at different preimplantation stages of development. Hum Reprod 5:203–208.

Kullander S, Sandahl B (1973): Fetal chromosome analysis after transcervical placental biopsy during early pregnancy. Acta Obstet Gynaecol Scand 52:355.

Kwok S, Higuchi R (1989): Avoiding false positives with PCR. Nature 339:237–238.

Leach P (1988): Why use the term "pre-embryo"? In The Third Report of the Voluntary Licensing Authority for Human In Vitro Fertilisation and Embryology. Sponsored by the Medical Research Council and the Royal College of Obstetricians and Gynaecologists, pp 22-23.

Leese HJ (1987a): Analysis of embryos by non-invasive methods. Hum Reprod 2:37–40.

Leese HJ (1987b): Non-invasive methods for assessing embryos. Hum Reprod 2:435–438.

Leese HJ, Hooper MAK, Edwards RG, Ashwood Smith MJ (1986): Uptake of pyruvate by early human embryos determined by a non-invasive technique. Human Reproduction 1, 181–182.

Lehn-Jensen H, Willadsen SM (1983): Deep freezing of cow "half" and "quarter" embryos. Theriogenology 19:49–54.

Li A, Gyllensten UB, Cui X, Saiki RK, Erlich HA, Arnheim N (1988): Amplification and analysis of DNA sequences in single human sperm and diploid cells. Nature 335:414–419.

Lichter P, Cremer T, Borden J, Manuelides L, Ward DC (1988a): Delineation of individual human chromosomes in metaphase and interphase cells by in situ suppression hybridisation using recombinant DNA libraries. Hum Genet 80:224–234.

Lichter P, Cremer T, Tang C-J, C, Watkins PC, Manuelides L, Ward DC (1988b): Rapid detection of human chromosome 21 aberrations by in situ hybridisation. Proc Natl Acad Sci USA 85:9664–9668.

Lindeman R, Lutjen J, O'Neill C, Trent RJ (1990): Exclusion of beta-thalassaemia by biopsy and DNA amplification in mouse pre-embryos. Prenat Diag 10:295–301.

Lopata A, Kohlman D, Johnston I (1983): The fine structure of normal and abnormal embryos developed in culture. In Beier HM, Lindner HR (eds): Fertilisation of the Human Egg In Vitro. Biological Basis and Clinical Application. Berlin: Springer-Verlag, pp 189–210.

Lowry OH, Passonneau JV (1972): A Flexible System of Enzymatic Analysis. New York: Academic Press.

Lucena E, Olivares R, Obando H, Uribe L, Lombana O, Davila A, Saa AM, Gomez M (1986): Pregnancies following transfer of human frozen-thawed embryos in Colombia, South America. Hum Reprod 1:383–385.

Magnuson T, Epstein CJ (1981): Genetic control of very early mammalian development. Biol Rev 56:369–408.

McLaren A (1985): Prenatal diagnosis before implantation: opportunities and problems. Prenat Diagn 5:85–90.

McLaren A (1986): Prelude to embryogenesis. In Human Embryo Research: Yes or No? Ciba Foundation. London: Tavistock Publications, pp 5–23.

Modlinski J (1970): The role of the zona pellucida in the development of mouse eggs in vivo. J Embryol Exp Morph 23:539–547.

Monk M (1978): Biochemical studies of X-chromosome activity in preimplantation mouse embryos. In Russell LB (ed). Genetic Mosaics and Chimeras in Mammals. New York: Plenum Press, pp 239–246.

Monk M (1987): Biochemical microassays for X-chromosome-linked enzymes HPRT and PGK. In Monk M (ed): Mammalian Development: A Practical Approach. Oxford: IRL Press, pp 139–161.

Monk M (1988): Preimplantation diagnosis. BioEssays 8:184–189.

Monk M, Handyside AH (1988): Sexing of preimplantation mouse embryos by measurement of X-linked gene dosage in a single blastomere. J Reprod Fertil 82:365–368.

Monk M, Handyside A, Hardy K, Whittingham D (1987): Preimplantation diagnosis of deficiency of hypoxanthine phosphoribosyltransferase in a mouse model for Lesch-Nyhan syndrome. Lancet 2:423–425.

Monk M, Handyside A, Muggleton-Harris A, Whittingham D (1990): Preimplantation sexing and diagnosis of hypoxanthine phosphoribosyl transferase deficiency in mice by biochemical microassay. Am J Med Genet 35:201–205.

Monk M, Harper M (1979): Sequential X chromosome inactivation coupled with cellular differentiation in early mouse embryos. Nature 281:311–313.

Monk M, Holding C (1990): Amplification of a β-haemoglobin sequence in individual human oocytes and polar bodies. Lancet 335:985–988.

Monk M, Muggleton-Harris AL, Rawlings E, Whittingham DG (1988): Pre-implantation diagnosis of HPRT-deficient male and carrier female mouse embryos by trophectoderm biopsy. Hum Reprod 3:377–381.

Moore KL (1977): The Developing Human: Clinically Orientated Embryology, 2nd ed. Philadelphia: WB Saunders.

Moore NW, Adams CE, Rowson LEA (1968): Developmental potential of single blastomeres of the rabbit egg. J Reprod Fertil 17:527–531.

MRC Working Party on Children Conceived by In Vitro Fertilisation (1990): Births in Great Britain resulting from assisted contraception, 1978–1987. British Medical Journal 300, 1229–1233.

Muggleton-Harris AL (1990): Proliferation of cells derived from the biopsy of pre-embryos. In Mashiach S (ed): Proceedings of the VIth World Congress on In Vitro Fertilization and Alternate Assisted Reproduction. Jerusalem, Israel: Plenum Press, pp 887–899.

Muggleton-Harris AL, Findlay I, Whittingham DG (1990): Improvement of the culture conditions for the development of human preimplantation embryos. Hum Reprod 5:217–220.

Mullis KB, Faloona FA (1987): Specific synthesis of DNA in vitro via a polymerase-catalysed chain reaction. Methods Enzymol 155:335–350.

Myerowitz R (1988): Splice junction mutation in some Ashkenazi Jews with Tay-Sachs disease: Evidence against a single defect within this ethnic group. Proc Natl Acad Sci USA 85:3955–3959.

Myerowitz R, Costigan FC (1988): The major defect in Ashkenazi Jews with Tay-Sachs disease is an insertion in the gene for the alpha-chain of beta-hexosaminidase. J Biol Chem 263: 18587–18589.

Nederlof PM, Robinson D, Wiegent J, Hopman AHN, Tanke HJ, Raap A (1989): Three color fluorescence for the simultaneous detection of multiple DNA targets. Cytometry 10:20–27.

Neveu S, Hedon B, Bringer J, Chinchole J-M, Arnal F, Humeau C, Cristol P, Viala J-L (1987): Ovarian stimulation by a combination of a gonadotropin-releasing hormone agonist and gonadotropins for in vitro fertilization. Fertil Steril 47:639–643.

Nichols J, Gardner RL (1989): Effect of damage to the zona pellucida on development of preimplantation embryos in the mouse. Hum Reprod 4:180–187.

Niemann H, Brem G, Scacher B, Smidt D, Krausslich H (1986): An approach to successful freezing of demi-embryos derived from 7-day bovine embryos. Theriogenology 25:519–524.

Nijs M, Camus M, Van Steirteghem AC (1988): Evaluation of different biopsy methods of blastomeres from 2-cell mouse embryos. Hum Reprod 3:999–1003.

Old JM (1986): Fetal DNA analysis. In Davies KE (ed): Human Genetic Diseases: A Pratical Approach. Oxford: IRL Press, pp 1–16.

Ozawa T, Yoneda M, Tonaka M, Ohno K, Sato W, Suzuki H, Nishikima M, Yamamoto M, Nonaka I, Horai S (1988): Maternal inheritance of deleted mitochondrial DNA in a family with mitochondrial myopathy. Biochem Biophys Res Commun 154:1240–1247.

Ozil JP (1983): Production of identical twins by bisection of blastocysts in the cow. J Reprod Fertil 69:463–468.

Papaioannou VE (1982): Lineage analysis of inner cell mass and trophectoderm using microsurgically reconstituted mouse blastocysts. J Embryol Exp Morphol 68:199–209.

Papaioannou VE, West JD, Bucher Th, Linke IM (1981): Non-random X-chromosome expression early in mouse development. Dev Genet 2:305–315.

Patrick AD (1983): Inherited metabolic disorders. Br Med Bull 39:378–385.

Patrick AD (1984): Prenatal diagnosis of inherited diseases. In Rodeck CH, Nicolaides KH (eds): Prenatal Diagnosis. Proceedings of the XIth Study Group of the Royal College of Obstetricians and Gynaecologists. Chichester: Wiley, pp 121–132.

Penketh RJA, Delhanty, JDA, Van den Berghe JA, Finklestone EM, Handyside AH, Malcolm S, Winston RML (1989): Rapid sexing of human embryos by non-radioactive in situ hybridization: potential for preimplantation diagnosis of X-linked disorders. Prenat Diagn 9:489–499.

Penketh R, McLaren A (1987): Prospects for prenatal diagnosis during preimplantation human development. In Rodeck CH (ed): Fetal Diagnosis of Genetic Defects. Baillieres Clinical Obstetrics and Gynaecology, International Practice and Research. London: Baillière Tindall, Volume 1, pp 747–764.

Pickering SJ, Johnson MH (1987): The influence of cooling on the organization of of the meiotic spindle of the mouse oocyte. Hum Reprod 2:207–216.

Pieters MHEC, Geraedts JPM, Meyer H, Dumoulin JCM, Evers JLH, Jongbloed RJE, Nederlof PM, van der Flier S (1990): Human gametes and zygotes studied by nonradioactive in situ hybridization. Cytogenet Cell Genet 53:15–19.

Piko L, Clegg KB (1982): Quantitative changes in total RNA, total poly (A), and ribosomes in early mouse embryos. Dev Biol 89:362–378.

Pinkel D, Landegent J, Collins C, Fuscoe J, Segraves R, Lucas J, Gray J (1988): Fluorescence in situ hybridization with human chromosome-specific libraries: detection of trisomy 21 and translocations of chromosome 4. Proc Natl Acad Sci USA 85:9138–9142.

Porter RN, Smith W, Craft IL, Abdulwahid NA, Jacobs HS (1984): Induction of ovulation for in-vitro fertilisation using Buserilin and gonadotropins. Lancet 2:1284–1285.

Quinn P, Kerin JFP (1986): Experience with the cryopreservation of human embryos using the mouse as a model to establish successful techniques. J In Vitro Fertil Embryo Transfer 3:40–45.

Rall WF, Fahy GM (1985): Ice free cryopreservation of of mouse embryos at −196°C by vitrification. Nature 313:573–575.

Rands GF (1985): Cell allocation in half- and quadruple-sized preimplantation mouse embryos. J Exp Zool 236:67–70.

Rands GF (1986): Size regulation in the mouse embryo. II The development of half embryos. J Embryol Exp Morphol 98:209–217.

Rieger R, Michaelis A, Green MM (1976): Glossary of Genetics and Cytogenetics, Classical and Molecular, 4th ed. Berlin: Springer-Verlag, 562.

Riis P, Fuchs F (1960): Antenatal determination of foetal sex in prevention of hereditary diseases. Lancet 2:180–182.

Roberts C, Lutjen J, Krzyminska U, O'Neill C (1990): Cytogenetic analysis of biopsied preimplantation mouse embryos: implications for prenatal diagnosis. Hum Reprod 5: 197–202.

Rutherford AJ, Subak-Sharpe RJ, Dawson KJ, Margara RA, Franks S, Winston RML (1988): Improvement in in vitro fertilisation after treatment with buserilin, an agonist of luteinising hormone releasing hormone. Br Med J 296:1765–1768.

Saiki RK, Arnheim N, Erlich HA (1985a): A novel method for the detection of polymorphic restriction sites by cleavage of oligonucleotide probes: Application to sickle-cell anemia. Biotechnology 3: 1008–1012.

Saiki RK, Bugawan TL, Horn GT, Mullis KB, Erlich HA (1986): Analysis of enzymatically amplified beta-globin and HLA-DQ alpha DNA with allele-specific oligonucleotide probes. Nature 324:163–166.

Saiki RK, Chang C-A, Levense CH, Warren TC, Boehm CD, Kazasian HH, Erlich HA (1988): Diagnosis of sickle cell anemia and beta-thalassaemia with enzymatically amplified DNA and nonradioactive allele-specific oligonucleotide probes. N Engl J Med 319:537–541.

Saiki RK, Scharf SJ, Faloona F, Mullis KB, Horn GT, Erlich HA, Arnheim N (1985b): Enzymatic amplification of beta-globin genomic sequences and restriction site analysis of diagnosis of sickle cell anemia. Science 230:1350–1354.

Salat-Baroux J, Cornet D, Alvarez S, Antoine JM, Tibi C, Mandelbaum J, Plachot M (1988): Pregnancies after replacement of of frozen-thawed embryos in a donation program. Fertil Steril 49:817–821.

Sathananthan H, Bongso A, Ng S-C., Ho J, Mok H, Ratnan S (1990): Ultrastructure of preimplantation human embryos co-cultured with ampullary cells. Hum Reprod 5:309–318.

Sauer MV, Anderson RE, Paulson RJ (1989): A trial of superovulation in ovum donors undergoing uterine lavage. Fertil Steril 51:131–134.

Serr DM, Sachs L, Danon M (1955): The diagnosis of sex before birth using cells from the amniotic fluid (a preliminary report). Bull Res Council Isr 5B:137–138.

Siebzehnrubl ER (1989): Cryopreservation of gametes and cleavage stage embryos. Hum Reprod [Suppl] 4:105–110.

Siebzehnrubl E, Trotnow S, Weigel M, Kniewald T, Harbermann PG, Kreuzer E, Hunlich T (1986): Pregnancy after in vitro fertilization, cryopreservation and embryo transfer. J In Vitro Fertil Embryo Transfer 3:261–263.

Speer A, Rosenthal A, Billwitz H, Hanke R, Forrest SM, Love, D, Davies KE Coutelle Ch (1989): DNA amplification of a further exon of Duchenne muscular dystrophy locus increase possibilities for deletion screening. Nucleic Acids Res 17:4892.

Steele MW, Breg WR (1966): Chromosome analysis of human amniotic-fluid cells. Lancet 1:383–385.

Strom CM, Verlinsky Y, Milayeva S, Evsikov S Cieslak J, Lifchez A, Valle, J., Moise J Ginsberg N. Applebau M (1990): Preconception diagnosis of cystic fibrosis. Lancet 336:306–307.

Sundstrom P, Nilsson O, Liedholm P (1981): Cleavage rate and morphology of early human embryos obtained after artificial fertilization and culture. Acta Obstet Gynecol Scand 60:109–120.

Summers PM, Campbell JM, Miller MW (1988): Normal in-vivo development of marmoset monkey embryos after trophectoderm biopsy. Hum Reprod 3:389–393.

Szollosi D (1965): The fate of sperm middle-piece mitochondria in the rat egg. J Exp Zool 159:367–378.

Testart J (1988): Results of in vitro fertilization with

embryo cryopreservation and a recommendation for uniform reporting. Fertil Steril 49:156–158.

Testart J, Lassalle B, Belaisch-Allart J, Hazout A, Forman R, Rainhorn JD, Frydman R (1986): High pregnancy rate after human embryo freezing. Fertil Steril 46:268–272.

Trounson A (1986): Preservation of human eggs and embryos. Fertil Steril 46:1–12.

Trounson AO (1990): Cryopreservation. Br Med Bull 46:695–708.

Trounson A, Mohr L (1983): Human pregnancy following cryopreservation, thawing and transfer of an eight-cell embryo. Nature 305:707–709.

Trounson A, Peura A, Freemann L, Kirby C (1988) Ultrarapid freezing of early cleavage stage human embryos and eight-cell mouse embryos. Fertil Steril 49:822–826.

Trounson A Peura A, Kirby C (1987): Ultrarapid freezing: A new low-cost and effective method of embryo cryopreservation. Fertil Steril 48:843–850.

Tsunoda Y, McLaren A (1983): Effect of various procedures on the viability of mouse embryos containing half the normal number of blastomeres. J Reprod Fertil 69:315–322.

Valenti C Schutta EJ, Kehaty T (1968): Prenatal diagnosis of Down's syndrome. Lancet 2:220.

Van Steirteghem AC, Van den Abbeel E, Camus M, Van Waesberghe L, Braeckmans P, Khan I, Nijs M, Smitz J, Staessen C, Wisanto A, Devroey P (1987): Cryopreservation of human embryos obtained after gamete intra-Fallopian transfer and/or invitro fertilization. Hum Reprod 2:593–598.

Verlinsky Y, Ginsberg N, Lifchez A, Valle J, Moise J, Strom CM (1990): Analysis of the first polar body: preconception genetic diagnosis. Hum Reprod 5:826–829.

Wales RG, Whittingham DG, Hardy K, Craft IL (1987): Metabolism of glucose by human embryos. J Reprod Fertil. 79:289–297.

Wallace DC (1989): Mitochondrial DNA mutations and neuromuscular disease. Trends Genet 5:9–13.

Wallace DC, Singh G., Lott MT, Hodge JA, Schurr TG, Lezza AMS, Elsas LJ, Nikoskelaonen EK (1988a): Mitochondrial DNA mutation associated with Leber's hereditary optic neuropathy. Science 242:1427–1430.

Wallace DC, Zheng X, Lott MT Shoffner JM, Hodge JA, Kelley RI, Epstein CJ, Hopkins LC (1988b): Familial mitochondrial encephalomyopathy (MERRF): genetic, pathological, and biochemical characterization of a mitochondrial DNA disease. Cell 55:601–610.

Warnock M (1984): Report of the Committee of Inquiry Into Human Fertilization and Embryology. Cmnd. 9314. London: Her Majesty's Stationery Office.

West JD (1990): Sexing the human conceptus by in situ hybridisation. In Harris N Wilkinson DG (eds): In Situ Hybrdisation: Application to Developmental Biology and Medicine. Society for Experimental Biology Seminar Series 40. Cambridge: Cambridge University Press, pp 205–239.

West JD (1991): Sexing the pre-embryo. In Chapman M, Grudzinskas, JG, Chard T (eds): The Embryo; Normal and Abnormal Development and Growth. Berlin: Springer-Verlag, pp 141–164.

West JD, Flockhart JH, Angell RR, Hillier SG, Thatcher SS, Glasier AE, Roger MW, Baird DT (1989a). Glucose phosphate isomerase activity in mouse and human eggs and pre-embryos. Hum Reprod 4:82–85.

West JD, Gosden JR, Angell RR, Hastie ND, Thatcher SS, Glasier AF, Baird DT (1987): Sexing the human pre-embryo by DNA-DNA in-situ hybridisation. Lancet 1:1345–1347.

West JD, Gosden JR, Angell RR, West KM, Glasier AF, Thatcher SS, Baird DT (1988): Sexing whole human pre-embryos by in situ hybridisation to a Y-chromosome specific DNA probe. Hum Reprod 3:1010–1019.

West JD, Green JF (1983): The transition from oocyte-coded to embryo-coded glucose phosphate isomerase in the early mouse embryo. J Embryol Exp Morphol 78:127–140.

West JD, Leask R, Green JF (1986): Quantification of the transition from oocyte-coded to embryo-coded glucose phosphate isomerase in mouse embryos. J Embryol Exp Morphol 97:225–237.

White TJ, Arnheim N, Erlich HA (1989): The polymerase chain reaction. Trends Genetics 5:185–189.

Whittingham DG, Leibo SP, Mazur P (1972): Survival of mouse embryos frozen to −196°C and −269°C. Science 178:411–414.

Willadsen SM (1979): A method for culture of micromanipulated sheep embryos and its use to produce monozygotic twins. Nature 277:298–300.

Willadsen SM (1981): The developmental capacity of blastomeres from 4- and 8-cell sheep embryos. J Embryol Exp Morphol 65:165–172.

Wilmut I (1972): The effect of cooling rate, warming rate, cryoprotective agent and stage of development on survival of mouse embryos during freezing and thawing. Life Sci 11:1071–1079.

Wilmut I, Rowson LEA (1973): Experiments on the low-temperature preservation of cow embryos. Vet Rec 92:686–690.

Wilton LJ, Trounson AO (1989): Biopsy of preimplantation mouse embryos: development of micromanipulated embryos and proliferation of single blastomeres in vitro. Biol Reprod 40:145–152.

Wilton LJ, Shaw JM, Trounson AO (1989): Successful single-cell biopsy and cryoprservation of preimplantation mouse embryos. Fertil Steril 51:513–517.

Wood MJ, Farrant J (1980): Preservation of mouse embryos by two-step freezing. Cryobiology 17:178–180.

Wood MJ, Whittingham DG, Rall WF (1987): The low temperature preservation of mouse oocytes and embryos. In Monk M (ed) Mammalian Development. A Practical Approach. Oxford: IRL Press, pp 255–280.

Wudl L, Chapman VM (1976): The expression of β-glucuronidase during preimplantation development of mouse embryos. Dev Biol 48:104–109.

Wudl L, Paigen K (1974): Enzyme measurements on single cells. Science 184:992–994.

Zeilmaker GH, Alberta ATh, van Gent I, Rijkmans CMPM, Drogendijk AC (1984): Two pregnancies following transfer of intact frozen-thawed embryos. Fertil Steril 42:293–296.

Zeviani M, Servidei S, Gellera C, Bertini E, Di Mauro S, DiDonato S (1989): An autosomal dominant disorder with multiple deletions of mitochondrial DNA starting at the D-loop region. Nature, 339:309–311.

ABOUT THE AUTHORS

JOHN R. GOSDEN is a Senior Scientist at the UK Medical Research Council's Human Genetics Unit in Edinburgh, where he carries out research in the general area of molecular cytogenetics. After receiving his B.A. from Cambridge University in 1959, he worked in the plantation industry in the Federation of Malaya (now Malaysia), where he qualified as Associate of the Incorporated Society of Planters. After returning to the United Kingdom, he received his Ph.D. from Edinburgh University. His thesis, under the direction of John Bishop, concentrated on the effect of mutations in the *gal* operon of *E. coli* on mRNA production. Dr. Gosden has worked for the MRC since 1971, initially on viral molecular biology. More recently his work has concentrated on the development of in situ hybridization methods for the investigation of human chromosome organization together with the analysis of chromosome abnormality and genetic disease. His research papers have appeared in such journals as *Nature, Cell, Chromosoma, Experimental Cell Research* and *Cytogenetics and Cell Genetics*. Dr. Gosden also has a Diploma in Animal Genetics from Edinburgh University, and is an executive editor for *Molecular Reproduction and Development*.

JOHN D. WEST is a non-clinical lecturer in the Department of Obstetrics and Gynaecology at the University of Edinburgh, Scotland, where he carries out research and teaches mammalian developmental biology and genetics. He received a B.Sc. degree from the University of East Anglia in 1970 and his Ph.D. from the University of Edinburgh in 1975. His Ph.D. work focused on the analysis of cell populations in chimaeric and mosaic mice, under the direction of Anne McLaren in the Department of Genetics. As a postdoctoral scientist with Verne Chapman in Buffalo, N.Y. he studied X chromosome inactivation in mouse embryos. He continued his interests in mammalian developmental biology and genetics, first with Richard Gardner in Oxford, then as a scientist in the Genetics Division of the Medical Research Council's Radiobiology Unit, Harwell until 1985. His subsequent involvement with the Edinburgh IVF programme led to a collaboration with John Gosden to test the feasibility of using DNA–DNA *in situ* hybridisation for sexing human pre-embryos. Dr West's current research interests includes confined mosaicism in mouse and human conceptuses. His research papers have appeared in such journals as *Nature, Cell, The Lancet, Genetical Research,* and *Development*.

Genes in Mammalian Reproduction: 131–171
© 1993 Wiley-Liss, Inc.

Epigenetic Interactions and Gene Expression in Peri-Implantation Mouse Embryo Development

Jean J. Latimer and Roger A. Pedersen

I. INTRODUCTION

One of the central questions in vertebrate developmental biology is how genetic and epigenetic information interact to form a three-dimensional embryo. Previous studies have established the general mechanisms of this process [Edelman, 1988, 1992; Nieuwkoop et al., 1985]: 1) The expression of certain developmentally important genes is regulated epigenetically and is position dependent. This expression depends on previously formed structures and is important for the subsequent interactions leading to the final pattern of the embryo. 2) The impetus for morphogenesis is cellular in origin, because it involves cell division, cell movement, cell adhesion, and cell death. In general, isolated molecular systems are not sufficient to provide the basis for patterning. 3) Cellular forces are linked to the nucleus via molecules such as proteins, peptide growth factors, hormones, receptors, and intracellular signal transduction cascades. The major tasks remaining for a resolution of this central question are to identify the specific gene products and control loops that link the tissue, cellular, and molecular levels of embryonic regulation.

The early morphogenesis of mammalian embryos emphasizes the differentiation of cell types that are essential for early embryonic nutrition within the maternal reproductive tract. Thus preimplantation development of eutherian embryos consists essentially of forming an outer layer of cells (trophectoderm) that is involved in attachment to the uterus [reviewed by Cruz and Pedersen, 1991]. Marsupial embryos also form an outer layer of cells (protoderm), but it is less invasive than the trophectoderm layer of eutherian embryos and generally does not develop into a chorioallantoic placenta [reviewed by Selwood, 1992]. Instead, marsupial embryos derive their nutrition through a vascularized yolk sac that becomes closely apposed to the uterine walls. Eutherian mammals, especially rodents, also utilize the yolk sac for early nutrition, before the chorioallantoic placenta is fully functional [Perry, 1981]. Thus mammals become equipped during their peri-implantation development for growth within the maternal uterine environment. By contrast, oviparous vertebrates are provisioned with nutritional resources by the yolk contained within the egg and proceed directly from cleavage to morphogenesis of the embryo proper.

Despite these differences in their strategies for embryonic nutrition, all vertebrates share the same basic body plan, i.e., they are bilaterally symmetrical quadripeds and might be expected to have similar embryonic strategies for achieving this form. Mechanisms involved in vertebrate body pattern formation are particularly evident during gastrulation, when mesoderm cells initially appear between ectoderm and endoderm layers at the prospective posterior end of the embryo, thus generating its anteroposterior axis. After gastrulation, neurulation occurs as the result of inductive interactions between mesoderm and the overlying ectoderm [Spemann, 1938; Cooke, 1985; Hamburger, 1988; Jones and Woodland, 1989]. Other dorsal mesoderm eventually forms the future notochord and somites, while ectoderm and endoderm develop into axial structures of the nervous system and the gut, respectively. The mesoderm becomes further subdivided as it moves toward the prospective head, trunk, and tail regions and mediates the specification of regional tissue identities as the anteroposterior structures are formed in the adjacent neural plate and endoderm. Morphogenetic mechanisms during this period of vertebrate development have been studied mainly in amphibian and bird embryos [Keller et al., 1992].

While these extensive experimental studies of morphogenetic mechanisms in other vertebrate classes have provided an understanding of the importance of epigenetic interactions in early development, relatively little is known about the cellular or molecular bases for these processes in mammals. This is due in part to the difficulty in obtaining sufficient amounts of preimplantation or peri-implantation material for biochemical or molecular analysis. However, the application of in situ hybridization and polymerase chain reaction (PCR) technologies has facilitated the analysis of gene expression at these early stages of development [Wilkinson and Green, 1990; Rappolee et al., 1989]. Although analysis of mammalian develop-

ment in vitro remains more cumbersome than that of amphibian or bird embryo systems, model cell systems have been developed for studying the differentiation of early cell lineages: These pluripotent cell types include both embryonal carcinoma (EC) cell lines and embryonic stem (ES) cell lines [Robertson, 1987]. Moreover, studies involving gene targeting and homologous recombination in ES cells are being used to produce genetically altered mice, thus providing a powerful approach for determining the role of specific molecules in mammalian development [Rossant and Joyner, 1989; Rossant and Hopkins, 1992].

In view of these advances in the field of mammalian developmental biology, we have undertaken a review of the role of epigenetic interactions in the differentiation of the extraembryonic and embryonic cell lineages that constitute the peri-implantation mouse embryo. In the first section, we review the differentiation and fate of the cell lineages that form during preimplantation mouse development (trophectoderm [TE], inner cell mass [ICM], and primitive endoderm and primitive ectoderm), emphasizing the evidence for epigenetic interactions in their development; we also consider the implications for epigenetic interactions of studies of model cell systems, EC and ES cells, and from genomic imprinting. In the second section we summarize observations of some potential epigenetic signalling molecules, or morphogens, that have been identified in mouse embryos, and we also consider the role of transcription factors that could be responsible for the influence of epigenetic signals on lineage-specific gene expression. Although evidence about the function of specific molecules in these early morphogenetic events is limited, we cite a combination of in vivo and in vitro studies to summarize the role of epigenetic interactions in mouse embryogenesis.

II. LINEAGE ANALYSIS IN PERI-IMPLANTATION MOUSE EMBRYOS

A. Differentiation of the Early Lineages

We now largely understand the fate of the primary cell types that form during the preimplantation development of eutherian mammals, but the molecular basis of their initial differentiation remains obscure. The earliest cell types, TE and ICM, arise as the embryo undergoes cleavage and compaction while it is being transported from the site of fertilization (the ampulla of the oviduct) to the uterine lumen, where implantation takes place. During this time, the cleaving blastomeres progressively increase their degree of adhesion with each other, until the outlines of individual blastomeres are not readily distinguishable. As this process of compaction continues, the outer surfaces of the blastomeres develop junctional complexes, an indication of incipient trophectoderm differentiation.

Beginning at the eight-cell stage, outer cells of mouse embryos form apical microvilli and other polarized characteristics that culminate in the differentiation of trophectoderm. The individual blastomeres of intact embryos have stereotypical fates that reflect their state of cytodifferentiation: Polarized, outer cells always contribute some descendants to TE, whereas apolar, inner cells contribute to the ICM [Ziomek and Johnson, 1982; Johnson and Ziomek, 1983]. Inner cells are first formed during the division from the 8-cell to 16-cell (morula) stage through cleavage of some blastomeres with a division plane parallel to the outer surface of the embryo; other blastomeres divide in a plane perpendicular to the surface of the embryo, generating two outer descendants at the morula stage [Sutherland et al., 1990]. During the progression from the morula to the early blastocyst (32-cell) stage, additional inner cells form, as the result of divisions of outer cells that yield an outer and an inner descendant

[Pedersen et al., 1986]. The number of cells recruited to the ICM at each of these two cleavage divisions (fourth and fifth cleavages) varies from embryo to embryo, through a regulative process that adjusts the number recruited in the fifth cleavage depending on the number of inner cells present after the fourth cleavage [Fleming et al., 1987]. Although the mechanism of this regulation is unknown, it has the consequence of providing a more consistent number of ICM cells and a more constant ratio of ICM to TE cells than would occur in the absence of such regulation. Thus, as early as preimplantation stages, there is evidence for a role of epigenetic mechanisms in development of the mammalian embryo.

The determination of the fate of individual blastomeres as TE or ICM also results from epigenetic interactions. The "inside–outside" hypothesis proposes that pluripotent cells of the compacting morula respond to differences in their microenvironments, outer cells respond to their environment by forming TE, and inner cells to theirs by forming ICM [Tarkowski and Wroblewska, 1967]. The two central predictions of the "inside–outside" hypothesis are that cleavage stage blastomeres would remain totipotent at least until the morula stage and that they would respond to changes in their environment by differentiating according to their inside or outside position. Studies of disaggregated blastomeres from each of the cleavage stages have shown that they are capable of forming both TE and ICM when aggregated with other genetically marked blastomeres [reviewed by Gardner, 1983; Pedersen, 1986]. This capacity has been demonstrated for blastomeres isolated from two-cell through late morula stages. Cells placed in outer positions of aggregation chimeras tend to form TE, while those placed inside tend to form ICM. Even the inner cells at the early blastocyst stage retain the capacity for differentiation into TE if placed in outer positions [see Pedersen,

1986, for review]. Therefore, the two essential predictions of the "inside–outside" hypothesis are fulfilled.

However, there is no evidence for specific microenvironmental factors produced by outer or inner cells that influence their path of differentiation as proposed by the "inside–outside" hypothesis. When eight-cell or morulae with intact zonae pellucidae are microinjected into the blastocyst cavity of giant, chimeric blastocysts, they develop as morphologically normal blastocysts, indicating that this microenvironment does not possess diffusible factors sufficient to induce ICM differentiation [Pedersen and Spindle, 1980]. Conversely, embryos exposed to the actin microfilament inhibitor cytochalasin D fail to undergo compaction and do not form morphologically normal blastocysts, yet they synthesize polypeptides characteristic of the ICM [Surani et al., 1980]. The absence of a clear case for the molecular basis of the "inside–outside" hypothesis leaves us unclear how to evaluate it further.

The "polarization" hypothesis proposes that TE differentiates in response to asymmetric cell contacts, while inner cells with their symmetric cell contacts develop as ICM [reviewed by Johnson and Maro, 1986]. This hypothesis also postulates epigenetic interactions for the first differentiative step in mouse embryogenesis. It is apparent that cell contacts are essential for polarization and compaction to occur, and that cell–cell adhesion is implicated in these processes, as proposed by the polarization hypothesis. Compaction depends on the close contact between adjacent blastomeres and on the calcium-dependent cell surface adhesion molecule known as uvomorulin [Hyafil et al., 1981]. Interference with uvomorulin function by calcium depletion or using antibodies blocks the polarized differentiation of mouse blastomeres, as in other cell culture systems [Shirayoshi et al., 1983; Vestweber et al., 1987; McNeill et al., 1990]. However, the

limited degree of polarized cytodifferentiation in rabbit morulae development raises the question of whether the polarized phenotype itself is a general phenomenon required for TE differentiation among eutherian mammals [Ziomek et al., 1990].

A genetic approach to the role of epigenetic interactions in TE and ICM differentiation would be helpful, provided that mutations affecting such early steps in mammalian development could be recovered. The earliest known mutations lead to developmental arrest during cleavage, without specifically affecting either TE or ICM [reviewed by Magnuson, 1986; Pedersen, 1988]. An informative mutation would be one that resulted in all cells of the embryo following a single differentiative pathway, which would be different from that followed by some cells under normal circumstances. Such a mutation would presumably render the mouse embryo "TE-less" or "ICM-less" and would fulfill the definition of a homeotic mutation, whether the gene was related to existing homeobox genes or had an independent molecular identity.

The next cell types to form in mammalian embryos are the primitive endoderm and primitive ectoderm, both derivatives of the ICM. Primitive endoderm differentiates on the blastocelic surface of the ICM just before the embryo attaches to the uterus at 4.0 days of gestation (dg) [Nadijca and Hillman, 1974; reviewed by Gardner, 1983]. The primitive ectoderm consists of the remaining undifferentiated cells in the core of the ICM [reviewed by Rossant, 1984]. Primitive endoderm appears to differentiate because of its outer position on the surface of the ICM. When ICMs were isolated from late blastocysts they formed an entire outer layer of primitive endoderm, whereas they would have formed endoderm on only one surface in the intact blastocyst [Rossant, 1975]. Thus the blastocelic surface is an outer environment conducive to primitive endoderm formation. Since early ICM cells have the capacity to form TE, why does the inner blastocele of the ICM differentiate into primitive endoderm rather than into TE? ICM cells retain the capacity for TE development until embryos reach the expanded blastocyst stage (3.5 dg). This is approximately the time of the sixth cleavage division, when the ICM gains the capacity for primitive endoderm formation [Nichols and Gardner, 1984; Chisholm et al., 1985; reviewed by Gardner, 1983; Rossant, 1986]. Thus, although the information conveyed by the "outside" environment of the blastocyst cavity may be similar to that surrounding the embryo, by 4.0 dg the ICM cells have undergone a transition to a new state, from which they can only differentiate into primitive endoderm or primitive ectoderm. There could be a role of the internal blastocyst environment and cell–cell contacts in suppressing trophectoderm differentiation before this time, since ICMs of early blastocysts do not differentiate into TE in the intact embryo. But after the transition, primitive endoderm differentiation is no longer suppressed by (nor does it require) the blastocyst environment, because ICMs from expanded blastocysts develop primitive endoderm either in situ or in isolation.

An alternative source of epigenetic information for primitive endoderm differentiation is through cell interactions. Enders et al. [1978] reported a compaction-like process just preceding endoderm differentiation by the ICM. However, there is no body of evidence to implicate uvomorulin as in TE and ICM differentiation. The amount of laminin present may be important in primitive endoderm differentiation, however, judging from the inhibitory effects of exogenous laminin or antilaminin antisera on endoderm differentiation in EC embryoid bodies [Grover et al., 1983]. In addition, studies on EC cells have implicated cell–cell and cell–matrix interactions in the differentiation of the primitive endoderm [Casanova and Grabel, 1988], and a similar

case can be made for the importance of these interactions in maintaining the differentiated state of visceral endoderm in embryos [Kalimi and Lo, 1989]. Further analysis of factors affecting primitive endoderm differentiation in isolated ICMs should be fruitful.

The differentiation of primitive ectoderm has not been studied in nearly such detail. Indeed, it is not clear from their morphology whether primitive ectoderm cells actually differentiate into a new phenotype or simply remain an undifferentiated stem cell population or an "embryoblast" at the late blastocyst stage (4.5 dg). The primitive ectoderm cells acquire a columnar phenotype and form an epithelial layer (also known as the embryonic ectoderm or epiblast) at the time the proamniotic cavity forms (5.5 dg). On the basis of culture studies using isolated cores of primitive ectoderm, it initially appeared that primitive ectoderm was capable of differentiating into primitive endoderm. However, Gardner [1985] demonstrated that pure cores of primitive ectoderm did not form primitive endoderm and attributed the previous observations to artifacts arising from a multilayered endoderm. Apparently the determination of ICM cells' capacity for primitive endoderm versus primitive ectoderm occurs at the late blastocyst stage (4.5 dg), with the innermost cells becoming primitive ectoderm [Gardner, 1985; reviewed by Gardner and Beddington, 1988]. The nature of the factors that regulate this developmental decision remains obscure.

B. Fate of the Early Lineages

The fate of the early cell lineages in mouse embryo—TE, ICM, primitive endoderm, and primitive ectoderm—has been studied by blastocyst micromanipulation and other methods for clonal analysis, leading to a coherent view of early allocation events [reviewed by Gardner, 1983; Rossant, 1984; 1986; 1987; Gardner and Beddington, 1988; Pedersen, 1988]. The TE develops into three trophoblast cell types, including trophoblast giant cells, extraembryonic ectoderm, and ectoplacental cone; all are extraembryonic and become part of the interface between conceptus and maternal tissues during in utero development. The ICM develops into primitive endoderm and primitive ectoderm, which have very different fates: Primitive endoderm subsequently forms visceral and parietal endoderm, which contribute strictly to extraembryonic tissues, while primitive ectoderm forms a wide variety of tissues, both embryonic and extraembryonic. The principal descendants of primitive ectoderm are the primary germ layers of the embryo proper—the embryonic mesoderm, endoderm, and ectoderm—which form during gastrulation. Other products are the amniotic ectoderm and the extraembryonic mesoderm, which contributes to the yolk sac, chorion, amnion, and allantois. Together, these descendants of the early lineages constitute the peri-implantation conceptus. In the following discussion, we will examine the evidence for the fate of TE, primitive endoderm, and primitive ectoderm, focusing on the role of epigenetic mechanisms in their development.

1. Trophectoderm. The diploid trophoblast descendants of TE—the extraembyronic ectoderm of the chorion and the diploid cells of the ectoplacental cone—arise through the continued proliferation of polar TE during peri-implantation development of the mouse conceptus [Gardner et al., 1973; Rossant and Tamura-Lis, 1979; Papaioannou, 1982]. The mural TE cells differentiate during early phases of implantation (4.0–4.5 dg) to become the primary trophoblast giant cells [Dickson, 1966; Barlow et al., 1972]. The nuclei of these cells enlarge through repeated cycles of DNA synthesis without cell division (endoreduplication), thus acquiring many copies of chromatids loosely resembling polytene chromosomes [Varmuza et al., 1988]. Differentiation of cells destined to

become primary trophoblast giant cells is altered, however, if a vesicle of isolated mural TE cells is injected with a donor ICM obtained from another embryo. In such "reconstituted blastocysts," the mural TE cells in contact with the donor ICM form diploid extraembyronic ectoderm and ectoplacental cone [Gardner et al., 1973; Rossant and Tamura-Lis, 1979; Papaioannou, 1982]. Both of these trophoblast cell types are able to differentiate into trophoblast giant cells under a variety of culture conditions when they are separated from ICM derivatives; cultured ectoplacental cone rapidly differentiates into giant cells, while extraembryonic ectoderm requires longer for this transition [Rossant and Tamura-Lis, 1981). The giant cells formed in culture thus resemble the "secondary giant cells" that surround the mouse conceptus during peri-implantation development.

The blastocyst reconstitution studies indicate that inductive interactions between ICM (and its derivatives) and the TE (and its derivatives) are responsible for maintaining the diploid phenotype in closely adjacent trophoblast cell types. Moreover, the trophoblast culture experiments indicate that the entire trophoblast lineage is capable of giant cell differentiation, which is thus the default pathway for trophoblast cells when inductive signals from the ICM are absent or are distant, as in the case of primary and secondary giant cells. There is no information about the possible identity of such inductive signals, although the giant cell formation by extraembryonic ectoderm cultured in conditions that preserve its tissue integrity suggests that substances derived from the ICM rather than embryonic organization per se are responsible for the inductive effect [Rossant and Tamura-Lis, 1981].

There has been considerable controversy over whether the ICM makes a significant cellular contribution to the trophoblast lineage, in addition to its evident inductive role. Blastocyst reconstitution studies demonstrated that with few exceptions proliferating mural TE was the source of the trophoblast cell types [Gardner et al., 1973; Rossant and Tamura-Lis, 1979; Papaioannou, 1982; Rossant and Croy, 1985]. But the cell population dynamics of the mouse blastocyst indicate that despite similar mitotic indices there is a gradual decline in the proportion of ICM cells relative to TE; this could be interpreted either as differential cell death or as ICM contribution to the polar TE [Handyside, 1978; Handyside and Hunter, 1986]. Because the polar TE was discarded in the blastocyst reconstitution studies, the possibility of ICM contribution to the TE lineage through polar TE had not been addressed. Analysis of the fate of marked polar TE cells in intact blastocysts revealed that they contributed descendants to the mural TE during subsequent blastocyst growth [Copp, 1979; Cruz and Pedersen, 1985]. Cruz and Pedersen [1985] interpreted the patterns of labeled mural TE descendants as indicating movement of labeled polar TE cells away from the embryonic pole and their replacement by unlabeled cells derived from the ICM. Movement of polar TE away from the embryonic pole occurs during development of the rabbit blastocyst and in other eutherian mammals [Williams and Biggers, 1990; reviewed by Cruz and Pedersen, 1991]. Its relevance to mouse blastocyst development, however, was questioned by Dyce et al. [1987], who concluded on the basis of labeling studies using tracer microinjection and fluorescent microparticles that labeled polar TE cells were passively displaced from the embryonic pole and that the ICM contributes descendants to the polar TE only rarely and in minimal numbers. Their conclusions contrast with those of Winkel and Pedersen [1988], who used microinjection of cell lineage tracers to mark progenitor cells in the ICM and concluded that there was a substantial contribution from ICM to TE. In a recent study Gardner and Nichols [1991] addressed this controversy by replacing some of the inner or outer cells of late morulae with donor cells from synchronous embryos. They found that donor inner cells rarely colonized the polar trophectoderm of recipient blastocysts, even when the donor cells were present throughout the period of polar TE development, and that donor outer cells regularly colonized the entire trophoblast lineage. This issue clearly deserves further study to resolve these divergent interpretations of TE origin. However, based on current information, the hypothesis that ICM is the predominant source of the definitive polar TE cells must be regarded with caution.

Taking into account all of these observations on lineage relationships within the trophoblast lineage, a model emerges (Fig. 1) for the fate of polar TE [Copp, 1979; Cruz and Pedersen, 1985; Rossant and Tamura-Lis, 1981]. Mural TE cells undergo endo-reduplication as implantation begins, but additional cells are recruited to the mural TE population through continued migration from the proliferating polar TE. After implantation, the polar trophectoderm becomes multilayered as ectoplacental cone and extraembryonic ectoderm form, thus increasing the distance between ICM derivatives and trophoblast. Peripheral ectoplacental cone cells, receiving diminishing inductive signals from the ICM, cease proliferating and begin endo-reduplication as secondary giant cells, continuing to accumulate with already-formed giant cells. According to the lineage model advanced by Rossant and Tamura-Lis [1981], extraembryonic ectoderm serves as a stem cell population for the trophoblast lineage, continuing proliferation throughout the peri-implantation period and contributing descendants to the ectoplacental cone that in turn form secondary giant cells (Fig. 1D). In summary, the trophoblast lineage develops in a highly organized process involving supporting interactions with the

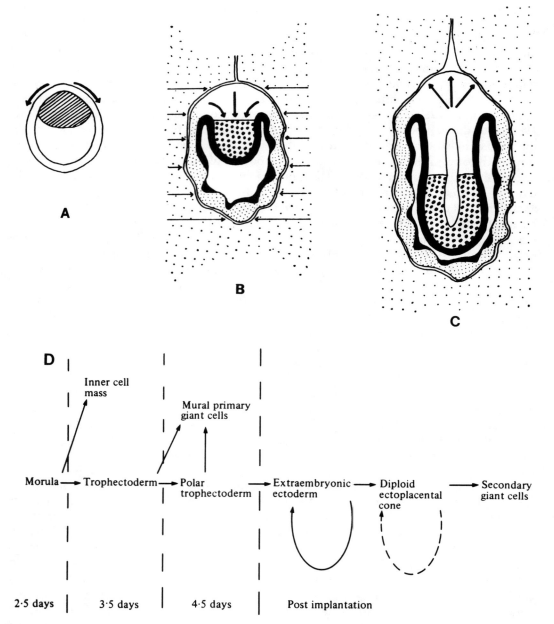

Fig. 1. *Model depicting the fate of cells in the trophoblast lineage during peri-implantation mouse embryo development.* **A:** *Polar trophectoderm cells (top) contribute descendants to mural trophectoderm (bottom) during preimplantation blastocyst growth (arrows).* **B:** *Polar trophectoderm cells contribute to extraembryonic ectoderm as trophectoderm proliferates (thick arrows), and trophoblast giant cells (finely stippled) undergo implantation in the uterine crypt (fine arrows); uterine decidual tissue is indicated by coarse stippling.* **C:** *Proliferating diploid trophoblast cells contribute to the elongating ectoplacental cone as implantation proceeds.* **D:** *Lineage map indicating the origins of the trophoblast lineage and modeling the relationships between polar trophectoderm, extraembryonic ectoderm, ectoplacental cone, and trophoblast giant cells. Extraembryonic ectoderm and diploid ectoplacental cone are envisioned as self-renewing stem cell populations for the trophoblast lineage during postimplantation development. (A–C, adapted from Copp, 1979, and D, adapted from Rossant and Tamura-Lis, 1981, with permission of the publisher.)*

ICM, progressive differentiation, and directional growth. Several of these morphogenetic events appear to involve epigenetic mechanisms, including effects of trophoblast derivatives on primitive endodermal derivatives.

2. Primitive endoderm. In studies of cell fate beginning at the late blastocyst stage (4.5 dg) Gardner and coworkers used blastocyst injections to analyze the fate of primitive endoderm, showing that descendants of this cell lineage populate the visceral extraembryonic endoderm and the parietal endoderm. In initial studies, visceral endoderm cells from 4.5 dg embryos were injected into expanded blastocysts (3.5 dg), and their descendants were analyzed at midgestation, showing that primitive endoderm cells contributed to yolk sac endoderm but not to definitive embryonic (gut) endoderm [Gardner and Rossant, 1979]. Subsequent studies led to the discovery that visceral endoderm was able to contribute to the parietal endoderm as well as yolk sac endoderm; parietal endoderm, on the other hand, had descendants only in the parietal endoderm population [Gardner, 1982; Cockroft and Gardner, 1987].

These results are consistent with other studies on 6.5 and 7.5 dg egg cylinders, indicating that the visceral endoderm layer has the capacity for differentiation into parietal endoderm. In experiments involving dissection of 6.5 dg mouse embryos, Hogan and Tilly [1981] found that visceral extraembryonic endoderm cells differentiated into parietal endoderm when they were cultured in contact with extraembryonic ectoderm undergoing the transition from diploid trophoblast to giant cells. They proposed that these transitions also occur in the intact embryo, where extraembryonic ectoderm may differentiate into trophoblast giant cells coordinately with the differentiation of visceral endoderm to parietal endoderm [Hogan and Tilly, 1981; Hogan and Newman, 1984]. Accordingly, their model (Fig. 2) envisions a movement

of visceral embryonic endoderm towards the extraembryonic region, concomitantly with movement of visceral extraembryonic endoderm toward the parietal endoderm regions and is consistent with the progenitor–descendant relationship between the visceral and parietal endoderm found in the blastocyst reconstitution studies [Gardner, 1982; Cockroft and Gardner, 1987].

This model has not been evaluated in the intact conceptus, because all experimental approaches that have been devised for studying cell lineage relationships in mouse embryos involve dissection of the peri-implantation embryo and culture in vitro. The studies by Gardner and coworkers reveal the fate of cells placed in a blastocyst environment, but do not directly address the fate of cells residing in the visceral endoderm of the intact egg cylinder. Analysis of lineage relationships within the primitive endoderm and trophectoderm lineages, in particular, would require an approach that allows cell marking after implantation to determine the fate of these cells. Creation of transgenic lines of mice possessing appropriate indicator genes under the regulation of transcriptional regulator genes such as *flp* might facilitate such experiments [O'Gorman et al., 1991]. Without such studies, we can only roughly estimate the limits of visceral endoderm contribution to the parietal endoderm population. Cockroft and Gardner [1987] found that the incidence of parietal endoderm colonization by descendants of visceral endoderm declined precipitously from 5.5 to 6.5 dg. These observations indicate that by 6.5 dg donor visceral endoderm no longer has the capacity to contribute substantially to the parietal endoderm, even though the visceral endoderm of the egg cylinder still contains many cells capable of parietal endoderm differentiation at this stage [Hogan and Tilly, 1981]. In studies of the fate of visceral embryonic endoderm cells of 6.5 and 7.5 dg embryos, Lawson and coworkers [Lawson et al., 1986; Lawson and Pedersen, 1987]

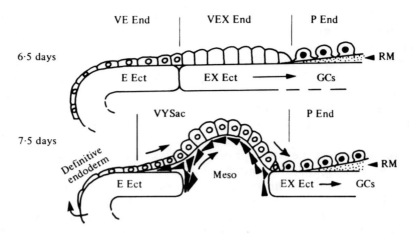

Fig. 2. *Model depicting the fate of cells in the primitive endoderm lineage during postimplantation development of the mouse egg cylinder. At 6.5 days (top), outer cell layer is composed entirely of primitive endoderm-derived cells, consisting of visceral embryonic endoderm (VE End), visceral extraembryonic endoderm (VEX End), and parietal endoderm (P End). α-Fetoprotein-synthesizing cells are indicated by open circles (VE End), and laminin and collagen IV–synthesizing cells are indicated by closed circles. Embryonic ectoderm (E Ect) and extraembryonic ectoderm (EX Ect), trophoblast giant cells (GCs) and Reichert's membrane (RM) are also indicated. At 7.5 days (bottom), extraembryonic mesoderm (Meso) has accumulated in the area formerly occupied by extraembryonic ectoderm, and definitive endoderm cells have accumulated at the distal tip of the egg cylinder (left) as a result of gastrulation. Visceral extraembryonic endoderm cells of the yolk sac (VYSac) synthesize α-fetoprotein at this stage. Also indicated is movement of cells from visceral embryonic endoderm to visceral extraembryonic endoderm, which is envisioned as the stem cell population for the parietal endoderm. (Reproduced from Hogan and Tilly, 1982, with permission of the publisher.)*

found that labeled descendants in yolk sac endoderm were spread in an arc roughly perpendicular to the anteroposterior embryonic axis and moreover tended to remain in coherent patches [Lawson and Pedersen, 1991], as described by Gardner [1984]. These results suggest that the contribution to parietal endoderm may diminish because the differentiation of visceral endoderm after the onset of gastrulation encourages coherent clonal growth rather than cell mixing or migration.

The synthetic pattern of α-fetoprotein (AFP) in the visceral endoderm layer shows that changing associations between early cell lineages alters their expression of lineage-specific genes. Dziadek and Adamson [1978] observed that AFP was first synthesized in the visceral embryonic endoderm of 6.5 dg embryos but was not synthesized in the visceral extraembryonic endoderm until later stages (7.5 dg). The onset of AFP synthesis in visceral extraembyronic endoderm coincided with the migration of extraembryonic mesoderm into space formerly occupied by extraembryonic ectoderm, leading to the close association of this mesoderm with the visceral extraembryonic endoderm in this region (Fig. 2). Further experimental studies showed that AFP synthesis by visceral endoderm was inhibited by the proximity of the extraembryonic ectoderm to the visceral extraembryonic endoderm. Visceral extraembryonic endoderm cells reassociated with embryonic ectoderm expressed AFP, whereas the same endoderm cells reassociated with extraembryonic ectoderm did not [Dziadek, 1978]. Moreover, extraembronic fragments that contained visceral extraembryonic endoderm cells in close association with extraembryonic ectoderm synthesized large amounts of collagen IV and laminin, the extracellular matrix components characteristic of parietal endoderm [Dziadek and Adamson, 1978]. Together, these observations suggest that epigenetic signals emanating from trophoblast lineages suppress the visceral extraembryonic endoderm phenotype and induce differentiation into parietal endoderm. The molecular identity of such morphogens is still unknown, but some ideas about the signalling pathways that can affect differentiation within the primitive endoderm lineage have emerged from studies of model cell systems, as discussed subsequently.

Because numerous studies of inductive interactions have demonstrated that there are mutual interactions between adjacent tissues, it is reasonable to expect that visceral endoderm cells also generate signals that affect the development of their neighbors. Visceral extraembryonic (yolk sac) endoderm has been shown to be necessary for blood island development in the extraembryonic mesoderm of chick embryos [Wilt, 1965; Miura and Wilt, 1969]. Similarly, the chick hypoblast layer (equivalent to visceral embryonic endoderm) appears to have a decisive role in the differentiation of the epiblast layer during gastrulation [Azar and Eyal-Giladi, 1981].

3. Primitive ectoderm. Blastocyst injection studies by Gardner and coworkers have demonstrated conclusively that the somatic and germ cell lineages of the fetus, as well as extraembryonic mesoderm and amniotic ectoderm, are descendants of the primitive ectoderm [Gardner and Papaioannou, 1975; Gardner and Rossant, 1979; reviewed by Rossant, 1984; Gardner, 1983]. This cell lineage is thus equivalent to the epiblast layer of the chick embryo, which gives rise to all fetal lineages, including the embryonic gut endoderm [Vakaet, 1985]. Mouse epiblast becomes recognizable as an embryonic epithelium upon formation of the proamniotic cavity at 5.5 dg [Snell and Stevens, 1966]. Thereafter the epiblast population expands by rapid proliferation [Snow, 1977] and contributes cells to the endodermal, mesodermal, and ectodermal lineages beginning at the onset of gastrulation, approximately 6.5 dg. The fate of mouse epiblast cells has been examined by microinjecting individual cells with lineage tracers at 6.7 dg [Lawson et al., 1991] and by transferring groups of donor epiblast cells labeled with gold-conjugated wheat germ agglutinin to unlabeled host embryos at 7.5 dg [Tam and Beddington, 1987; Tam, 1989], followed by culture for 1 to 2 days.

For the purpose of analyzing the fate of epiblast cells, Lawson et al.[1991] divided the cup-like embryonic portion of the mouse egg cylinder (containing the epiblast) into 11 arbitrary sectors (Fig. 3A). The first tier of sectors (I–V) passes from the anterior of the epiblast to its posterior, the base of the primitive streak. The second tier of sectors also passes from anterior to posterior, from sector VI to the tip of the primitive streak in sector X. The distal end of the egg cylinder forms sector XI. The analysis of epiblast cell fates and allocation to the primary germ layers and organ rudiments was performed by microinjecting a single progenitor cell of each embryo with a combination of horseradish peroxidase and rhodamine-conjugated dextran. The distribution of descendants in the neuroectoderm shows that this tissue is derived primarily from the anterior axis of the epiblast (sectors VI and XI). A smaller contribution was derived from sectors I and VII.

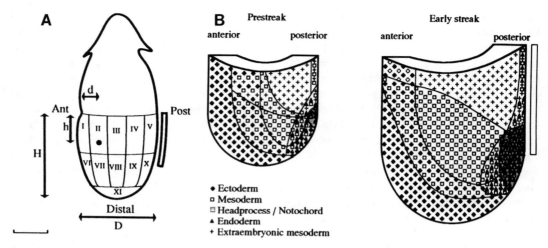

Fig. 3. A: *Diagram of pregastrula/early gastrula stage mouse egg cylinder (6.7 day), identifying sectors (I–XI) marked by lineage tracer microinjection. The anterior of the prospective craniocaudal axis is oriented to the left, the posterior to the right. The typical extent of the primitive streak in early gastrula stage embryos is indicated as a vertical rectangular box (right). The position of the injected epiblast cell is estimated along the H axis (dorsoventral height) and D axis (egg cylinder diameter) as the distance (h) from the embryonic/extraembryonic border and distance (d) from the anterior margin. Bar-0.1 mm. **B:** Fate map of the epiblast of the prestreak and early streak stages of the mouse embryo (6.7 days) showing the derivation of the primary germ layers, extraembryonic mesoderm, and amniotic ectoderm when cultured from prestreak to midstreak or from early streak to neural plate stages. (Adapted from Lawson et al., 1991, with permission of the publisher.)*

The distribution of descendants in the embryonic gut endoderm indicates that this germ layer is derived from a region extending from sector VIII to sector X. This region overlaps the region that gives rise to the head process, which contributes to the notochord of the embryo [Lawson and Pedersen, 1986]. The descendants in the embryonic mesoderm are derived from the lateral and posterior regions (including sectors II–V and VII–X), but not from the anterior-most sectors (sectors I and VI) or from the tip of the epiblast (sector XI). The distribution of descendants in yolk sac mesoderm indicates that this tissue is derived mainly from the posterior epiblast region near the border of the embryonic and extraembryonic portions of the conceptus (sectors III–V). The allantois was derived mainly from sector II, in the anterior epiblast, with an occasional contribution from sector V, in the posterior epiblast. Epiblast progenitors that gave rise to the blood islands in the yolk sac also produced other yolk sac mesoderm cells. This indicates that the hematopoietic stem cell progenitors are derived from a common ancestor with extraembryonic mesoderm (KA Lawson and RA Pedersen, unpublished observations). Together with other examples of single epiblast cells that contributed to two or more primary germ layers, this observation indicates that neither determination of cell fate nor cell allocation in the epiblast of mouse embryos is complete before epiblast cells enter the primitive streak [Beddington, 1983; Lawson et al., 1991]. This conclusion is not consistent with the prediction from a study on gastrulating chick embryos describing the epiblast as a mosaic of pure clones destined to generate ectoderm, mesoderm, or endoderm [Stern and Canning,

1990]. Rather, germ layer fate appears to be determined in mouse embryos by epigenetic signals that reach the epiblast cells during or after the process of gastrulation.

Studies of the gastrula fate map of the mouse embryo have revealed striking topological similarities to the avian gastrula fate map. Not only do the migratory movements of mouse epiblast converge on the primitive streak as in the chick embryo, but also the germ layer precursors are arrayed in the same successive order of convergence on the streak in the two species (Fig. 3B): Precursors of notochord and gut endoderm are the first to enter the streaks, followed by those for embryonic and extraembryonic mesoderm, then for neuroectoderm and surface ectoderm, which do not leave the epiblast layer [Lawson et al., 1991; Vakaet, 1984]. There is also a strong similarity in the pattern of surface ectoderm and brain segment precursors in the mouse epiblast as compared with the chick, with anterior epiblast forming nonneural ectoderm as well as telencephalon and diencephalon [Tam, 1989; Couly and LeDouarin, 1987]. These similarities imply that fundamental mechanisms of organ formation are conserved among vertebrates. Thus, in searching for epigenetic mechanisms in mammalian development, it is important to take into account evidence from other vertebrate models, in addition to mammalian embryos and model cell systems.

C. Model Cell Systems

EC cell lines established from teratocarcinomas have been used extensively as models for the early cell lineages of mouse embryos [reviewed by Martin, 1980; Hogan et al., 1983]. After in vitro exposure to retinoic acid, these cells differentiate into derivatives of primitive endoderm [Strickland and Mahdavi, 1978]. Treatment with cyclic AMP shifts their differentiation toward a parietal endoderm phenotype, while culture in suspension induces embryoid bodies to form an outer layer with a visceral endoderm phenotype [Strickland et al., 1980; Hogan et al., 1981]. These observations may suggest a role for protein kinase A in the signal transduction pathway leading to parietal endoderm differentiation. While the regulated differentiation of EC cells provides a useful model for analyzing the primitive endoderm lineage, they have been less satisfactory models for primitive ectoderm derivatives. The cells that remained at the core of embryoid bodies of the F9 line of EC cells did not differentiate into embryonic gut endoderm or mesoderm, although they could form neural-like cells [Kuff and Fewell, 1980; Hogan et al., 1983]. Other EC lines varied markedly in their capacity to contribute to somatic tissues of blastocyst injection chimerae, but rarely generated germline chimerism [Papaioannou and Rossant, 1983]. The isolation of pluripotent human EC cell lines has raised expectations that they would provide similar models for studying early differentiative processes in human development [Andrews, 1984; Pera et al., 1989; Wiles, 1988].

The greatest progress in developing model cell systems for early mouse embryogenesis has been establishment of ES cells by direct culture of inner cell masses from mammalian embryos [Martin, 1981; Evans and Kaufman, 1981; reviewed by Robertson, 1987]. These cell lines have been shown to differentiate spontaneously in culture and must in fact be prevented from differentiating, using the cytokine leukemia inhibitory factor differentiation inhibitory activity (LIF/DIA). The endoderm that forms around the embryoid bodies of these cell lines has been shown to express genes characteristic of visceral extraembryonic endoderm, although only in a subset of the total endodermal cells [JJ Latimer, CA Burdsal, and RA Pedersen, unpublished observations]. The other cells in the endoderm layer probably represent primitive and parietal endoderm. Furthermore, these cell lines show distinct patterns of differentiation when allowed to differentiate spontaneously rather than under the control of retinoic acid in aggregation culture. The spontaneous differentiation of the ES cells is more progressive and less uniform than that of the retinoic acid–induced differentiation and differs significantly from that seen in F9 cultures induced to differentiate with retinoic acid (JJ Latimer, CA Burdsal, and RA Pedersen, unpublished observations). Interestingly, the expression of certain related gene families such as the globin gene family is induced in these differentiated cells in the correct temporal pattern (i.e., with fetal globins actively expressed first and adult globin last), although the early genes do not seem to be switched off at the time the adult genes become active [Lindenbaum and Grosveld, 1990] ES cells cultured in medium supplemented with methylcellulose and peptide growth factors can undergo hematopoiesis [Wiles and Keller, 1991], and embryoid bodies cultured for prolonged periods can form structures resembling blood islands [Risau et al., 1988; Wang et al., 1992]. Embryoid bodies derived from ES cells can also transcribe genes that characterize cardiac muscle [Robbins et al., 1990; Sanchez et al., 1991]. A more compelling observation, however, is the absence of axial development in ES as well as EC embryoid bodies, indicating that gastrulation does not typically occur. These data suggest that although a more progressive type of differentiation is manifested by ES cells in culture, some of the proper signalling required for axial formation is missing and the resultant differentiation is consequently disorganized. Future studies involving the definition of serum-free media that contain the appropriate growth factors and cytokines for specific types of differentiation need to be performed. Growth on defined extracellular matrix components may also be required to induce differentiation of specific cell types.

Because of their high rates of contribution to both somatic and germ line lineages in chimerae, ES cells are being used extensively as vehicles for introducing transgenes into mice [see Rossant and Hopkins, 1992, for review]. In this context, it has been shown that the cells must be maintained in undifferentiated form to retain their pluripotency. When LIF/DIA is added to culture medium it prevents differentiation, although culture of ES cells on feeder layers of embryonic fibroblasts or STO cells is also useful in maintaining their pluripotency [Robertson, 1987].

ES cell lines have also been derived from both parthenogenetic and androgenetic blastocysts [Robertson et al., 1983; Stewart, et al., 1991] (KS Sturm, JJ Meneses, and RA Pedersen, unpublished observations) as well as normally fertilized diploid blastocysts. These ES cells have been valuable for studying the mechanism and consequences of genomic imprinting, a perturbation of epigenetic mechanisms that provides additional models for studying early developmental events.

D. Genomic Imprinting

Experimental evidence with mice indicates that normal prenatal development of mammalian embryos requires both maternal and paternal genetic contributions [reviewed by Surani, 1986; Solter, 1988]. This specialization of egg and sperm nuclei, defined as genomic imprinting, is thought to occur during gametogenesis. Imprinting seems to act through the expression of specific genes, which are located on six different mouse autosomes [reviewed by Cattanach and Beechey, 1990]. The morphological and developmental phenotypes of parthenogenetic (diploid maternal genotype) and androgenetic (diploid paternal genotype) embryos have been studied extensively, and the identities of several imprinted genes are now known [see Pedersen et al., 1992 for review].

Diploid parthenogenotes are capable of preimplantation development, but die at midgestation stages with extensive perturbations in the differentiation of their extra-embryonic and embryonic cell lineages [Sturm et al., 1992, reviewed in Pedersen et al., 1992]. The most conspicuous abnormalities are the tendency for trophoblast to differentiate into giant cells, with few diploid cells capable of proliferation in this lineage, and for the primitive endoderm lineage to differentiate completely into parietal endoderm, with few morphologically recognizable visceral extraembryonic endoderm cells. Parthenogenesis also causes defects in the embryo proper, including abnormalities of mesodermal and ectodermal lineages ranging from their complete absence to slight morphological defects in the heart, somites, and other organ rudiments [Sturm et al., 1992]. When diploid parthenogenotes are combined with normally fertilized embryos as aggregation chimerae, they contribute extensively to many tissues of adult mice, including the germline, but are notably absent from skeletal muscle; descendants of parthenogenetic cells are selectively eliminated from the trophoblast and primitive endoderm lineages during early embryonic development from the trophoblast and primitive endoderm lineages [Fundele et al., 1989; Nagy et al., 1989; Clarke et al., 1988; Thomson and Solter, 1989]. The deficiencies in parthenogenetic embryos thus seem to be of two kinds: 1) cell autonomous defects in extraembryonic lineages, which are not rescued by combination with normal cells in aggregation chimerae; and 2) defects in embryonic lineages (other than skeletal muscle) that involve epigenetic signals, because they are largely correctable by normal embryonic cells in aggregation chimerae. Chimerae produced by reconstituting blastocysts so they have normal trophoblast and primitive endoderm cells, but have parthenogenetic primitive ectoderm, also die at midgestation stages, although they reach slightly more advanced stages than pure parthenogenotes [Gardner et al., 1990]. These observations further indicate

that embryonic lineages are perturbed by parthenogenesis and suggest that they are deficient for diffusible factor(s) normally generated either by embryonic sources or by extraembryonic sources that are not accessible to the parthenogenetic embryo, even in such chimerae.

Although less extensive studies have been done on androgenetic embryos, descendants of their cells also can be sustained in chimerae beyond the midgestation time when pure androgenotes die, suggesting perturbation of some diffusible component(s) in androgenotes [Barton et al., 1991; Mann and Stewart, 1991]. In these cases, however, the chimerae containing descendants of androgenote embryos or ES cells display skeletal defects, suggesting that androgenotes may generate an excess of diffusible factors responsible for these and other abnormalities [Mann et al., 1990; Mann and Stewart, 1991; Barton et al., 1991].

The genes that have been identified as imprinted endogenous genes include growth and differentiating-regulating factors and their receptors. The first endogenous gene to be identified was insulin-like growth factor (IGF)-II [DeChiara et al., 1991], followed by the IGF-II/mannose-6-phosphate receptor [Barlow et al., 1991] and the H19 gene, with no known function [Bartolomei et al., 1991]. Recently, Rappolee et al. [1992] observed that the IGF-I receptor gene is not transcribed in preimplantation mouse parthenogenotes, suggesting that its expression is affected by imprinting at early stages of parthenogenetic development. The identities of these genes, taken together, provide evidence that genomic imprinting can act through perturbations of growth and differentiation, as suggested by Surani et al. [1988] and more recently by Haig and Graham [1991] and Sturm et al. [1992]. However, deficiencies in the known genes cannot account for all the abnormalities observed in parthenogenotes, androgenotes, or other maternal duplication/paternal deficiency and paternal duplication/maternal deficiency phenotypes; therefore, other imprinted endogenous genes must be involved in generating the biological effects of imprinting observed during mouse development [see Cattanach and Beechey, 1990; Pedersen et al., 1992, for review]. The rationale for continued examination of these model systems is that they will reveal the identities of additional genes that are integral components of epigenetic mechanisms that act during early mammalian development.

III. GENE EXPRESSION IN THE PERI-IMPLANTATION EMBRYO

Some of the seminal epigenetic interactions that drive inductive developmental processes have been described morphologically in the previous sections. Although the subject of this section is gene expression in the peri-implantation embryo that contributes to these cell–cell interactions, a discussion of extracellular matrix (ECM)–involved interactions has been omitted. This subject has been recently reviewed [Edelman, 1988; Jessel and Melton, 1992]. We have chosen instead to restrict our discussion to the expression of *trans*-active effector molecules that impact more directly on the nucleus either as mitogenic agents or transcriptional regulators.

Although a discussion of the morphogenetic properties of retinoic acid is included in the section on mammalian homeotic genes, we have not included a section on retinoic acid receptor molecules. These have been recently reviewed [Linney, 1992; Ruberte et al., 1991; Dolle et al., 1989b].

A. Selected Cytokines

Several cytokines have been identified as having a role in mammalian embryonic development. Among these is a cytokine known as leukemia inhibitory factor (LIF) or differentiation inhibitory factor (DIA). This cytokine was originally purified, characterized and cloned because of its ability to induce the differentiation as well as suppress the clonogenicity of mammalian myeloid leukemia cell lines [Tomida et al., 1990; Maekawa and Metcalf, 1989]. However, LIF/DIA ligand, like certain other factors active with myeloid cells (interleukin-6, [IL-6], granulocyte colony stimulating factor [G-CSE]; and macrophage colony stimulating factor [M-CSF]), seems to be involved in the hepatic acute phase reaction as it is released into the serum by the liver in response to acute inflammation [Metcalf, 1988; Baumann and Wong, 1989; Kordula et al., 1991]. In addition to these functions, LIF/DIA promotes bone resorption in vitro [Abe et al., 1986] and is identical to the melanoma-derived lipoprotein lipase inhibitor [Mori et al., 1989] and cholinergic neuronal differentiation factor from heart cells [Yamamori et al., 1989]. LIF/DIA receptors are present on monocyte macrophage cells of the hematopoietic system although their biological significance is not clear.

Of greatest significance to early mouse development is the differentiation inhibitory activity of LIF/DIA on ES cells derived from the inner cell mass of the mammalian blastocyst [Williams et al., 1988; Gough et al., 1989]. This functional role coupled with the fact that expression of the *Lif* gene varies at different stages of embryogenesis [Murray et al., 1990] suggests that it plays an important role in embryogenesis. In fact *Lif* expression has been shown to coincide with blastocyst formation and always precedes implantation [Bhatt et al., 1991]. These results may indicate that one of the principal functions of LIF/DIA is to regulate blastocyst growth and initiate implantation.

Although there is a single copy of the *Lif* gene in both mouse and humans, two independently transcribed forms of this mRNA exist, one that is secreted and diffusible, and one that is extracellular matrix-bound [Rathjen et al., 1990a,b]. In view of the pleiotropic effects of this molecule in many different organ systems, such differential transcription could provide at least two distinct types of cellular regulation.

LIF/DIA and IL-6 are both cytokines that have pleiotropic effects on many cell types and systems [Hilton and Gough, 1991; Allan et al., 1990; Gough and Williams, 1989; Reid et al., 1990]. Although in many cases the activities of these two molecules seem to be similar, the structures of the two proteins are unrelated. However, the cDNA of the *Lif* receptor has recently been isolated and characterized [Gearing et al., 1991] and has been shown to be related to the gp130 signal-transducing component of the IL-6 and G-CSF receptors. The transmembrane and cytoplasmic regions of the LIF/DIA receptor and gp130 are most closely related. These structural similarities may suggest a common signal transduction pathway existing for the two receptors and may explain the similarities in biological function. Furthermore, it has now been shown that both LIF/DIA and IL-6 inhibit the differentiation of mouse EC F9 cells when induced to differentiate by retinoic acid alone or in combination with dibutyryl cAMP [Hirayoshi et al., 1991].

B. Peptide Growth Factors in Development

The concept of induction in experimental embryology was illustrated elegantly by Spemann and Mangold [1924] when they demonstrated that the dorsal lip of the blastopore of a gastrula stage amphibian embryo had special *trans*-active properties; i.e., when taken from one gastrula and transplanted into the ventral side of another, this piece of dorsal mesoderm could induce a secondary neural plate, leading to a twinned tadpole. It is now known that many factors operate in *trans* to induce proliferation of specific cell types as well as

differentiation of others in the vertebrate embryo. These secreted proteins include peptide growth factors such as the growing family of fibroblast growth factors (FGFs; aFGF, bFGF, WNT-2, HST/k-FGF, FGF-5, FGF-6), transforming growth factors (TGFs), and TGF-related factors, i.e. the activins and the oncogenes of the WNT family; and the insulin-like growth factors (IGFs) [for review, see Nilsen-Hamilton, 1990]. The patterns of expression of these genes as well as those of the genes of their receptors help to elucidate the complex nature of their respective mechanisms during development. Increasing evidence suggests that peptide growth factors have crucial roles in development [Davidson, 1990; Whitman and Melton, 1989]. Polypeptide growth factors mediate many cell–cell interactions that presumably occur during mammalian development [Mercola and Stiles, 1988].

FGF is involved in various developmental events in amphibia, including mesoderm induction, induction of homeobox genes, establishment of anteroposterior polarity [Ruiz i Altaba and Melton, 1989], neuronal differentiation and survival [Morrison et al., 1988], and angiogenesis [Folksman and Klagsbrun, 1987]. Originally FGF was isolated in two forms in mammals, acidic (aFGF) and basic (bFGF). Both forms act via the same cell surface receptor and are potent fibroblast mitogens [Caday et al., 1990]. The ability of both forms to interact with a common receptor presumably enables aFGF and bFGF to exert similar effects on a wide range of mesodermal and neuroectodermal cell types in amphibia to control their proliferation and differentiation [Gospodarowicz, 1987].

The best known effect of bFGF is its ability to act as a mitogen for cells of mesenchymal origin [Gospodarowicz et al., 1987]. Both aFGF and bFGF exert a wide range of effects on different adult mammalian cell types in vitro. These effects include stimulation of endothelial cell migration and proliferation, neurite outgrowth, and enhanced nerve cell survival and differention [Gospodarowicz et al., 1987]. A possible role for the FGF family in pathological processes such as cancer has become evident with the identification of oncogenes that encode proteins having a 40%–50% sequence homology to aFGF and bFGF [Dickson and Peters, 1987]. These oncogenes include *Wnt-2, Hst, Fgfk,* and *Fgf,* and they are discussed in detail below. Another member of this family, keratinocyte growth factor (KGF), is a specific mitogen for epithelial cells [Finch et al., 1989].

In mammalian embryos the presence of FGF protein and mRNA during development has been detected in several ways including a fibroblast mitogenesis assay [Caday et al., 1990], immunoprecipitation [Seed et al., 1988], immunocytochemistry [Gonzalez et al., 1990] and Northern blotting analysis [Hebert et al., 1990]. Recently, Gonzalez et al. [1990] reported the protein localization of bFGF in 18-day rat fetus. The distribution of this protein was found to be widespread, suggesting multiple functions in various developing organ systems.

In an attempt to understand further the roles of FGF in development, another study involving identification of FGF-responsive or FGF receptor (FGF-R)–bearing cells was done to determine spatial and temporal patterns of expression. Transcripts were localized with varying intensities to all organs, with the possible exception of the liver [Wanaka et al., 1991]. Relatively strong signals were located in the mesoderm-derived tissues such as perichondrium, metanephros, prevertebral column, and myotome. Since the purification and cloning of the chicken b*Fgf*-r [Lee et al., 1989], several laboratories have isolated *Fgf*-r cDNAs [Reid, 1990; Mansukhani et al., 1990]. This work suggests that the FGF-R belongs to a family of tyrosine kinase receptors based on highly conserved tyrosine kinase domains

and relatively unconserved extracellular domains.

In amphibia FGF is probably a vegetalizing factor, working in combination with another growth factor, transforming growth factor-β (TGF-β) to induce ventral mesenchyme induction in early embryos [Slack et al., 1987]. In the chick it may have a similar function [Mitrani et al., 1990a]. Functional testing of FGF in *Xenopus* has been performed using a dominant negative approach to prove that it functions in mesoderm formation [Amaya et al., 1991]. TGF-β has been shown to be capable of positively or negatively regulating aFGF or bFGF, depending on the cell type involved [Frater-Schoder et al., 1986].

The FGFs are expressed differentially in cell culture systems [Hebert et al., 1990]. Since EC and ES cell lines are believed to be model systems of the preimplantation mammalian embryo, it may be possible to draw some limited conclusions about expression of these genes in the early embryo from these types of studies. These conclusions can now be verified using PCR or in situ hybridization on embryos. PSA-1 cells can be grown as aggregates that differentiate into simple embryoid bodies after 3–6 days in culture or allowed to progress to cystic embryoid bodies after 12–15 days in culture. Finally, embryoid bodies that have been replated in tissue culture dishes for an additional 10 days have also been analyzed for *Fgf* expression. The most restricted pattern of expression was shown by *Fgfk,* as this gene was expressed only in undifferentiated EC or ES cells or in early differentiation stages of PSA-1 cells. Transcripts from the three other genes *Fgfb, Fgfa* and *Fgf5* were detectable by Northern blot analysis from whole embryos or various embryonic tissues ranging from 10.5 to 15.5 days. In PSA-1 cells, which were assumed to represent earlier cell types, *Fgfb, Fgfa,* and *Fgfk* were expressed in embryoid bodies, regardless of their state of differentiation. *Fgf5* mRNA expression underwent a dramatic 15-fold increase in steady-state levels when PSA-1 cells differentiated into embryoid bodies. Interestingly, the level of *Fgf5* did not increase when PSA-1 cells were specifically differentiated into visceral or parietal endoderm, so the changes in *Fgf5* mRNA upon differentiation of these cells were presumed to be due to changes in the inner core cells of the embryoid bodies rather than the endoderm cells themselves [Hebert et al., 1990]. Although EC cells are not analogous to preimplantation embryos in the sense that they are transformed, and also lack the three-dimensional cell–cell contacts present in the embryo, studies like these are useful in providing clues as to the cell type exclusivity of expression one might be able to expect of a gene within the mammalian embryo.

Another gene that shares homology with *Fgf* is the *Wnt-2* gene, which has no homology with the *Wnt-1* gene. The WNT-2 protein has an atypical hydrophobic leader sequence and may be proteolytically processed at dimers of basic residues [Nusse, 1988]. In the mouse embryo expression of the *Wnt-2* gene is found in the primitive streak from 7.5 to 9.5 days [Wilkinson et al., 1988]. Since the embryonic mesoderm arises from the primitive streak, this is the area where one would expect to find a response to a mammalian mesoderm-inducing factor, although at this point there is no direct evidence that *Wnt-2* is such a factor in mammals. At later stages of development *Wnt-2* mRNA is detectable in the endoderm of the first three pharyngeal pouches, the hindbrain adjacent to developing otocyst, and the mesenchyme of developing teeth. Although these are not areas associated with mesoderm formation, they are places where differentiation is modulated by epithelial mesenchymal interactions. Therefore *Wnt-2* may function as an extracellular inducer at this time in development [Wilkinson et al., 1988, 1989].

The TGF-β family of proteins has also been implicated as playing a role in mammalian development. Originally in vitro studies showed that TGF-βs 1–3 were mitogenic for cells derived from supporting tissues such as bone and cartilage. They have, however, also been shown to inhibit proliferation of many cell types. The TGF-βs have a similarly complex regulation of differentiation, having a stimulatory effect on certain cell types and an inhibitory effect on others. In addition, TGF-βs stimulate extracellular matrix deposition, are chemotactic for particular cells, and in the amphibian embryo induce mesoderm formation [Nilsen-Hamilton, 1990].

The TGF-βs 1–3, like the FGFs, are involved in complex cell–cell interactions occurring in the adult as well as the embryo [Akhurst et al., 1991]. Although five cDNA clones have been isolated in the *Tgf*-β family, only TGF-βs 1–3 have been found in mammals. TGF-βs 1–3 are similar in that each polypeptide is synthesized as a precursor monomeric protein that is subsequently cleaved to form a 112 amino acid polypeptide. This polypeptide remains associated with the latent "pro" portion of the molecule [Lyons and Moses, 1990; Miller et al.1990a,b]. Biological activity in all three forms of TGF-β is associated with dimerization of the monomers, generally as homodimers, and subsequent release of the latent "pro" peptide. The fully processed form of the TGF-β 3 protein has approximately 80% identity with the analogous regions of both β 1 and β 2, although there is only 27% identity at the amino-terminal regions [ten Dijke et al., 1988; Derynck et al., 1988]. In general, TGF-β 1–3 have qualitatively similar activities when added to cells in culture and seem to interact with the same cell surface molecules [Graycar et al., 1989], although some differences in their biological activities have been reported [Merwin et al., 1991; Rosa et al., 1988].

Recent studies have shown that expression patterns of the *Tgf*-β mRNAs in the murine embryo are overlapping but distinct [Pelton et al., 1990]. Furthermore, these patterns were shown to change over the course of development and were most often found to be present in tissues that were in the process of undergoing morphogenetic alterations, such as in the developing whisker follicle [Lyons et al., 1990]. All three *Tgf*-βs were expressed similarly in the mature follicle, whereas they were all different in the immature follicle. Similar trends in expression have been seen in the human embryo [Gatherer et al., 1990; Sandberg et al., 1988]. In terms of the protein expression of these genes, a study utilizing isoform-specific antibodies has shown that the *Tgf*-βs 1–3 are expressed in unique temporal and spatial patterns relative to one another in a wide range of tissues [Pelton et al., 1991]. These results may suggest that all three mammalian forms of TGF-β act through both paracrine and autocrine mechanisms.

Activins are also members of the TGF-β family. Activins were first described as factors that stimulate the release of follicle-stimulating hormone (FSH) from the anterior pituitary [Mason et al., 1985; Ling et al., 1986]. They have been purified in vivo as a homodimer of two βA chains (activin A) or a heterodimer of βA and βB chains (activin AB) [Vale et al., 1986]. Activin A has also been shown to stimulate erythroid differentiation [Yu et al., 1987]. Within the activin family is a related molecule, inhibin, that blocks the FSH-releasing capacity of pituitary cells and consists of an inhibin-specific A chain paired with either of the B chains [Mason et al., 1985].

Both mammalian activin A and recombinant *Xenopus* activin B induce a variety of dorsal anterior mesodermal and neural structures, including eyes, in *Xenopus*. *Xenopus* animal caps cultured in the presence of activins will differentiate into structures that possess well-defined head structures, notochord, and muscle [Asashima et al., 1990a,b; Sokol et al., 1990]. These re-

sults suggest that activins induce presumptive ectoderm to form mesoderm dorsoanterior structures [Sokol et al., 1991]. Another piece of evidence supporting this hypothesis is that activin mRNA injected into *Xenopus* embryos causes the formation of a partial second dorsal axis [Thomsen et al., 1990]. However, evidence against the latter proposal includes the fact that activins (βA and βB) are not expressed at the stage of embryonic development in vivo (32–64 cell stage) when the mesoderm-inducing capacity is present (both activins βA and βB are expressed later in development) and the fact that the endogenous inducer (but not activins) is capable of respecifying the fate of ventral ectoderm, resulting in the formation of a dorsal axis on the ventral side of the embryo [Ogi, 1967; Nieuwkoop, 1969]. In addition, vegetal blastomeres, which contain an endogenous dorsal inducer, are capable of rescuing axial formation (of neural or dorsal mesoderm tissues) after embryos have been UV irradiated. Activins are not able to achieve either of the two latter feats [Sokol and Melton, 1991].

The timing of activin gene expression suggests that in *Xenopus* activin may play an important role in mesoderm induction and axial patterning [Thomsen et al., 1990; Smith et al., 1990]. Furthermore, activins have also been shown to induce axial formation and been shown to be expressed in the hypoblast of the chick [Mitrani et al., 1990b]. At present there is no published evidence that activins are analogous in function with those of *Xenopus* in the context of the mammalian embryo.

Some studies have been performed on EC cells in culture using combinations of activins and other growth factors. In one study, activin A and bFGF were shown to prevent the death of P19 cells, which normally die within 2 days grown in serum-free N2 medium on tissue culture plastic [Shuber and Kimura, 1991]. Furthermore, the substratum to which these cells were exposed was capable of mediating their mitogenic response to growth factors, i.e., when P19 cells were cultured on extracellular matrix components laminin and fibronectin, they responded to activin A and bFGF by actively dividing. This study may indicate that, to evaluate the role of activins in mammalian cells, investigators will have to provide more complex culture conditions that include extracellular matrix components that may be present during embryonic development in utero. This idea may also serve to improve and extend embryo culture [Copp and Cockroft, 1990].

Another group of peptide growth factors that is involved in mammalian embryogenesis is the insulin and insulin-like growth factor (IGF) family. The IGFs, like the growth factors previously discussed, also participate in a wide variety of biological responses. One of the most intriguing observations on the effects of the insulin family in the preimplantation period of embryogenesis is that the treatment of blastocysts in culture with insulin leads to the formation of mice that are larger than controls after transfer to a pseudopregnant foster mother [Gardner and Kaye, 1991]. During preimplantation development, the major responses to insulin develop when there are numerous receptors on the cell surface during the morula and blastocyst stages. It has been shown that at these times insulin stimulates the incorporation of glucose into nonglycogen macromolecules and increases the incorporation of labeled precursors into RNA, DNA, and protein [Rao et al., 1990]. The presence of transcripts for *Igf* I, *Igf* II, and their receptors in early mouse embryos provides additional evidence for an autocrine role in growth regulation during peri-implantation stages [Rappolee et al., 1992; Telford et al., 1990].

The IGFs are mitogenic growth factors whose activities are modulated by the pres-ence of accessory binding molecules. Both IGF I and IGF II are held on binding proteins in serum that increase their half lives and alter their access to cells. These high-molecular-weight complexes are unable to pass through the walls of capillaries and therefore limit the access of IGFs to tissues. These binding complexes do not appear, however, until late in fetal development, and are therefore not involved in embryogenesis, although they may restrict IGF access to the placenta [Daughaday, and Rotwein, 1989; Froesch et al., 1985].

Although the IGFs are encoded by single-copy genes in mammals, there is considerable heterogeneity in their expression at the level of the proteins. All of the members of this family are produced as precursor prohormones that are proteolytically cleaved to their mature size by peptidases. Different proteins are produced by alternative splicing of mRNAs or posttranslational processing [Gammeltoft, 1989].

The IGFs share receptors with insulin. The receptor for IGF I with highest affinity is IGF I receptor. IGF II interacts with the mannose-6-phosphate receptor as well as the IGF I receptor with lower affinity. The mannose-6-phosphate receptor also has the ability to bind glycosylated TGF-β 1 precursor [Clairmont and Czech, 1989].

IGF II has recently been shown to play a role in embryonic growth [Gray et al., 1987]. The *Igf* II gene uses at least three different promoters and expresses several transcripts in many tissues during embryogenesis. Abundant IGF II mRNA expression is found in all the trophectoderm derivatives after implantation [Lee et al., 1990]. Later it is detected in the extraembryonic mesoderm at the early primitive streak stage. IGF II mRNA is also expressed transiently in the primitive endoderm, disappearing after yolk sac formation. Expression in the embryo proper appeared first at the late primitive streak/neural plate stage in lateral mesoderm and in the anteroproximal cells located between the visceral endoderm and the most cranial region of the embryonic ectoderm.

The first direct evidence that IGF II plays a role in embryonic development comes from experiments by DeChiara et al., in which disruption by gene targeting was performed in ES cells. Subsequent reintroduction into mouse host blastocysts gave rise to heterozygous mutant mice that were smaller than wild-type ES-derived littermates, leading to the identification of *Igf* II as an endogenous imprinted gene. These growth-deficient mice were apparently normal and fertile [DeChiara et al., 1990, 1991]. The *Igf* II receptor gene has also been shown to be imprinted [Barlow et al., 1991].

C. Oncogenes

It has been assumed for some time that mechanisms of carcinogenesis could be related to control of normal embryogenesis. Indeed, a tumor resembles an undifferentiated, rapidly proliferating embryonic cell, and, conversely, an embryo placed into certain ectopic nonuterine sites eventually develops into a tumor [Silver, 1983]. Many tumors display embryonic antigens such as the carcinoma embryonic antigens [Huang, 1990b]. Some forms of cancer are thus postulated to be the consequence of aberrant expressions of genes whose normal function is to modulate proliferation during development.

Immortalized and fully transformed cells frequently transcribe genes that are expressed in, and presumably influence, normal mammalian development [Ruddon, 1987]. In some cases these oncofetal genes do not appear to contribute to the neoplastic phenotype; for example, AFP is expressed by trophoblast cells and by many tumor cells. In other cases developmentally regulated genes play a primary role in the conversion of cells to the transformed phenotype; for example, the protooncogenes: *c-Myc, c-Src, c-Fos,* and *c-Fms* are all ex-

pressed during embryonic development and have been shown to regulate developmental steps in vitro [Adamson, 1987]. Another example of these types of genes is illustrated by the gene *Pem,* derived from a T lymphoma, which is not expressed in adult tissue but is first expressed in early embryos and subsequently in extraembryonic tissues.

Many protooncogenes are expressed during specific stages of mammalian development. These include *Wnt-1* (formerly *int-1*), *Wnt-2, Wnt-3, c-Fms, c-Myc, N-Myc, c-Jun,* and *jun*B which are all expressed in particular temporal and spatial patterns in the developing mammalian embryo and presumably play a role in embryonic cell growth or differentiation [Wilkinson et al., 1987, 1988, 1989; Regenstreif and Rossant, 1989; Schmid et al., 1989; Mugrauer et al., 1989]. It is generally assumed that these genes do not play an exclusive role in control of development, but are performing the same essential function in the embryo or fetus as they do in the adult.

Oncogenes can be divided into several categories. Certain oncogenes are growth factors. One of the best examples of oncogenes of this type is *v-Sis,* which is nearly identical to the β-chain of platelet-derived growth factor (PDGF). Cells overexpressing the cellular homolog of PDGF can grow in the absence of PDGF in the medium [Owen et al., 1984]. Other examples include the *Wnt-2* gene, which was identified by its activation in mouse mammary tumors after insertion of a proviral DNA [Moore et al., 1986]. This gene has been shown to have 40%–60% amino acid identity with the aFGF and bFGF proteins [Dickson and Peters, 1987]. Further examples are provided by the *Hst*/ks (*Fgfk*) oncogene, which shares similar amino acid identity with the 120 amino acid core characteristic of the FGFs [Delli-Bovi et al., 1987], and *Fgf5,* originally identified as a human oncogene [Zhan et al., 1988].

The *Wnt-1* gene encodes a protein with a hydrophobic leader sequence that is presumably a signal peptide. Furthermore, it is cysteine rich, a fact that is consistent with its putative identity as a growth factor or receptor, although there is no direct functional evidence for such extracellular action or cell surface properties [Nusse, 1988]. The *Wnt-1* gene is 54% identical with the *Drosophila wingless* gene, which plays a role in segmentation. It is expressed at the mRNA level in the mouse embryo between 8 and 14 days, within the developing nervous system, specifically in areas of the neural plate, anterior head folds, neural tube, and spinal cord [Shackleford and Varmus, 1987; Wilkinson et al., 1987]. The spatial and temporal patterns of expression of this gene in the mouse suggested that it may be involved in neural tube maturation, although it is now known that targeted disruption of both alleles of this gene gives rise to mice with severe abnormalities in the development of the mesencephalon and cerebellum and that the recessive mutation in mice known as *swaying* is a mutant allele of the *Wnt-1* gene [Thomas and Capecchi, 1990; Thomas et al., 1991; McMahon and Bradley, 1990].

The *Wnt-1* and *Xwnt-8 (Xenopus wnt-8)* genes have been shown to be important in *Xenopus* axial formation [see Jessell and Melton, 1992, for review]. In *Xenopus*, mesoderm is induced in an area known as the marginal zone of the blastula stage embryo. This marginal zone is located at the border of the animal and the vegetal hemispheres, destined to become ectoderm and endoderm, repectively. Factors responsible for the induction of mesoderm apparently arise from the vegetal cells [Nieuwkoop, 1973; Boterenbrood and Nieuwkoop, 1973; Gimlich and Gerhart, 1984; Dale et al., 1985] and act upon the marginal zone. This vegetal dorsalizing region is also known as the Nieuwkoop center [Gerhart et al., 1989], which acts to stimulate the exposed marginal zone cells to form dorsal meso-

derm. This dorsal mesoderm, known as the Spemann organizer [Spemann and Mangold, 1938], forms the dorsal lip of the blastopore, a place where gastrulation occurs earlier and more extensively than on the ventral side of the embryo [Gerhart and Keller, 1986].

Recent studies have shown that the *Wnt* oncogene families and the activins can induce certain aspects of dorsal axis formation in *Xenopus* embryos. When *Xenopus* embryos are injected with *Wnt-1* and *Xwnt-8* mRNAs, dorsal axial structure bifurcation occurs. This result resembles that observed in *Xenopus* embryos injected with activin mRNAs. The hypothesis formed from these results was that the ectopically expressed WNT proteins were interfering with the Spemann organizer and subsequently causing it to split [Sokol et al., 1991]. However, in a recent study by Smith and Harland the vegetal blastomeres exposed to exogenous *Xwnt-8* mRNA contributed progeny to the endoderm rather than the induced dorsal axis, indicating that the *Xwnt-8* mRNA may instead cause cells to act as a Nieuwkoop center rather than a Spemann organizer. Furthermore, both *Wnt*-related mRNAs were shown to have the ability to rescue UV-treated axis-deficient embryos, whereas activins could provide only partial rescue [Sokol et al., 1991; Smith and Harland, 1991]. Recently an additional insertion site for MMTV proviruses has been identified, *Wnt-3*. This gene encodes a protein that is 47% identical to WNT-1 at the amino acid level. The expression of this gene has not yet been characterized at the biochemical or molecular level in the embryo [Roelink et al., 1990].

Another class of oncogenes includes that of growth factor receptors. Especially interesting are the receptors with tyrosine kinase activity that include epidermal growth factor (EGF) and HER 2/*Neu,* both of which can become oncogenes if properly

activated, i.e., a point mutation in the transmembrane domain of HER2/*Neu* receptor
causes this receptor to become oncogenic
[Bargmann et al., 1986]. Other receptor-de-
rived oncogenes possess different structural lesions such as point mutations in the
cytoplasmic regions or deletions and carboxy-terminal truncations that appear to
enhance and modulate the transforming
signal [Studzinski, 1989].

D. Homeobox Domain Genes

The homeobox is a highly conserved 180
bp sequence that was first described in
genes controlling pattern formation in *Drosophila melanogaster* [Akam, 1987; Gehring, 1987]. This sequence is normally located close to the C terminus of the protein,
encoding a 60 amino acid homeodomain
containing a helix–turn–helix motif that
function as a sequence-specific DNA-binding domain. More than 30 murine
homeobox genes (*Hox*) have been isolated
by homology to *Antennapedia* or other
members of this highly conserved gene
family. In the mouse four clusters of these
genes have been identified: *Hox-1* (Chrm.
6), *Hox-2* (Chrm.11), *Hox-3* (Chrm.15) and
Hox-4 (Chrm.2) [Colberg-Poley, 1985]. The
identical order of paralogous genes
[Schughart, 1988] in all four clusters suggests that they emerged by complete or
partial duplication of a common ancestral
cluster. Alignment can be extended to the
Drosophila ANT-C and *BX-C* complexes
[Gaunt, 1988]. A strong correlation exists
between the position of a gene within the
cluster and its expression pattern along the
anteroposterior (A–P) axis in both *Drosophila* and vertebrates [Akam, 1989]. In
the mouse, this is most obvious in the central nervous system where the distinct anterior boundaries of expression are exhibited [Gaunt, 1988].

The *Hox-5* gene complex exhibits an interesting pattern of expression in murine
limb buds after the ninth day of gestation
[Dolle et al., 1989a], although no function

for the encoded genes has been defined.
There are few other genetic markers known
that characterize cells in early and intermediate stages of murine development; for
example, no gene has been associated with
the initial process of segmentation which
occurs at 8dg in the mouse [Rossant and
Joyner, 1989].

Many of the known murine homeobox
genes encode multiple transcripts during
embryogenesis [Odenwald et al., 1987]. Tissue-specific differences are seen in the sizes
of transcripts produced by respective
homeobox genes such as *Hox-3.2* and *Hox-
2.6*. It is possible that alternative splicing
plays an important role in tissue-specific
regulation of many of the vertebrate *Hox*
genes [Graham et al., 1989].

Most murine homeobox genes can be
induced during differentiation of F9 EC
cells by retinoic acid. Furthermore, the
genes of the human *Hox-2* cluster are differentially activated by retinoic acid in the
EC line NT2/D1, depending on their position in the cluster [Simeone et al., 1990],
with genes at the 5′ end requiring much
higher levels of retinoic acid (up to 10^{-5}M)
for activation than genes in the 3′ part of
the cluster (10^{-8}M). These results are remarkably similar to the colinear patterns
detected in the mouse embryo. An analogous result was achieved in a similar study
using *Xenopus* embryos treated with
retinoic acid [Papalopulu et al., 1990, 1991].
Although a series of comparable experiments has not been performed in F9 EC
cells, it has been shown that the *Hox-3.2*
gene lying in the 5′ region of the cluster is
not induced during differentiation of F9
cells, whereas the next 3′ gene, *Hox-3.1*, is
activated by retinoic acid (5×10^{-7}M) in
combination with 10^{-3}M cAMP [Breier et
al., 1986].

The colinearity of response to retinoic
acid and gene order of *Hox-2* genes
could arise from a graded morphogen
signal in the embryo. This morphogen
could then induce a differential homeo-

box gene response. As has been cited, retinoic acid stimulates expression of many of the homeobox genes in teratocarcinoma cells. Experimental evidence in chicken embryos suggests that retinoic acid is important in patterning in the limb [Brockes, 1989]. Therefore, the differential response of *Hox-2* genes to retinoic acid could be one of the means that is used during embryogenesis to set up a gradient of homeobox expression and ordered, partially overlapping domains of expression.

The patterns of expression of homeobox domain genes suggest that they play a role in positional specification in vertebrate development, particularly in axis specification, where they may establish an "address" or identity through an overlapping "combinatorial code" [Sham et al., 1991]. In the absence of naturally occurring mouse mutants, homologous recombination and gene targeting will presumably address the problems of assigning function to each of these genes and determining functional redundancy where homologies between different *Hox* clusters is apparent. Indeed such studies have already begun with targeted disruption of the mouse homeobox gene *Hox-1.5* [Chisaka and Capecchi, 1991], which has led to the formation of homozygous mice that exhibit reduced thyroid and submaxillary tissue and a wide range of throat abnormalities, but no apparent failure of peri-implantation embryogenesis.

E. POU/Oct Genes

The octamer motif (ATGCAAAT) is a *cis*-active promoter element required for both ubiquitous and cell-specific type expression of certain histone and immunoglobulin light and heavy chain genes. The obvious paradox in the same element being required for two different types of expression was explained by the result that two different proteins were able to interact with this sequence in the different cell types tested [Kemler and Schaffner, 1990]. Addi-

tional proteins that bind directly to this octamer motif have now been identified at specific developmental stages and in various model systems of development. These proteins all belong to the POU family, as defined by the sequence homology with three mammalian transcriptional factors and one nematode regulatory protein (Pit-1/GHF-1, Oct-1, Oct-2, and Unc-86) [Herr et al., 1988].

POU proteins have two conserved domains: a domain specified by the POU-specific domain and a homeodomain that is distantly related to the prototype *Antennapedia* homeodomain and nearby on the amino-terminal side. Outside of these two domains the amino acid sequences are highly divergent and contain sequences required for transcriptional activation. The POU domain is 81 amino acids long and contains two subdomains of high sequence homology. Each of these subdomains shares two features: a cluster of basic amino acids in the center and a predicted α-helix at the carboxy-terminal end. The "recognition helix" or the most conserved part of the POU domain contains 11 invariant residues. The RVFCN motif within this recognition helix has a cysteine residue at position 9 that is specific for the POU family, whereas the other residues are extremely well conserved in all homeodomain proteins.

The homeodomain found in POU proteins is 60 amino acids in length. It contains a DNA-binding domain as well as three well-defined helices, two of which form helix–turn–helix motifs. An intact homeodomain is required for DNA binding of all POU proteins, whereas the contribution of the POU-specific domain varies, depending on the DNA-binding site and on the identity of the POU protein.

Comparison of all POU homeodomain and POU subdomain sequences divides the POU family into five classes. This classification is based on the similarity of the variable linker sequences that flank the

POU-specific domain and the POU homeodomain. This linker area contains 14–26 amino acids.

Most POU genes are differentially expressed throughout embryogenesis (see Table I). However, unlike the *Hox* and the *Pax* genes, their expression patterns are not regular and not explainable with any sort of unifying hypothesis.

Two of the POU genes are expressed in the early embryo, *Oct-4* (also called *Oct-3* or *NF-A3*) and *Oct-6*. *Oct-4* expression seems to be correlated with an undifferentiated or stem cell population. It is first detectable in the totipotent and pluripotent stem cells of the pregastrula stage embryo, is subsequently downregulated during differentiation of these cells, and eventually becomes confined to the germ cell lineage [Okamoto et al., 1990]. In ES and EC cells, both *Oct-4* and *Oct-6* are downregulated when the cells are induced to differentiate with retinoic acid. *Oct-6* is also expressed later in development in specific neurons of the developing and adult brain and also in testis. *SCIP/tst-1* is the rat homologue of *Oct-6,* which is expressed transiently during the period of rapid cell division separating the premyelinating and myelinating phases of Schwann cell differentiation. This gene may therefore play a role in the progressive determination of Schwann cells [Monuki et al., 1989, 1990].

Although most murine developmental control genes that have been identified by homology with *Drosophila* genes are expressed at postimplantation stages, several of the *Oct* genes are expressed at preimplantation stages. The analysis of POU genes in model systems of early embryo ES and EC cells may indicate their putative roles in early development. Activation and repression via the octamer motif have been documented in EC cells, depending on the regulatory elements present and on the amount of *Oct* factors present [Lenardo et al.,1989; Scholer et al., 1989]. Several developmental control genes contain octamer motifs in their promoters and are developmentally expressed after *Oct-4* and *Oct-6* (e.g., *Hox-1.3*), and these may be acted upon by the octamer genes [Scholer et al., 1991].

In summary, based on the present evidence, POU domain genes seem to be involved in cell proliferation. Some POU genes are clearly more highly expressed during increased cellular proliferation, sug-

TABLE I. Summary Of Octamer Proteins

OCT protein*	Embryonic expression	Adult expression
OCT-1	Ubiquitous	Ubiquitous
OCT-2	Neural tube, entire brain except telencephalon	Lymphoid cells, nervous system, intestine, testis, kidney
N-OCT-2	Nervous system	Nervous system: astrocytes, glioblastoma, and neuroblastoma cell lines
MiniOCT-2	Nervous system, developing nasal neuroepithelium	Nervous system, primary spermatids
N-OCT-3	Nervous system	Nervous system, glioblastoma, and neuroblastoma cell lines
OCT-4A, OCT-4B, OCT-5	Totipotent stem cells: pregastrulation embryonic ectoderm, primordial germ cells, testis, ovaries	Oocytes
OCT-6	Blastocyst, ES and EC cells, brain	Nervous system, testis
OCT-7, 8	Nervous system	Nervous system

*OCT-9 and OCT-10 proteins have also been isolated, but their expression patterns are presently unknown.

gesting a possible role in replication (e.g., *SCIP* (*tst-1/Oct-6*). The importance of *Oct-4* (*Oct-3*) in particular at early embryonic stages has been demonstrated by injection of antisense *Oct-4* oligonucleotides into fertilized oocytes, resulting in an inhibition of DNA synthesis and arrest of the embryo at the one-cell stage [Rosner et al., 1991]. Future studies using this technique and others will elucidate the role of these interesting gene products.

IV. DISCUSSION

It is apparent from this perspective on the cell lineage relationships during early mouse embryogenesis that epigenetic interactions play a substantial role in peri-implantation mammalian development. We summarize below the time, place, and direction of these interactions; we also consider the nature of disturbances to these epigenetic interactions that may be involved in the aberrant development of model cell systems and as a result of imprinting. Finally, we evaluate the possible roles of specific peptide growth factors and other agents in the epigenetic interactions that occur during early mouse embryogenesis.

A. Sites of Epigenetic Interactions in Normal Development

The cell lineage relationships described above provide evidence for several distinct epigenetic interactive processes in peri-implantation mouse embryogenesis (Fig. 1): 1) outer cells at the morula stage interact with the inner cells, suppressing their tendencies for trophectoderm differentiation or preventing exposure to the outside environment; 2) ICM cells interact with the overlying polar TE of the early blastocyst, suppressing their tendencies for mural TE differentiation and sustaining the proliferation of diploid cells of the trophectoderm lineage; 3) newly differentiated primitive endoderm cells of the late blastocyst inter-

act with the remaining core cells of the inner cell mass (primitive ectoderm), suppressing their tendencies for primitive endoderm differentiation or preventing exposure to the blastocele environment; 4) trophoblast giant cells of the late blastocyst and later stages interact with adjacent cells of the primitive endoderm lineage, inducing their differentiation as parietal endoderm; 5) visceral embryonic endoderm cells (primitive endoderm) of the early egg cylinder interact with the adjacent epiblast (primitive ectoderm) layer, inducing their differentiation into mesoderm, leading to formation of the primitive streak; 6) extraembryonic ectoderm of the early egg cylinder interacts with the adjacent visceral extraembryonic endoderm, suppressing the synthesis of AFP; and 7) visceral extraembryonic endoderm of the late egg cylinder interacts with the adjacent extraembryonic mesoderm, inducing the differentiation of blood islands. There are undoubtedly other epigenetic interactions between other early cell lineages of the mouse embryo, and some of these interacting tissues may have bidirectional signalling. However, these examples are sufficient to establish the importance of such signaling mechanisms even at the earliest stages of mammalian development.

The evidence for such epigenetic interactions in normal mouse embryogenesis may provide clues for understanding the etiology of the developmental perturbations observed in model cell systems. The earliest stages of mouse embryogenesis are not applicable to EC and ES differentiation in vitro, because neither of these cell systems generates trophectoderm in isolation. However, ES cells transferred to mouse blastocysts were able to contribute to trophectoderm and primitive endoderm populations at a low frequency [Beddington and Robertson, 1989], showing that some cells were capable of responding to signals for such differentiative events. Although EC cells do not spontaneously differentiate to form primitive endoderm cell types, these are induced by exposure to retinoic acid. The tendency for retinoic acid–induced EC cells to differentiate into parietal endoderm after exposure to cyclic AMP indicates that they are capable of responding to their molecular signals and suggests that such signals could act through the protein kinase A signal transduction pathway. The capacity that is conspicuously absent from most EC cells is mesodermal differentiation, and neither EC nor ES cells establish an embryonic axis by primitive streak formation at gastrulation despite their capacity for expression of cardiac muscle-specific genes [Robbins et al., 1990; Sanchez et al., 1991]. This may reflect a deficiency in signals responsible for mesoderm induction, which (by analogy with chick and amphibian embryos) probably arise from primitive endoderm cells at the posterior of the prospective embryonic axis. The capacity for EC and ES cells to participate in axial development when transferred to the blastocyst microenvironment indicates that their capacity to respond to such signals is intact. In mice, the source of these signals for axial development may originate in the relationship between the primitive ectoderm and the extraembryonic lineages trophectoderm and primitive endoderm, which are absent from embryoid bodies. Moreover, the site of primitive streak formation in the mouse embryo in utero appears to reflect either the contact point of the trophectoderm with the uterine wall, or a precocious differentiation of trophoblast that presages anteroposterior axis formation and determines the orientation of implantation [Smith, 1985].

Another feature in the development of embryoid bodies is their failure to form well-differentiated blood vessels, despite the capacity of ES cells to differentiate into hematopoietic cells [Doetschman et al., 1985; Wiles and Keller, 1991; Lindenbaum and Grosveld, 1990; Wang et al., 1992]. This may also reflect a deficiency in signals involved in angiogenesis [Klagsbrun and D'Amore, 1991].

The abnormalities of mouse isoparental embryos (parthenogenotes and androgenotes) that arise from genomic imprinting may also be partially accounted for as deficiencies in epigenetic interactions. Both parthenogenotes and androgenotes are capable of forming morphologically normal preimplantation embryos and thus appear competent in generating signals required for early lineage differentiation. However, the morphological abnormalities of parthenogenotes may indicate deficiencies in the epigenetic signals that regulate polar trophectoderm differentiation, since parthenogenotes form an excess of giant cells and retain few diploid cells in the trophoblast lineage. Similarly, there may be deficiencies in epigenetic mechanisms regulating the differentiation of primitive endoderm, because parthenogenotes form an excess of parietal endoderm [Sturm et al., 1992; reviewed by Pedersen et al., 1992]. While parthogenotes were capable of forming both embryonic and extraembryonic mesoderm, they often lacked an embryonic axis, suggesting eviatic function of the epigenetic processes leading to axial mesoderm induction. On the other hand, the more advanced parthenogenotes were fully capable of forming blood islands, indicating that they could generate any epigenetic signals necessary for this process.

B. Role of Specific Factors in Epigenetic Interactions

A wide variety of scenarios can be envisioned for the inductive interactions between the various cell lineages of mammalian embryos. General properties of epigenetic signalling are the synthesis of ligand(s) by an inducing cell population and the reception of such signals by the responding cell population. The ligand may be secreted as a diffusible factor or may be retained in a membrane-bound form. The capacity for numerous ligands to interact with extracellular matrix materials provides additional possibilities of reservoirs, barriers, or concentration of such factors and blurs the distinction between diffusible and static factors [see Jessell and Melton, 1992, for review]. Because many of the potential ligands known to be synthesized in early embryos share these properties and because other ligands may well be produced by the interacting tissues, it is probably not fruitful to speculate at this point on the identity of ligands responsible for each of the epigenetic interactions that characterize early mouse development. However, further consideration of the effects of certain ligands or receptors allows us to exclude a role for the epigenetic interactions during peri-implantation mouse embryo lineage differentiation. The relevance of specific peptide growth factors and receptors to mouse development can be evaluated on the basis of 1) direct evidence for a function or, more commonly, 2) circumstantial evidence indicating the synthesis of the transport or protein in appropriate tissues and stages of development or a biological response to exogenously added agent.

We can conclude on the basis of direct functional evidence that certain genes are essential for early events of lineage differentiation and fate in mouse embryos. The role of stem cell factor [*Steel* locus; Copeland et al., 1990; Zsebo et al., 1990; Huang et al., 1990a], *c-Kit* [*W* locus; Chabot et al., 1988; Geissler et al., 1988], PDGF-alpha [*Patch* locus; Mercola et al., 1990], IGF-II [De Chiara et al., 1990] and *Wnt-1* [McMahon and Bradley, 1990; Thomas and Capecchi, 1990] have been examined experimentally by gene targeting using homologous recombination in ES cells or by analysis of existing mutations. In every case, the phenotype of the homozygous mutant shows effects later in fetal development rather than at peri-implantation stages. A possible explanation for the lack of an effect of these mutations is that there are other redundant functions that mask

their consequences. Such explanations must be invoked to account for the lack of effect of these mutations in tissues that express the gene in order to reconcile this with the circumstantial evidence for a developmental role. An example of this situation is stem cell factor, which is strongly transcribed in the embryonic endoderm at 7.5 dg and the ectoplacental cone at 9.0 dg [Motro et al., 1991], but shows no mutational effects until germ cell and hematopoietic development begins.

The stage and tissue-specific expressions of peptide growth factors and cytokines and their receptors have been interpreted as circumstantial evidence for a developmental role in mammals. Genes in this category include *c-Fms, Egf, Fgfs, Igf*-I, *Lif, Pdgf*-A, *Tgf*-α, and *Tgf*-β. The effect of exogenous ligand on in vitro embryonic development provides another line of indirect evidence for developmentally active epigenetic signals. Genes in this category include b-*Fgf*, insulin, *Pdgf*-A, and others. A thorough evaluation of the epigenetic role of each of these products will require analysis by antisense oligonucleotides, dominant negative transgenics, function-perturbing antibodies, gene targeting, or other novel methods for disrupting the function of the endogenous gene and its products.

A similar evaluation of specific transcription factors, including *Hox* genes and POU genes, suggests on the basis of their transcriptional patterns that they function during early mouse development, but there is no decisive evidence for their role in lineage differentiation and fate. We conclude that, despite the evidence for numerous epigenetic processes in early mammalian development, there is no documented role of known signalling factors in the emergence of the tissues that comprise the midgestation mouse conceptus. The most extensive analysis of epigenetic processes during early vertebrate development have been carried out using amphibian embryos. While these studies have implicated members of the *Tgf*-β and *Wnt* families as inducers of mesoderm, only b-*Fgf* has been shown to have a role by disrupting the endogenous gene product, leading to abnormalities in axial development [Amaya et al., 1991; Jessell and Melton, 1992; see Slack and Tannahill, 1992, for review]. Thus there is still no consensus on the identity of the factors responsible for vertebrate epigenetic interactions. Because of their unique strategy of devoting early stages to differentiation of extraembryonic membranes, mammalian embryos may have expression patterns and roles for these genes that are distinct from those found in amphibian embryos. Further work is therefore needed to determine which, if any, of the gene products known to be synthesized in early mammalian embryos generate the epigenetic phenomena summarized here.

ACKNOWLEDGMENTS

The work carried out by the authors was supported by NIH grant HD26732 and by USDOE/OHER contract No. DE-AC03-76-SF01012. We thank Liana Hartanto for assistance with the manuscript and Drs. Rik Derynck and Stephen Grant for their comments.

REFERENCES

Abe T, Murakami M, Sato T, Kajiki M, Ohno M, Kodaira R (1989): Macrophage differentiation inducing factor from human monocytic cells is equivalent to murine leukemia inhibitory factor. J Biol Chem 264:8941–8945.

Adamson (1987): Expression of proto-oncogenes in the placenta. Rev Placenta 8:449–466.

Akam M (1987): The molecular basis for metameric pattern in the *Drosophila* embryo. Development 101:1–22.

Akam M (1989): *Hox* and HOM: Homologous gene clusters in insects and vertebrates. Cell 57:347–349.

Akhurst RJ, FitzPatrick DR, Gatherer D, Lehnert SA, Millan FA (1990): Transforming growth factor betas in mammalian embryogenesis. Prog Growth Factor Res 2(3):153–168.

Akhurst RJ, Lehnert SA, Faissner A, Duffie E (1990): TGF beta in murine morphogenetic processes: the early embryo and cardiogenesis. Development 108:645–656.

Allan EH, Hilton DJ, Brown MA, Evely RS, Yumita S, Metcalf D, Gough NM, Ng KW, Nicola NA, Martin TJ (1990): Osteoblasts display receptors for and responses to leukemia-inhibitory factor. J Cell Physiol 145:110–119.

Amaya E, Musci TJ, Kirschner MW (1991): Expression of a dominant negative mutant of the FGF receptor disrupts mesoderm formation in *Xenopus* embryos. Cell 66:257–270.

Andrews PW (1984): Retinoic acid induces neuronal differentiation of a cloned human embryonal carcinoma cell line *in vitro.* Dev Biol 103:285–293.

Asashima M, Nakano H, Shimada K, Kinoshita K, Ishii K, Shibai H, Ueno N (1990a): Mesodermal induction in early amphibian embryos by activin A (erythroid differentiation factor). Roux Arch Dev Biol 198:330–335.

Asashima M, Nakano H, Uchiyama H, Davids M, Plessow S, Loppnow-Blinde B, Hoppe P, Dau H, Tiedemann H (1990b): The vegetalizing factor belongs to a family of mesoderm-inducing proteins related to erythroid differentiation factor. Naturwissenschaften 77(8):389–391.

Azar Y, Eyal-Giladi H (1981): Interaction of epiblast and hypoblast in the formation of the primitive streak and the embryonic axis in the chick, as revealed by hypoblast rotation experiments. J Embryol Exp Morphol 61:133–144.

Balinsky BI (1981): An Introduction to Embryology, 5th ed. Philadelphia: WB Saunders.

Bargmann CI, Hung M-C, Weinberg RA (1986): Multiple independent activations of the *neu* oncogene by a point mutation altering the transmembrane domain of p185. Cell 45:649–657.

Barlow DP, Stoger R, Herrmann BG, Saito K, Schweifer N (1991): The mouse insulin-like growth factor type-2 receptor is imprinted and closely linked to the *Tme* locus. Nature 349:84–87.

Barlow P, Owen DAJ, Graham C (1972): DNA synthesis in the preimplantation mouse embryo. J Embryol Exp Morphol 27:431–445.

Bartolomei MS, Zemel S, Tilghman SM (1991): Parental imprinting of the mouse *H19* gene. Nature 351:153–155.

Barton SC, Ferguson-Smith AC, Fundele R, Surani MA (1991): Influence of paternally imprinted genes on development. Development 113:679–688.

Baumann H, Wong GG (1989): Hepatocyte-stimulating factor III shares structural and functional identity with leukemia-inhibitory factor. J Immunol 143(4):1163–1167.

Beddington RSP (1983): Histogenetic and neoplastic potential of different regions of the mouse embryonic egg cylinder. J Embryol Exp Morphol 75:189–204.

Beddington RS, Robertson EJ (1989): An assessment of the developmental potential of embryonic stem cells in the midgestation mouse embryo. Development 105:733–737.

Bhatt H, Brunet LJ, Stewart CL (1991): Uterine expression of leukemia inhibitory factor coincides with the onset of blastocyst implantation. Proc Natl Acad Sci USA 88(24):11408–11412.

Boterenbrood EC, Nieuwkoop PD (1973): The formation of the mesoderm in Urodelan amphibians. V. Its regional induction by the endoderm. Rouxs Arch 173:319–332.

Breier G, Bucan M, Francke U, Colberg-Poley AM, Gruss P (1986): Sequential expression of murine homeobox genes during F9 EC cell differentiation. EMBO J 5:2209–2215.

Brockes JP (1989): Retinoids, homeobox genes, and limb morphogenesis. Neuron 2:1285–1294.

Caday CG, Klagsbrun M, Fanning PJ, Mirzabegian A, Finklestein SP (1990): Fibroblast growth factor (FGF) levels in the developing rat brain. Dev Brain Res 52(1–2):241–246.

Casanova JE, Grobel LB (1988): The role of cell interactions in the differentiation of teratocarcinoma-derived parietal and visceral endoderm. Dev Biol 129:124–139.

Cattanach BM, Beechey CV (1990): Autosomal and X-chromosome imprinting. Development (Suppl):63–72.

Chabot B, Stephenson DA, Chapman VM, Besmer P, Bernstein A (1988): The proto-oncogene c-kit encoding a transmembrane tyrosine kinase receptor maps to the mouse W locus. Nature 335:88–89.

Chisaka O, Capecchi MR (1991): Regionally restricted developmental defects resulting from targeted disruption of the mouse homeobox gene hox-1.5. Nature 350:473–479.

Chisholm JC, Johnson MH, Warren PD, Fleming TP, Pickering SJ (1985): Developmental variability within and between mouse expanding blastocysts and their ICMs. J Embryol Exp Morphol 86:311–336.

Clairmont KB, Czech MP (1989): Chicken and Xenopus mannose 6-phosphate receptors fail to bind insulin-like growth factor II. J Biol Chem 264:16390–16392.

Clarke HJ, Varmuza S, Prideaux VR, Rossant J (1988): The developmental potential of parthenogenetically derived cells in chimeric mouse embryos: implications for action of imprinted genes. Development 104(1):175–182.

Cockroft DL, Gardner RL (1987): Clonal analysis of the developmental potential of 6th and 7th day visceral endoderm cells in the mouse. Development 101:143–155.

Colberg-Poley AM, Voss SD, Chowdhury K, Gruss P (1985a): Structural analysis of murine genes containing homeobox sequences and their expression in embryonal carcinoma cells. Nature 314:713–718.

Cooke J (1985): Embryonic origins of the nervous system and its pattern. Trends Neurosci. 00:58–63.

Copeland NG, Gilbert DJ, Cho BC, Donovan PJ, Jenkins NA, Cosman D, Anderson D, Lyman SD, Williams DE (1990): Mast cell growth factor maps near the steel locus on mouse chromosome 10 and is deleted in a number of steel alleles. Cell 63:175–183.

Copp AJ (1979): Interaction between inner cell mass and trophectoderm of the mouse blastocyst. II. The fate of the polar trophectoderm. J Embryol Exp Morphol 51:109–120.

Copp AJ, Cockroft DL (1990): Postimplantation Mammalian Embryos: A Practical Approach. Oxford: IRL Press.

Couly GF, LeDouarin NM (1987): Mapping of the early neural primordium in quail-chick chimeras. II. The prosencephalic neural plate and neural fold: implications for the genesis of cephalic human congenital abnormalities. Dev Biol 120:198–214.

Cruz YP, Pedersen RA (1985): Cell fate in the polar trophectoderm of mouse blastocysts as studied by microinjection of cell lineage tracers. Dev Biol 112:73–83.

Cruz YP, Pedersen RA (1991): Origin of embryonic and extraembryonic cell lineages in mammalian embryos. In: Animal Applications of Research in Mammalian Development, Cold Spring Harbor, NY: Cold Spring Harbor Laboratory Press.

Dale L, Smith JC, Slack JMW (1985): Mesoderm induction in Xenopus laevis: A quantitative study using a cell lineage label and tissue-specific antibodies. J Embryol Exp Morphol 89:289–312.

Daughaday WH, Rotwein P (1989): Insulin-like growth factors I and II. Peptide, messenger ribonucleic acid and gene structures, serum and tissue concentrations. Endocr Rev 10:68–91.

Davidson EH (1990): How embryos work: A comparative view of diverse modes of cell fate specification. Development 108(3):365–389.

DeChiara TM, Efstratiadis A, Robertson EJ (1990): A growth-deficiency phenotype in heterozygous mice carrying an insulin-like growth factor II gene disrupted by targeting. Nature 345:78–80.

DeChiara TM, Robertson EJ, Efstratiadis A (1991):

Parental imprinting of the mouse insulin-like growth factor II gene. Cell 64:849–859.

Delli Bovi P, Curatola AM, Kern FG, Greco A, Ittmann M, Basilico C (1987): An oncogene isolated by transfection of Kaposi's sarcoma DNA encodes a growth factor that is a member of the FGF family. Cell 50(5):729–737.

Derynck R, Lindquist PB, Lee A, Wen D, Tamm J, Graycar JL, Rhee L, Mason AJ, Miller DA, Coffey RJ, Moses HL, Chen EY (1988): A new type of transforming growth factor-b, TGF-β3. EMBO J 7:3737–3743.

Dickson AD (1966): The form of the mouse blastocyst. J Anat 100:335–348.

Dickson C, Peters G (1987): Potential oncogene product related to growth factors. Nature 326(6116):833.

Doetschman TC (1991): Gene targeting in embryonic stem cells. Biotechnology 16:89–101.

Doetschman TC, Eistetter H, Katz M, Schmidt W, Kemler R (1985): The in vitro development of blastocyst-derived embryonic stem cell lines: formation of visceral yolk sac, blood islands and myocardium. J Embryol Exp Morphol 87:27–45.

Dolle P, Duboule D (1989a): Two gene members of the murine Hox-5 complex show regional and cell-type specific expression in developing limbs and gonads. EMBO J 8:1507–1515.

Dolle P, Izpisua-Belmonte JC, Falkenstein H, Renucci A, Duboule D (1989b): Coordinate expression of the murine Hox-5 complex homoeobox-containing genes during limb pattern formation. Nature 342(6251):767–772.

Dolle P, Ruberte E, Kastner P, Petkovich M, Stoner CM, Gudas LJ, Chambon P (1989): Differential expression of genes encoding alpha, beta and gamma retinoic acid receptors and CRABP in the developing limbs of the mouse. Nature 342(6250):702–705.

Dyce J, George M, Goodall H, Fleming TP (1987): Do trophectoderm and inner mass cells in the mouse blastocyst maintain discrete lineages? Development 100:685–698.

Dziadek M (1978): Modulation of alphafetoprotein synthesis in the early postimplantation mouse embryo. J Embryol Exp Morph 46:135–146.

Dziadek M, Adamson E (1978): Localization and synthesis of alphafoetoprotein in post-implantation mouse embryos. J Embryol Exp Morph 43:289–313.

Edelman GM (1988): Topobiology: An Introduction to Molecular Embryology. New York: Basic Books.

Edelman GM (1992): Morphoregulation. Dev Dynam 193:2–10.

Erselius JR, Goulding MD, Gruss P (1990): Structure and expression pattern of the murine Hox-3.2 gene. Development 110:629–642.

Evans MJ, Kaufman MH (1981): Establishment in culture of pluri-potential cells from mouse embryos. Nature 292:154–155.

Finch PW, Rubin JS, Miki T, Ron D, Aaronson SA (1989): Human KGF is FGF-related with properties of a paracrine effector of epithelial cell growth. Science 245: 752–755.

Fleming TP (1987): A quantitative analysis of cell allocation to trophectoderm and inner cell mass in the mouse blastocyst. Dev Biol 119:520–531.

Fleming TP, Johnson MH (1988): From egg to epithelium. Annu Rev Cell Biol 4:459–485.

Folkman J, Klagsbrun M (1987): Vascular physiology: A family of angiogenic peptides. Nature 329(6141):671–672.

Frater-Schoder M, Risau W, Hallmann R, Gautschi P, Bohlen P (1987): Tumor necrosis factor type alpha, a potent inhibitor of endothelial cell growth in vitro, is angiogenic in vivo. Proc Natl Acad Sci USA 84:5277–5281.

Froesch ER, Schmid C, Schwander J, Zapf J (1985): Actions of insulin-like growth factors. Annu Rev Physiol 47:443–467.

Frohman MA, Boyle M, Martin GR (1990): Isolation of the mouse Hox-2.9 gene: Analysis of embryonic expression suggests that positional information along the anterior-posterior axis is specified by mesoderm. Development 110:589–607.

Fundele R, Norris ML, Barton SC, Reik W, Surani MA (1989): Systematic elimination of parthenogenetic cells in mouse chimeras. Development 106:29–35.

Gammeltoft S (1989): Insulin-like growth factors and insulin: gene expression, receptors and biological actions. In Martinez J (ed): Peptide Hormones as Prohormones. New York: Halsted Press, pp 176–210.

Gardner RL (1982): Investigation of cell lineage and differentiation in the extraembryonic endoderm of the mouse embryo. J Embryol Exp Morph. 68:175–198.

Gardner RL (1983): Origin and differentiation of extraembryonic tissues in the mouse. Int Rev Exp Pathol 24:63–133.

Gardner RL (1984): An in situ cell marker for clonal analysis of development of the extraembryonic endoderm in the mouse. J Embryol Exp Morphol 80:251–288.

Gardner RL (1985): Regeneration of endoderm from primitive ectoderm in the mouse embryo: fact or artifact? J Embryol Exp Morphol 88:303–326.

Gardner RL (1986): Cell mingling during mammalian embryogenesis. J Cell Sci Suppl 4:337–356.

Gardner RL, Beddington RSP (1988): Multi-lineage

stem cells in the mammalian embryo. J Cell Sci [Suppl] 10:11–27.

Gardner RL, Nichols J (1991): An investigation of the fate of cells transplanted orthotopically between morulae/nascent blastocysts in the mouse. Hum Reprod 6:25–35.

Gardner RL, Papaioannou VE (1975): Differentiation in the trophectoderm and inner cell mass. In Balls M, Wild AE (eds): The Early Development of Mammals. Cambridge: Cambridge University Press, pp 107–132.

Gardner RL, Papaioannou VE, Barton SC (1973): Origin of the ectoplacental cone and secondary giant cells in mouse blastocysts reconstituted from isolated trophoblast and inner cell mass. J Embryol Exp Morph, 30:561–572.

Gardner RL, Rossant J (1979): Investigation of the fate of 4.5 day post-coitum mouse inner cell mass cells by blastocyst injection. J Embryol Exp Morphol 52:141–152.

Gardner RL, Barton SC, Surani MAH (1990): Use of triple tissue blastocyst reconstitution to study the development of diploid parthenogenetic primitive ectoderm in combination with fertilization-derived trophectoderm and primitive endoderm. Genet Res Comb 56:209–222.

Gardner HG, Kaye PL, (1991): Insulin increases cell numbers and morphological development in mouse pre-implantation embryos in vitro. Reprod Fertil Dev 3:79–91.

Gatherer D, Ten Dijke P, Baird DT, Akhurst RJ (1990): Expression of TGFβ isoforms during first trimester human embryogenesis. Development 110:445–460.

Gaunt SJ (1988): Mouse homeobox gene transcripts occupy different but overlapping domains in embryonic germ layers and organs: A comparison of Hox-3.1 and Hox-1.5. Development 103:135–144.

Gaunt SJ, Singh PB (1990): Homeogene expression patterns and chromosomal imprinting. Trends Genet 6(7):208–212.

Gearing DP, Thut CJ, VandeBos T, Gimpel SD, Delaney PB, King J, Price V, Cosman D, Beckmann MP (1991): Leukemia inhibitory factor receptor is structurally related to the IL-6 signal transducer gp130. EMBO J 10:2839–2848.

Gehring WJ (1987): Homeoboxes in the study of development. Science 236(4806):1245–1252.

Geissler EN, Ryan MA, Housman DE (1988): The dominant-white spotting (W) locus of the mouse encodes the c-kit proto-oncogene. Cell 55:185–192.

Gerhart JC, Danilchik M, Doniach T, Roberts S, Browning B, Stewart R (1989): Cortical rotation of the Xenopus egg: Consequences for the anteroposterior pattern of embryonic dorsal development. Development 107:37–51.

Gerhart, Keller RE (1986): Region-specific cell activities in amphibian gastrulation. Annu Rev Cell Biol 2:201–229.

Ghulam H, Kalimi, Lo CW (1989): Gap junctional communication in the extraembryonic tissues of the gastrulating mouse embryo. J Cell Biol 109:3015–3026.

Gillespie LL, Paterno GD, Slack JM (1989): Analysis of competence: receptors for fibroblast growth factor in early Xenopus embryos. Development 106:203–208.

Gimlich RL, Gerhart JC (1984): Early cellular interactions promote embryonic axis formation in Xenopus laevis. Dev Biol 104:117–130.

Gonzalez AM, Buscaglia M, Ong M, Baird A (1990): Distribution of basic fibroblast growth factor in the 18-day rat fetus: localization in the basement membranes of diverse tissues. J Cell Biol 110(3):753–765.

Gospodarowicz D (1987): Isolation and characterization of acidic and basic fibroblast growth factor. Method Enzymol 147:106–119.

Gospodarowicz D (1990): Fibroblast growth factor. Chemical structure and biologic function. Clin Orthop Rel Res 257:231–248.

Gospodarowicz D, Ferrara N, Schweigerer L, Neufeld G (1987): Structural characterization and biological functions of fibroblast growth factor. Endocr Rev 8(2):95–114.

Gospodarowicz D, Neufeld G, Schweigerer L (1987b): Fibroblast growth factor: Structural and biological properties. J Cell Physiol [Suppl] 5:15–26.

Gough NM, Williams RL (1989): The pleiotropic actions of leukemia inhibitory factor. Cancer Cells 1:77–80.

Gough NM, Williams RL, Hilton DJ, Pease S, Wilson TA, Stahl J, Gearing DP, Nicola NA, Metcalf D (1989): LIF: A molecule with divergent actions on myeloid leukaemic cells and embryonic stem cells. Reprod Fertil Dev 1:281–288.

Graham A, Papalopulu N, Krumlauf R (1989): The murine and Drosophila homeobox gene complexes have common features of organization and expression. Cell 57:367–378.

Gray A, Tam AW, Dull TJ, Hayflick J, Pintar J, Cavanee WK, Koufos A, Ullrich A (1987): Tissue-specific and developmentally regulated transcription of the insulin-like growth factor 2 gene. DNA 6:283–295.

Graycar JL, Miller DA, Arrick BA, Lyons RM, Moses HL, Derynck R (1989): Human tranforming growth factor-β3: recombinant expression, purification and biological activities in comparison with transforming growth factors-β1 and β2. Mol Endocrinol 7:1977–1986.

Grover A, Andrews G, Adamson ED (1983): Role of laminin in epithelium formation by F9 aggregates. J Cell Biol 97:137–144.

Haig D, Graham C (1991): Genomic imprinting and the strange case of the insulin-like growth factor II receptor. Cell 64:1045–1046.

Hamburger V (1988): The Heritage of Experimental Embryology. New York: Oxford University Press.

Handyside AH (1978): Time of commitment of inside cells isolated from preimplantation mouse embryos. J Embryol Exp Morphol 45:37–53.

Handyside AH, Hunter S (1986): Cell division and death in the mouse blastocyst before implantation. Roux's Arch Dev Biol 195:519–526.

Heath JK, Paterno GD, Lindon AC, Edwards DR (1989): Expression of multiple heparin-binding growth factor species by murine embryonal carcinoma and embryonic stem cells. Development 107(1): 113–122.

Hebert JM, Basilico C, Goldfarb, Martin G (1990): Isolation of cDNAs encoding four mouse *Fgf* family members and characterization of their expression patterns during embryogenesis. Dev Biol 138:454–463.

Hebert JM, Boyle M, Martin G (1991): mRNA localization studies suggest that murine *FGF-5* plays a role in gastrulation. Development 11:407–415.

Herr W, Sturm RA, Clerc RG, Corcoran LM, Baltimore D, Sharp PA, Ingraham HA, Rosenfeld MG, Finney M, Ruvkun G (1988): The POU domain: A large conserved region in the mammalian *pit-1, oct-1, oct-2,* and *Caenorhabditis elegans unc-86* gene products. Genes Dev 2(12a):1513–1516.

Hilton DJ, Gough NM (1991): Leukemia inhibitory factor: A biological perspective. J Cell Biochem 46:21–26.

Hirayoshi K, Tsuru A, Yamashita M, Tomida M, Yamamoto-Yamaguchi Y, Yasukawa K, Hozumi M, Goeddel DV, Nagata K (1991): Both D factor/LIF and IL-6 inhibit the differentiation of mouse teratocarcinoma F9 cells. FEBS LETT 282(2):401–404.

Hogan BLM, Newman R (1984): A scanning electron microscope study of the extraembryonic endoderm of the 8th-day mouse embryo. Differentiation 26:138–143.

Hogan BLM, Taylor A, Adamson ED (1981): Cell interactions modulate embryonal carcinoma cell differentiation into parietal or visceral endoderm. Nature 291:235–237.

Hogan BLM, Tilly R (1981): Cell interactions and endoderm differentiation in cultured mouse embryos. J Embryol Exp Morphol 62:379–394.

Huang E, Nocka K, Beier DR, Chu TY, Buck J, Lahm HW, Wellner D, Leder P, Besmer P (1990a): The hematopoietic growth factor KL is encoded by the *Sl* locus and is the ligand of the *c-kit* receptor, the gene product of the *W* locus. Cell 63:225–233.

Huang JQ, Turbide C, Daniels E, Jothy S, Beauchemin N (1990b): Spatiotemporal expression of murine carcinoembryonic antigen (CEA) gene family members during mouse embryogenesis Development 110:573–588.

Hyafil F, Morello D, Babinet C, Jacob F (1981): Cell–cell interaction in early embryogenesis: A molecular approach to the role of calcium. Cell 26:447–454.

Jessell TM, Melton DA (1992): Diffusible factors in vertebrate embryonic induction. Cell 68:257–270.

Johnson MH, Maro B (1986): Time and space in the mouse early embryo: A cell biological approach to cell diversification. In Rossant J, Pedersen RA (eds): Experimental Approaches to Mammalian Embryonic Development. Cambridge: Cambridge University Press, pp 35–65.

Johnson MH, Ziomek CA (1983): Cell interactions influence the fate of mouse blastomeres undergoing the transition from the 16- to the 32-cell stage. Dev Biol 95:211–218.

Jones EA, Woodland HR (1989): Spatial aspects of neural induction in *Xenopus laevis.* Development 107(4): 785–791.

Keller R, Clark WH Jr, Griffin F (1991): Gastrulation: Movements, Patterns and Molecules. Bodega Marine Laboratory, Marine Science Series. New York: Plenum Press.

Kemler I, Schaffner W (1990): Octamer transcription factors and the cell type-specificity of immunoglobulin gene expression. FASEB J 4(5):1444–1449.

Kimelman D, Kirschner M (1987): Synergistic induction of mesoderm by FGF and TGF-β and the identification of an mRNA coding for FGF in the early *Xenopus* embryo. Cell 51:869–877.

Klagsburn M, D'Amore PA (1991): Regulators of angiogenesis. Annu Rev Physiol 53:217–239.

Kordula T, Rokita H, Koj A, Fiers W, Gauldie J, Baumann H (1991): Effects of interleukin-6 and leukemia inhibitory factor on the acute phase response and DNA synthesis in cultured rat hepatocytes. Lymphocyte Res 10:23–26.

Kuff EL, Fewell JW (1980): Induction of neural-like cells and acetylcholinesterase activity in cultures of F9 teratocarcinoma treated with retinoic acid and dibutyryl cyclic adenosine monophosphate. Devel Biol 77:103–115.

Lawson KA, Meneses JJ, Pedersen RA (1986): Cell fate and cell lineage in the endoderm of the presomite mouse embryo, studied with an intracellular tracer. Dev Biol 115:325–339.

Lawson KA, Meneses JJ, Pedersen RA (1991): Clonal analysis of epiblast fate during germ layer formation in the mouse embryo. Development 113:891–911.

Lawson KA, Pedersen RA (1987): Cell fate, morphogenetic movement and population kinetics of embryonic endoderm at the time of germ layer formation in the mouse. Development 101:627–652.

Lee JE, Pintar J, Efstratiadis A (1990): Pattern of the insulin-like growth factor II gene expression during early mouse embryogenesis. Development 110:151–159.

Lee PL, Johnson DE, Cousens LS, Fried VA, Williams LT (1989): Purification and complementary DNA cloning of a receptor for basic fibroblast growth factor. Science 245(4913):57–60.

Lenardo MJ, Staudt L, Robbins P, Kuang A, Mulligan RC; Baltimore D (1989): Repression of the IgH enhancer in teratocarcinoma cells associated with a novel octamer factor. Science 243(4890):544–546.

Lindenbaum MH, Grosveld F (1990): An in vitro globin gene switching model based on differentiated embryonic stem cells. Genes Dev 4:2075–2085.

Ling N, Ying S-Y, Ueno N, Shimasaki S, Esch F, Hotta M, Guillemin R (1986): Pituitary FSH is released by a heterodimer of the β-subunits from the two forms of inhibin. Nature 321:779–782.

Linney E (1992): Retinoic acid receptors: Transcription factors modulating gene regulation, development and differentiation. Curr Top Dev Biol 27 (in press).

Lyons KM, Pelton RW, Hogan BLM (1990): Organogenesis and pattern formation in the mouse: RNA distribution patterns suggest a role for *Bone Morphogenetic Protein-2A (BMP-2A)*. Development 109: 833–844.

Lyons RM, Moses HL (1990): Transforming growth factors and the regulation of cell proliferation. Eur J Biochem 187(3):467–473.

Maekawa T, Metcalf D (1989): Clonal suppression of HL60 and U937 cells by recombinant human leukemia inhibitory factor in combination with GM-CSF or G-CSF. Leukemia 3(4):270–276.

Magnuson T (1986): Mutations and chromosomal abnormalities: How are they useful for studying genetic control of early mammalian development? In Rossant J, Pedersen RA, (eds): Experimental Approaches to Mammalian Embryonic Development. New York: Cambridge University Press, pp 437–474.

Mann JR, Gadi I, Harbison ML, Abbondanzo SJ, Stewart CL (1990): Androgenetic mouse embryonic stem cells are pluripotent and cause skeletal defects in chimeras: Implications for genetic imprinting. Cell 62:251–260.

Mann JR, Stewart CL (1991): Development to term of mouse androgenetic aggregation chimeras. Development 113:1325–1333.

Mansukhani A, Moscatelli D, Talarico D, Levytska V, Basilico C (1990): A murine fibroblast growth factor (FGF) receptor expressed in CHO cells is activated by basic FGF and Kaposi FGF. Proc Natl Acad Sci USA (11):4378–4382.

Martin GR (1980): Teratocarcinomas and mammalian embryogenesis. Science 209:676–678.

Martin GR (1981): Isolation of a pluripotent cell line from early mouse embryos cultured in medium conditioned by teratocarcinoma stem cells. Proc Natl Acad Sci USA 78:7634–7638.

Mason AJ, Hayflick JS, Ling N, Esch F, Ueno N, Ying SY, Guilleman R, Niall H, Seeburg PH (1985): Complementary DNA sequences of ovarian follicular fluid inhibin show percursor structure homology with transforming growth factor-β. Nature 318:659–663.

McMahon AP, Bradley A (1990): The *Wnt-1 (int-1)* proto-oncogene is required for development of a large region of the mouse brain. Cell 62(6): 1073–1085.

McNeill H, Ozawa M, Kemler R, Nelson WJ (1990): Novel function of the cell adhesion molecule uvomorulin as an inducer of cell surface polarity. Cell 62:309–316.

Mercola M, Deininger PL, Shamah SM, Porter J, Wang CY, Stiles CD (1990): Dominant-negative mutants of a platelet-derived growth factor gene. Genes Dev 4:2333–2341.

Mercola M, Stiles CD (1988): Growth factor superfamilies and mammalian embryogenesis. Development 102:451–460.

Merwin JR, Roberts A, Kondaiah P, Tucker A, Madri J (1991): Vascular cell responses to TGF-beta 3 mimic those of TGF-beta 1 in vitro. Growth Factors 5(2):149–158.

Metcalf D (1990): Haemopoietic growth factors (1988): Med J Aust 148(10):516–519.

Miller DA, Pelton RW, Derynck R, Moses HL. (1990a): Transforming growth factor-beta. A family of growth regulatory peptides. Ann NY Acad Sci 593:208–217.

Miller DA, Pelton RW, Derynck R, Moses HL (1990b): Transforming growth factor-beta. A family of growth regulatory peptides. Ann NY Acad Sci 593:208–217.

Mitrani E, Gruenbaum Y, Shohat H, Ziv T (1990a): Fibroblast growth factor during mesoderm induction in the early chick embryo. Development 109:387–93.

Mitrani E, Ziv T, Thomsen J, Shimoni Y, Melton DA, Bril A (1990b): Activin can induce the formation of axial structures and is expressed in the hypoblast of the chick. Cell 63:495–501.

Miura Y, Wilt FH (1969): Tissue interaction and the formation of the first erythroblasts of the chick embryo. Dev Biol 19:201–211.

Monuki ES, Kuhn R, Weinmaster G, Trapp BD, Lemke G (1990): Expression and activity of the POU transcription factor SCIP. Science 249(4974):1300–1303.

Monuki ES, Weinmaster G, Kuhn R, Lemke G (1989): SCIP: A glial POU domain gene regulated by cyclic AMP. Neuron 2:783–793.

Moore, R, Casey G, Brookes S, Dixon M, Peters G, Dickson C (1986): Sequence, topography and protein coding potential of mouse int-2: A putative oncogene activated by mouse mammary tumour virus. EMBO J. 5:919–924.

Mori M, Yamaguchi K, Abe K (1989): Purification of a lipoprotein lipase-inhibiting protein produced by a melanoma cell line associated with cancer cachexia. Biochem Biophys Res Commun 160:1085–1092.

Morrison RS, Keating RF, Moskal JR (1988): Basic fibroblast growth factor and epidermal growth factor exert differential trophic effects on CNS neurons. J Neurol Res 21:71–79.

Motro B, Van Der Kooy D, Rossant J, Reith A, Bernstein A (1991): Contiguous patterns of c-kit and steel expression: Analysis of mutations at the W and Sl loci. Development 113:1207–1221.

Mugrauer G, Alt FW, Ekblom P (1988): N-myc protooncogene expression during organogenesis in the developing mouse as revealed by in situ hybridization. J Cell Biol 107:1325–1335.

Murray R, Lee F, Chiu CP (1990): The genes for leukemia inhibitory factor and interleukin-6 are expressed in mouse blastocysts prior to the onset of hemopoiesis. Mol Cell Biol 10(9):4953–4956.

Nadijcka M, Hillman N (1974): Ultrastructural studies of the mouse blastocyst substages. J Embryol Exp Morphol 32:675–695.

Nagy A, Gocza E, Diaz EM, Prideaux VR, Ivanyi E, Markkula M, Rossant J (1990): Embryonic stem cells alone are able to support fetal development in the mouse. Development 110(3):815–821.

Nagy A, Sass M, Markkula M (1989): Systematic nonuniform distribution of parthenogenetic cells in adult mouse chimaeras. Development 106:321–324.

Nichols J, Gardener RL (1984): Heterogenous differentiation of external cells in individual isolated early mouse inner cell masses in culture. J Embryol Exp Morphol 80:225–240.

Nieuwkoop PD (1969): The formation of mesoderm in early urodelean amphibians. I. Induction by the endoderm. Rouxs Arch Dev Biol 162:341–373.

Nieuwkoop PD (1973): The "organization center" of the amphibian embryo: Its origin, spatial organization, and morphogenetic action. Adv Morphol 10:1–39.

Nilsen-Hamilton M (1990): Growth factors and development. Curr Top Dev Biol 24 (in press).

Nusse R (1988): The int genes in mammary tumorigenesis and in normal development. Trends Genet 4(10):291–295.

O'Gorman S, Fox DT, Wahl GM (1991): Recombinase-mediated gene activation and site-specific integration in mammalian cells. Science 251:1351–1355.

Odenwald WF, Taylor CF, Palmer-Hill FJ, Friedrich V Jr, Tani M, Lazzarini RA (1987): Expression of a homeo domain protein in noncontact-inhibited cultured cells and postmitotic neurons. Genes and Dev 1:482–496.

Ogi KI (1967): Determination in the development of the amphibian embryo. Sci Rep Tohuku Univ Ser IV (Biol) 33:239–247.

Okamoto K, Okazawa H, Okuda A, Sakai M, Muramatsu M, Hamada H (1990): A novel octamer binding transcription factor is differentially expressed in mouse embryonic cells. Cell 60(3):461–472.

Owen AJ, Pantazis, Antoniades HN (1984): Simian virus transformed cells secrete a mitogen identical to platelet derived growth factor. Science 225:54–56.

Papaioannou VE (1982): Lineage analysis of inner cell mass and trophectoderm using microsurgically reconstituted mouse blastocysts. J Embryol Exp Morphol 68: 199–209.

Papaioannou VE, Rossant J (1983): Effects of the embryonic environment on proliferation and differentiation of embryonal carcinoma cells. Cancer Surveys 2:165–183.

Papalopulu N, Hunt P, Wilkinson D, Graham A, Krumlauf R (1990): Hox-2 homeobox genes and retinoic acid: Potential roles in patterning the vertebrate nervous system. Neural Regen Res 60:291–308.

Papalopulu N, Lovell-Badge R, Krumlauf R (1991): The expression of murine Hox-2 genes is dependent on the differentiation pathway and displays a collinear sensitivity to retinoic acid in F9 cells and Xenopus embryos. Nucleic Acids Res 19:5497–5506.

Paterno GD, Gillespie LL (1989): Fibroblast growth factor and transforming growth factor beta in early embryonic development. Prog Growth Factor Res 1(2):79–88.

Paterno GD, Gillespie LL, Dixon MS, Slack JMW, Heath JK (1989): Mesoderm-inducing properties of INT-2 and kFGF: two oncogene-encoded growth factors related to FGF. Development 106:79–83.

Pedersen RA (1986): Potency, lineage, and allocation in preimplantation mouse embryos. In Rossant J, Pedersen RA (eds): Experimental Approaches to Mammalian Embryonic Development. Cambridge: Cambridge University Press, pp 3–33.

Pedersen RA, Spindle AI (1980): Role of the blastocoel microenvironment in early mouse embryo differentiation. Nature 284:550–552.

Pedersen RA, Wu K, Balakier H (1986): Origin of the

inner cell mass in mouse embryos: Cell lineage analysis by microinjection. Dev Biol 117:581–595.

Pedersen RA (1988): Early mammalian embryogenesis. In Knobil E, Neill JD, Markert CL (eds): The Physiology of Reproduction, Vol 1. New York: Raven Press, pp 187–230.

Pedersen RA, Sturm KS, Rappolee DA, Werb Z (1992): Effects of imprinting on early development of mouse embryos. In Proceedings of the Serono Symposium on Preimplantation Embryo Development, Boston, MA, August 15–18, 1991 (in press).

Pelton RW, Dickinson ME, Moses HL, Hogan BLM (1990): In situ hybridization analysis of TGFβ3 RNA expression during mouse development: Comparative studies with TGFβ1 and β2. Development 110:609–620.

Pelton RW, Saxena B, Jones M, Moses HL and Gold LI (1991): Immunohistochemical localization of TGFβ1, and TGFβ2, and TGFβ3 in the mouse embryo: Expression patterns suggest multiple roles during embryonic development. J Cell Biol 115:1091–1105.

Pera MF, Cooper S, Mills J, Parrington JM (1989): Isolation and characterization of a multipotent clone of human embryonal carcinoma cells. Differentiation 42:10–23.

Perry JS (1981): The mammalian fetal membranes. J Reprod Fertil 62:321–335.

Pierce GB, Arechaga J, Muro C, Wells RS (1988): Differentiation of ICM cells into trophectoderm. Am J Pathol 132(2):356–364.

Pratt HPM (1989): Marking time and making space: Chronology and topography in the early mouse embryo. Int Rev Cytol 117: 99–130.

Rao, LV, Wikarczak ML, Heyner S (1990): Functional roles of insulin and insulin-like growth factors in preimplantation mouse embryo development. In Vitro Cell Devl Biol 26:1043–1048.

Rappolee DA, Wong A, Mark D, Werb Z (1989): Novel method for studying mRNA phenotypes in single or small numbers of cells. J Cell Biochem 39:1–11.

Rappolee DA, Sturm KS, Behrendsten O, Schultz GA, Pedersen RA, Werb Z (1992): Insulin-like growth factor II acts through an endogenous growth pathway regulated by imprinting in early mouse embryos. Genes Dev 6:939–952.

Rathjen PD, Nichols J, Toth S, Edwards DR, Heath JK, Smith AG (1990): Developmentally programmed induction of differentiation inhibiting activity and the control of stem cell populations. Genes Dev 4(12B):2308–2318.

Rathjen PD, Toth S, Willis A, Heath JK, Smith AG (1990): Differentiation inhibiting activity is produced in matrix-associated and diffusible forms that are generated by alternate promoter usage. Cell 62(6):1105–1114.

Regenstreif LJ, Rossant J (1989): Expression of the c-fms proto-oncogene and of the cytokine, CSF-1, during mouse embryogenesis. Dev Biol 133(1):284–294.

Reid L (1990): From gradients to axes, from morphogenesis to differentiation. Cell 63(5):875–882.

Reid LR, Lowe C, Cornish J, Skinner SJ, Hilton DJ, Willson TA, Gearing DP, Martin TJ (1990): Leukemia inhibitory factor: A novel bone-active cytokine. Endocrinology 126(3):1416–1420.

Rickords LF, White KL (1982): Electrofusion-induced intracellular Ca^{2+} flux and its effect on murine oocyte activation. Mol Reprod Dev 31:152–159.

Risau W, Sariola H, Zerwes HG, Sasse J, Ekblom P, Kemler R, Doetschman T (1988): Vasculogenesis and angiogenesis in embryonic-stem-cell-derived embryoid bodies. Development 102:471–478.

Robbins J, Gulick J, Sanchez A, Howles P, Doetschman T (1990): Mouse embryonic stem cells express the cardiac myosin heavy chain genes during development in vitro. J Biol Chem 265:11905–11909.

Robertson EJ (1991): Using embryonic stem cells to introduce mutations into the mouse germ line. Biol Reprod 44:238–245.

Roelink H, Wagenar E, da Silva SL, Nusse R (1990): Wnt-3, a gene activated by proviral insertion in mouse mammary tumors, is homologous to int-1/Wnt-1 and is normally expressed in mouse embryos and adult brain. Proc Natl Acad Sci USA 87:4519–4523.

Rosa F, Roberts AB, Danielpour D, Dart LL, Sporn MB, Dawid IB (1988): Mesoderm induction in amphibians: the role of TGF-beta 2-like factors. Science 239:783–785.

Rosner MH, De Santo RJ, Arnheiter H, Staudt LM (1991): Oct-3 is a maternal factor required for the first mouse embryonic division. Cell 64:1103–1110.

Rossant J (1975b): Investigation of the determinative state of the mouse inner cell mass. II. The fate of isolated inner cell masses transferred to the oviduct. J Embryol Exp Morphol 33:991–1001.

Rossant J (1984): Somatic cell lineages in mammalian chimeras. In LeDouarin, N and McLaren A (eds): Chimeras in Developmental Biology. London: Academic Press, pp 89–109.

Rossant J (1986): Development of extraembryonic cell lineages in the mouse embryo. In Rossant J Pedersen RA (eds): Experimental Approaches to Mammalian Embryonic Development. Cambridge: Cambridge University Press, pp 97–120.

Rossant J, Croy BA (1985): Genetic identification of tissue of origin of cellular populations within the mouse placenta. J Embryol Exp Morphol 86:177–189.

Rossant J, Hopkins N (1992): Of fin and fur: Mutational analysis of vertebrate embryonic development. Genes Dev 6:1–13.

Rossant J, Joyner AL (1989): Towards a molecular–genetic analysis of mammalian development. Trends Genet 5(8):277–283.

Rossant J, Tamura-Lis W (1979): The possible dual origin of the ectoderm of the chorion in the mouse embryo. Dev Biol 70: 249–254.

Rossant J, Tamura-Lis W (1981): Effect of culture conditions on diploid to giant-cell transformation in postimplantation mouse trophoblast. J Embryol Exp Morphol 62:217–227.

Rossant J (1987): Cell lineage analysis in mammalian embryogenesis. Curr Top Dev Biol 23: 115–146.

Ruberte, E, Dolle P, Chambon P, Morris-Kay G (1991): Retinoic acid receptors and cellular retinoid binding proteins. Development 111:45–60.

Ruddon RW (1987): Cancer Biology. New York: Oxford, pp 431–436.

Ruiz i Altaba A, Melton DA (1989): Interaction between peptide growth factors and homeobox genes in the establishment of antero-posterior polarity in frog embryos. Nature 341(6237):33–38.

Sanchez A, Jones WK, Gulick J, Doetschman T, Robbins J (1991): Myosin heavy chain gene expression in mouse embryoid bodies: An in vitro developmental study. J Biol Chem 266:22419–22426.

Sandberg M, Vuorio T, Hirvonen H, Alitalo K, Vuorio E (1988): Enhanced expression of TGF-β and c-fos mRNAs in the growth plates of developing human long bones. Development 102:461–470.

Schmid P, Cox D, Bilbe G, Maier R, McMaster GK (1991): Differential expression of TGF β1, β2 and β3 genes during mouse embryogenesis. Development 111:117–130.

Scholer HR, Hatzopoulos AK, Balling R, Suzuki N, Gruss P (1989): A family of octamer-specific proteins present during mouse embryogenesis: evidence for germline-specific expression of an Oct factor. EMBO J 8:2543–2550.

Scholer HR (1991): Octamania: The POU factors in murine development. Trends Genet 10:323–329.

Schubert D, Kimura H (1991): Substratum-growth factor collaborations are required for the mitogenic activities of activin and FGF on embryonal carcinoma cells. J Cell Biol 114:841–846.

Schughart K, Kappen C, Ruddle FH (1988): Mammalian homeobox-containing genes: genome organization, structure, expression and evolution. Br J Cancer 58:9–13.

Seed J, Olwin BB, Hauschka SD (1988): Fibroblast growth factor levels in the whole embryo and limb bud during chick development. Dev Biol 128(1):50–57.

Shackleford GM, Varmus HE (1987): Expression of the proto-oncogene Int-1 is restricted to postmeiotic male germ cells and the neural tube of mid-gestational embryos. Cell 50:89–95.

Sham MH, Nonchev S, Whiting J, Papalopulu N, Marshall H, Hunt P, Muchamore I, Cook M, Krumlauf R (1991): Hox-2: Gene regulation and segmental patterning in the vertebrate head. In Gerhart J (ed): Cell–Cell Interactions in Early Development. New York: Wiley-Liss, pp 129–143.

Shirayoshi Y, Okada TS, Takeichi M (1983): The calcium dependent cell–cell adhesion system regulates inner cell mass formation and cell surface polarization in early mouse development. Cell 35:631–638.

Silver LM, Martin GR, Strickland S (1983): Teratocarcinoma Stem Cells, Vol 10. Cold Spring Harbor, NY: Cold Spring Harbor Laboratory.

Simeone A, Acampora D, Arcioni L, Andrews PW, Boncinelli E, Mavilio F (1990): Sequential activation of HOX2 homeobox genes by retinoic acid in human embryonal carcinoma cells. Nature 346(6286):763–766.

Slack JM (1990): Growth factors as inducing agents in early Xenopus development. J Cell Sci[Suppl] 13:119–30.

Slack JM (1991): From Egg to Embryo, 2nd ed. Cambridge: Cambridge University Press.

Slack JMW, Tannahill D (1992): Mechanism of anteroposterior axis specification in vertebrates. Development 114:285–302.

Slack JMW, Darlington BG, Heath JK, Godsave SF (1987): Mesoderm induction in early Xenopus embryos by heparin-binding growth factors. Nature 326:197–200.

Smith LJ (1985): Embryonic axis orientation in the mouse and its correlation with blastocyst relationships to the uterus. II. Relationships from 4 1/2 to 9 1/2 days. J Embryol Exp Morphol 89:15–35.

Smith R, Peters G, Dickson C (1988): Multiple RNAs expressed from the int-2 gene in mouse embryonal carcinoma cell lines encode a protein with homology to fibroblast growth factors. EMBO J 7:1013–1022.

Smith JC, Price BMJ, Van Nimmen KV, Huylebroeck D (1990): Identification of a potent Xenopus mesoderm-inducing factor as a homologue of activin A. Nature 345:729–731.

Smith WC, Harland RM (1991): Injected Xwnt-8 RNA acts early in Xenopus embryos to promote formation of a vegetal dorsalizing center. Cell 67:753–765.

Snell GD, Stevens LC (1966): Early embryology. In Green EL (ed): Biology of the Laboratory Mouse, 2nd ed. New York: McGraw Hill, pp 205–245.

Snow MHL (1977): Gastrulation in the mouse: growth and regionalization of the epiblast. J Embryol Exp Morphol 42:293–303.

Sokol S, Christian JL, Moon RT, Melton DA (1991): Injected *Wnt* RNA induces a complete body axis in *Xenopus* embryos. Cell 67(4):741–752.

Sokol S, Melton D (1991): Pre-existent pattern in *Xenopus* animal pole cells revealed by induction with activin. Nature 351(6325):409–411.

Sokol S, Wong GG, Melton DA (1990): A mouse macrophage factor induces head structures and organizes a body axis in *Xenopus*. Science 249(4968):561–564.

Solter D (1988): Differential imprinting and expression of maternal and paternal genomes. Annu Rev Genet 22:127–146.

Spemann H, Mangold H (1924): Induction of embryonic primordia by implantation of organizers from different species. Rouxs Arch Dev Biol 100: 555–638.

Spemann H (1938): Embryonic Development and Introduction. New Haven, CT: Yale University Press.

Stern CD, Canning DR (1990): Origin of cells giving rise to mesoderm and endoderm in chick embryo. Nature 343:273–275.

Strickland S, Mahdavi V (1978): The induction of differentiation in teratocarcinoma stem cells by retinoic acid. Cell 15:393–403.

Strickland S, Smith KK, Marotti KR (1980): Hormonal induction of differentiation in teratocarcinoma stem cells: generation of parietal endoderm by retinoic acid and dibutyryl cAMP. Cell 21:347–355.

Strickland S (1983): In Silver LM, Martin GR, Strickland S (eds): Teratocarcinoma Stem Cells, Vol 10. Cold Spring Harbor, NY: Cold Spring Harbor Laboratory.

Studzinski GP (1989): Oncogenes, growth and the cell cycle. Cell Tissue Kinet 22:405–424.

Sturm, KS, Flannery ML, Pedersen RA (1992): Abnormal development of embryonic and extraembryonic cell lineages in parthenogenetic mouse embryos. Submitted.

Surani MAH, Barton SC, Burling A (1980): Differentiation of 2-cell and 8-cell mouse embryos arrested by cytoskeletal inhibitors. Exp Cell Res 125:275–286.

Surani MAH (1986): Evidences and consequences of differences between maternal and paternal genomes during embryogenesis in the mouse. In Rossant J, Pedersen RA (eds): Experimental Approaches to Mammalian Embryonic Development. New York: Cambridge University Press, pp 401–435.

Surani MA, Barton SC, Howlett SK, Norris ML (1988): Influence of chromosomal determinants on development of androgenetic and parthenogenetic cells. Development 103:171–178.

Sutherland AE, Speed TP, Calarco PG (1990): Inner cell allocation in the mouse morula: the role of

oriented division during fourth cleavage. Dev Biol 137:13–25.

Tam PPL, Beddington RSP (1987): The formation of mesodermal tissues in the mouse embryo during gastrulation and early organogenesis. Development 99:109–126.

Tam, PPL (1989): Regionalisation of the mouse embryonic ectoderm: Allocation of prospective ectodermal tissues during gastrulation. Development 107:55–67.

Tarkowski AK, Wroblewska J (1967): Development of blastomeres of mouse eggs isolated at the 4- and 8-cell stage. J Embryol Exp Morphol 18:155–180.

Telford NA, Hogan A, Franz CR, Schultz GA (1990): Expression of genes for insulin and insulin-like growth factors in early postimplantation mouse embryos and embryonal carcinoma cells. Mol Reprod Dev 27:81–92.

ten Dijke P, Hansen P, Iwata KK, Pieler C, Foulkes JG (1988): Identification of another member of the transforming growth factor type beta gene family. Proc Natl Acad Sci USA 85(13):4715–4719.

Thomas KR, Capecchi MR (1990): Targeted disruption of the murine *int-1* proto-oncogene resulting in severe abnormalities in midbrain and cerebellar development. Nature 346:847–850.

Thomas KR, Musci TS, Neumann PE, Capecchi MR (1991): Swaying is a mutant allele of the proto-oncogene *Wnt-1*. Cell 67:969–976.

Thomsen G, Woolf T, Whitman M, Sokol S, Vaughan J, Vale W, Melton DA (1990): Activins are expressed early in *Xenopus* embryogenesis and can induce axial mesoderm and anterior structures. Cell 63:485–493.

Thomson JA, Solter D (1989): Chimeras between parthenogenetic or androgenetic blastomeres and normal embryos: Allocation to the inner cell mass and trophectoderm. Dev Biol 131:580–583.

Tomida M, Yamamoto-Yamaguchi Y, Hozumi M, Holmes W, Lowe DG, Goeddel DV (1990): Inhibition of development of Na(+)-dependent hexose transport in renal epithelial LLC-PK1 cells by differentiation-stimulating factor for myeloid leukemic cells/leukemia inhibitory factor. FEBS Lett 268(1):261–264.

Travali S, Koniecki J, Petralia S, Baserga R (1990): Oncogenes in growth and development. FASEB J 4:3209–3214.

Vakaet L (1984): Early development of birds. In LeDouarin N, McLaren A (eds): Chimaeras in Developmental Biology. New York: Academic Press, pp 71–88.

Vakaet L (1985): Morphogenetic movements and fate maps in the avian blastoderm. In Edelman GM (ed): Molecular Determinants of Animal Form. New York: Alan R Liss, pp 99–109.

Vale W, Rivier J, Vaughan J, McClintock R, Corrigan

A, Wilson W, Karr D, Spiess J (1986): Purification and characterization of an FSH releasing protein from porcine ovarian follicular fluid. Nature 321:776–779.

van den Eijnden-van Raaij AJM, van Zoelent EJJ, van Nimmen K, Koster CH, Snoel GT, Durston AJ, Huylebroek D (1990): Activin-like factor from a *Xenopus laevis* cell line responsible for mesoderm induction. Nature 345:732–734.

Varmuza S, Prideaux V, Kothary R, Rossant J (1988): Polytene chromosomes in mouse trophoblast giant cells. Development 102:127–134.

Vestweber D, Gossler A, Boller K, Kemler R (1987): Expression and distribution of cell adhesion molecule uvomorulin in mouse preimplantation embryos. Dev Biol 124:451.

Wanaka A, Milbrandt J, Johnson EM Jr (1991): Expression of FGF receptor gene in rat development. Development 111:455–468.

Wang R, Clark R, Bautch VL (1992): Embryonic stem cell-derived cystic embryoid bodies form vascular channels: An *in vitro* model of blood vessel development. Development 114:303–316.

Whitman M, Melton DA (1989): Growth factors in early embryogenesis. Annu Rev Cell Biol 5:93–117.

Wiles MV (1988): Isolation of differentially expressed human cDNA clones: similarities between mouse and human embryonal carcinoma cell differentiation. Development 104:403–413.

Wiles MV, Keller G (1991): Multiple hematopoietic lineages develop from embryonic stem (ES) cells in culture. Development 111:259–267.

Wiley LM (1978): Apparent trophoblast giant cell production *in vitro* by core cells isolated from cultured mouse inner cell masses. J Exp Zool 206:13–16.

Wilkinson DG, Bailes JA, McMahon AP (1987): Expression of the proto-oncogene *int-1* is restricted to specific neural cells in the developing mouse embryo. Cell 50:79–88.

Wilkinson DG, Bailes JA, Champion JE, McMahon AP (1987): A molecular analysis of mouse development from 8 to 10 days *post coitum* detects changes only in embryonic globin expression. Development 99(4):493–500.

Wilkinson DG, Bhatt S, McMahon AP (1989): Expression pattern of the FGF-related proto-oncogene *int-2* suggests multiple roles in fetal development. Development 105:131–136.

Wilkinson DG, Green J (1990): *In situ* hybridization and the three-dimensional reconstruction of serial selections. In Copp AJ, Cockroft DL (eds): Postimplantation Mammalian Embryos. London: IRL Press, pp 155–171.

Wilkinson DG, Peters G, Dickson C, McMahon AP (1988): Expression of the FGF-related proto-oncogene *int-2* during gastrulation and neurulation in the mouse. EMBO J 7: 691–695.

Williams BS, Biggers JD (1990): Polar trophoblast (Rauber's layer) of the rabbit blastocyst. Anat Rec 227:211–222.

Williams RL, Hilton DJ, Pease S, Willson TA, Stewart CL, Gearing DP, Wagner EF, Metcalf D, Nicola NA, Gough NM (1988): Myeloid leukaemia inhibitory factor maintains the developmental potential of embryonic stem cells. Nature 336(6200):684–687.

Wilt FH (1964): Erythropoiesis in the chick embryo: the role of endoderm. Science 147:1588–1590.

Winkel GK, Pedersen RA (1988): Fate of the inner cell mass in mouse embryos as studied by microinjection of lineage tracers. Dev Biol 127:143–156.

Yamamori T, Fukada K, Aebersold R, Korsching S, Fann MJ, Patterson PH (1989): The cholinergic neuronal differentiation factor from heart cells is identical to leukemia inhibitory factor. Science 46.(4936):1412–1416.

Yu J, Shao L, Lemas V, Yu AL, Vaughan J, Rivier J, Vale W (1987): Importance of FSH-releasing protein and inhibin in erythrodifferentiation. Nature 330:765–767.

Zhan X, Bates B, Hu XG, Goldfarb M (1988): The human *FGF-5* oncogene encodes a novel protein related to fibroblast growth factors. Mol Cell Biol 8(8):3487–3495.

Ziomek C, Chatot CL, Manes C (1990): Polarization of blastomeres in the cleaving rabbit embryo. J Exp Zool 256:84–91.

Ziomek CA, Johnson MA (1982): The roles of phenotype and position in guiding the fate of 16-cell mouse blastomeres. Dev Biol 91:440–447.

Zsebo KM, Williams DA, Geissler EN, Broudy VC, Martin FH, Atkins HL, Hsu RY, Birkett NC, Okino KH, Murdock DC (1990): Stem cell factor is encoded at the *Sl* locus of the mouse and is the ligand for the *c-kit* tyrosine kinase receptor. Cell 63:213–224.

ABOUT THE AUTHORS

JEAN J. LATIMER is a postdoctoral fellow in the laboratory of Roger Pedersen of the Laboratory of Radiobiology and Environmental Health at the University of California, San Francisco. Dr. Latimer received her B.A. in Cell Biology from Cornell University in 1982, and her Ph.D. in Molecular and Cellular Biology from the State University of New York at Buffalo for her work with Heinz Baumann at the Roswell Park Cancer Institute. Her thesis involved characterization of the molecular evolution of the α_1-antitrypsin gene in the mouse genus, particularly the unusually abundant renal expression of this gene in the Asian wild mouse species *Mus caroli*. Her current work in the field of developmental biology is focused on the establishment and utilization of embryonic stem (ES) cell lines as models of early mouse development. Besides establishing such cell lines herself, Dr. Latimer has used ES cells to define molecular markers of early differentiation and to isolate pure populations of differentiated cell types via fluorescence-activated cell sorting. She is currently using these cells to create mouse models of human DNA repair-deficiency syndromes using the technique of targeted homologous recombination, beginning with *ERCC*1, a gene implicated in repair of lesions induced in DNA by ultraviolet light.

ROGER A. PEDERSEN is Professor of Radiology and Anatomy at the University of California at San Francisco, where he teaches developmental biology and mammalian embryology. He received his B.A. degree from Stanford University in 1965 and his Ph.D. under Clement Markert at Yale University in 1970. He pursued postdoctoral research with John Biggers at the Johns Hopkins University, beginning his studies using the mouse embryo as a model system for mammalian embryonic development. Since 1971, Dr. Pedersen has been a member of the Laboratory of Radiobiology and Environmental Health at the University of California. His research has involved mechanisms of embryotoxicity and repair in early mammalian development, analysis of cell potency and fate in pre- and postimplantation mouse embryos, and mechanisms of genomic imprinting in mammals. He has written numerous reviews on early mouse development, and co-produced an instrumental videotape on the use of mice in transgenic research. Since 1991 he has served as Series Editor of *Current Topics in Developmental Biology*. His research papers have appeared in *Development, Developmental Biology, Genes and Development, Nature, Science*, and other journals.

Genes in Mammalian Reproduction: 173–184
© 1993 Wiley-Liss, Inc.

Imprinting and Methylation

Christine Pourcel

I. DEMONSTRATION OF THE PARENTAL EFFECT

A. The Role of Genomic Imprinting in Mouse Development

The observations that some mutations are phenotypically expressed only when transmitted by one parent suggest that the paternal and maternal contributions to mouse development are not equivalent. The *hairpintail* (*Thp*) mutation described by Johnson in 1975 deletes the *Tme* locus on chromosome 17. It is lethal in heterozygotes that receive the mutated chromosome from their mother but is viable in those receiving it from the father. An imprinted gene was recently identified at this locus (see Identification of Imprinted Genes, below). Other evidence has come from observations in F_1 mice produced by reciprocal cross between inbred strains [reviewed by Shire, 1989]. The DDK mice provide one striking example of such an effect with variable phenotypic expression, depending on the strain [Renard and Babinet, 1986].

More direct evidence of the complementarity of the parental chromosomes in the mouse came from pronuclear transplantation experiments conducted by McGrath and Solter [1984] and by Surani et al. [1984] demonstrating that neither androgenones (two male pronuclei) nor gynogenones (two female pronuclei) were able to develop to term. In addition, Surani et al.

[1987] reported that chimerae made by aggregation of parthenogenetic and androgenetic cells were not viable, thus showing that the maternal and paternal chromosomes must be present in the same cell.

The contribution of the parental chromosomes to different stages in development was demonstrated using chimerae made from normal and parthenogenetic or androgenetic embryos. During early development, paternal genes contribute to the formation of trophectoderm-derived tissues, whereas maternal genes contribute to the development of the embryo and some extraembryonic tissues [reviewed by Surani, 1986; Solter, 1988]. Postnatal development of parthenogenetic chimerae has been observed, but growth was reduced in proportion to the contribution of parthenogenetic cells to the organs [Paldi et al., 1989]. Androgenetic embryonic chimerae usually do not develop to term. However, recent work showed that chimerae can be formed using androgenic embryonic stem cells and that these cells contribute to many tissues. When the stem cells are injected into adult mice, they form tumors composed of striated muscle, and in chimerae skeletal abnormalities are observed postnatally. This suggests that a specific class of gene involved in bone and muscle formation might be imprinted [Mann et al., 1990].

B. Evidence of Imprinting in Humans

In humans difference between maternal and paternal chromosomes come from naturally occurring malformations such as the hydatidiform mole, a placental malformation without embryonic tissues found during pregnancy. The mole results from the presence of two paternal genomes and the absence of a maternal genome. Other evidence comes from the observation of several genetic disorders in which the phenotypic expression of the disease depends on the parental origin of the mutation. For example, Angelman syndrome is associated with the deletion of band q11–q13 on maternal chromosome 15, whereas in the Prader-Willi syndrome the same deletion of the paternal chromosome is found. Parental origin effects on penetrance and expressivity of dominant disorders such as Huntington have also been reported. Finally, there is a preferential loss of maternal chromosomes in embryonal tumors such as Wilms tumor and osteosarcoma, suggesting that antitumor genes may be differentially imprinted on the two parental alleles [Reviewed by Reik, 1989; Hall, 1990].

C. Only Part of the Chromosomes Is Imprinted

The use of robertsonian translocations and reciprocal translocations in the mouse has allowed the definition of chromosomes or chromosome regions that are involved in the parental effect. The presence of two chromosomes (or chromosome regions) from one parent in the absence of the corresponding chromosome from the other parent can be lethal or lead to abnormalities expressed at different times during development. The results have shown that only certain regions of certain chromosomes are imprinted (Fig. 1) [Cattanach and Beechey, 1990a,b]. Interestingly the reciprocal duplication/deficiency of some regions produce opposite effects on development. Maternal disomy/paternal nullisomy

of the proximal region of chromosome 11 caused lower birth weight and reduced growth rate. The paternal disomy/maternal nullisomy results in the opposite phenotype [Cattanach and Kirk, 1985].

It is possible to predict homologous areas of the human and mouse genome. Some genes suspected of demonstrating imprinting in humans have already been shown to lie within these homologous areas. However, it was not always possible to assign the mutations involved in the human diseases to a chromosome region of the mouse known to be imprinted [Cattanach and Beechey, 1990a].

D. Imprinting and Sex Determination X-Chromosome Inactivation

In some insects, sex determination depends on the specific loss of the paternal set of chromosomes and/or the differential imprinting of the chromosomes at the time of fertilization [Chandra and Brown, 1975]. Although imprinting of chromosomes in insects is a necessary feature of sex determination, parthenogenesis is possible, showing that maternal chromosomes alone can conduct normal development.

In animals in which sex is genetically determined, imprinting is restricted to the X chromosome and is responsible for nonrandom inactivation. In marsupials the X inactivation is paternal rather than random and is incomplete in some tissues [Marshall-Graves, 1987].

In mammals the extraembryonic tissues only have kept such a specific X inactivation. In the trophectoderm and the primitive endoderm of female embryos (XX), the paternal X is inactivated. This inactivation is different from the random complete inactivation in the embryo since it can be reversed [Kaslow and Migeon, 1987].

Marshall-Graves [1987] suggested that the evolution of an inner cell mass with its complete segregation from the intrauterine environment, a process that occurred only in the eutherian lineage, was accompanied

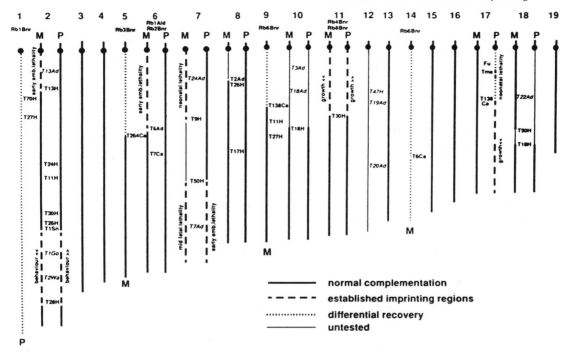

Fig. 1. *Autosomal and X-chromosome imprinting. (Reproduced from Cattanach and Beechey, 1990b, with permission of the publisher.)*

by evolution of random, complete inactivation, which may, then, depend for its signals on different positional information.

E. Imprinting and Recombination

There are sex-specific differences in genetic recombination in many organisms reflected by differences in the male and female genetic maps. This can be explained either by differences in the physiology of the meiotic male and female gamete or by differential expression of the gene involved in the recombination event. Alternatively, it has been suggested that actively transcribed sequences were more often likely to undergo recombination. If differential imprinting in the male and female germ cells results in gene activation at the time of meiosis, then these genomic regions might

be potentially recombinogenic [Thomas and Rothstein, 1991].

Another observation concerns the appearance of neomutations at minisatellite loci. Two types of loci with high mutation rates have been described in man. In the first class, mutations are of paternal or maternal origin at equal rates [Jeffreys et al., 1988]. In the second class, mutations are predominantly of paternal origin [Vergnaud et al., 1991]. In such sequences, variation of the number of tandem repeats may be due to different mechanisms, including anomaly in replication, unequal recombination between sister chromatids, or gene conversion. Some minisatellites might be located in imprinted regions and thus be subjected to differential variations in male and female gametes.

II. IDENTIFICATION OF IMPRINTED GENES

Recent observations with transgenic mice have suggested that transgenes might be useful to target imprinted genes. In transgenic mice, the transgene integrates at random but analyses on the rate of insertional mutagenesis suggest that active chromatin domains containing genes are prefered sites of integration [Palmiter and Brinster, 1986]. The transgene expression is often influenced by the surrounding genomic DNA in particular when weak promoters are used to drive the expression of the injected gene. The position effect sometimes results in the inactivation of the transgene or on variable levels of expression. This phenomenon can be used to localize genomic regulatory sequences by the enhancer/promoter trap technique [Allen et al., 1988; Kothary et al., 1988]. DeLoia and Solter [1990] have described a transgenic mouse strain in which transmission of the transgene through the paternal lineage is associated with limb and skull deformities in the offspring, whereas maternal transmission results in phenotypically normal mice. The integration site maps to mouse chromosome 5, which is not imprinted according to Cattanach and Beechey [1990a,b]. However, the cloning of the insertion site may lead to the identification of an imprinted locus. In a following paragraph we will discuss some examples of differential transgene methylation that may be due to the effect of an imprinted locus. Using different approaches, two imprinted genes have been now identified: the insulin-like growth factor 2 (IGF-2) receptor gene on mouse chromosome 17 and its ligand the IGF-2 gene on mouse chromosome 7. The IGF-2 receptor gene is expressed only from the maternal allele [Barlow et al., 1991], which is in agreement with the imprinting map (Fig. 1). It lies near the *Tme* locus (deleted in the *Thp* mutation) and is a good candidate for this gene.

In mice in which the IGF-2 gene is mutated by homologous recombination, the remaining allele is shown to be normally expressed only when paternally transmitted [de Chiara et al., 1991]. The observation by de Chiara et al., [1990] suggest that the IGF-2 gene is an imprinted gene necessary for growth. It lies in a chromosomal region subjected to imprinting (Fig. 1). Surprisingly, the H19 gene, tightly linked to IGF-2 on chromosome 7, is expressed only from the maternal allele [Bartolomei et al., 1991].

III. MECHANISMS OF IMPRINTING

The precedent observations clearly indicate that the two copies of some genes carried by homologous chromosomes are not identically transcribed. In some cases the level of expression may be different as suggested by the experiments of Cattanach and Kirk [1985], or one of the alleles may be totally silenced. The imprinting of genes that are differentially expressed during development must be maintained during multiple cycles of cell division, and it must be erased during subsequent passage in the germ line.

Holliday [1990] suggested that the basis for genomic imprinting might be allelic exclusion, reducing genes to functional haploidy. This regulation is known for immunoglobulin genes and, in yeast, is possibly involved in the segregation of daughter cells different from their progenitor [Klar, 1990]. For imprinted genes the allelic exclusion would not be at random but would be programmed in a sex-specific fashion. The site for the reversible DNA modifications is probably the gamete in which meiosis leads to genome haploidy.

A. Epigenetic Modifications: DNA Methylation and Chromatin Arrangement

1. Regulation of gene expression. The studies on gene expression have shown that transcription activation requires the interactions of proteins with regulatory sequences usually but not exclusively located upstream from the coding regions. Housekeeping genes are thought to be constitutively active in response to ubiquitous factors. Tissue-specific genes are activated during development by specific factors that must be first turned on in the differentiating cell. In addition, other factors (hormone receptors, negative factors, growth factors) are known to modulate gene expression.

The interaction of the transcriptional factors with their target is affected by modifications of the genome such as the chromatin structure, cytosine methylation of the DNA, DNA replication, and nuclear organization. The modification of the gene expression that is not due to a change in DNA base sequence has been termed an *epimutation* by Holliday [1987]. These different modifications are thought to evolve in particular from the germ cells to the egg and thereafter during cellular differentiation and probably during the cell cycle.

2. Methylation and active chromatin domain. The most studied epigenetic modifications of genes relative to their transcriptional activity are cytosine methylation and chromatin structure, as reflected by the sensitivity of the chromatin to DNase digestion. Active promoters coincide with nuclease hypersensitive sites that are sequences of DNA with a particular chromatin arrangement that makes them more available to several enzymatic probes and especially to DNaseI [Weintraub and Groudine, 1976]. Hypersensitive sites can be defined around active genes. They are different from the binding sites for transcriptional factors and in some instances can confer the independence of chromosomal position to a reporter gene [Reitman et al., 1990].

There is a clear correlation between the activity of the chromatin and the methylation status of some genes [Navey-Many and Cedar, 1981]. Cytosine methylation of CG dinucleotides is a reversible DNA modification involved in the regulation of plant and animal gene expression. Change in the methylation status of genes accompanies changes in transcription during development or in tumors and cell cultures. In many instances, hypomethylation of 5' domains and/or all the gene was shown to be associated with expression. Hypermethylation results in the inactivation of the genes both in vitro and in vivo [Cedar, 1988]. Specific de novo methylation is not believed to be the primary event that turns off genes. Many tissue-specific regulatory factors are now known, and it is not necessary to imagine the existence of specific methylases to regulate gene expression. However, Szyf et al. [1989] described the nucleotide sequence-specific de novo methylation of the mouse steroid 21-hydroxylase, independent of gene transcription.

The binding of transcriptional factor is influenced by methylation of regulatory sequences. Hypermethylation has been shown to prevent the binding of activators [Becker et al., 1987]. In addition, the binding of specific proteins that recognize methylcytosine may be responsible for the formation of the inactive chromatin [Meehan et al., 1989]. Methylation of a DNA strand is stably propagated to the daughter cell during replication by maintenance methyltransferase [Holliday, 1987]. However, de novo methylation does not necessitate replication [Paroush et al., 1990].

B. Methylation as a Way to Maintain a Regulatory State

It is clear from many observations that cytosine methylation is not the primary event in gene inactivation but follows this inactivation. Its role is rather to maintain a repressed state in the absence of the factors that initiated it. This is clearly the case for some genes on the inactive X chromosome [Monk, 1986]. Following inactivation of gene transcription, DNA sequences may be nonspecifically de novo methylated by maintenance methylase if they are no longer protected by proteins. In vertebrates the bulk of DNA is methylated (60%–90%), but particular sequences rich in restriction sites containing CGs regularly spaced on the chromosomes are completely unmethylated both in the germ line and in all somatic cells [Bird, 1987]. These sequences, called CpG islands, have been found 5' to all the known housekeeping genes and many testis-specific genes [Ariel et al., 1991] in a region corresponding to the promoter and the first exon. Tissue-specific genes are almost fully methylated in sperm and probably the female germ line and in the somatic tissues in which they are not expressed.

The egg genome is strikingly undermethylated compared with the male germ cells and somatic cells [Sanford et al., 1987]. A de novo methylation is seen after the blastocyst stage and proceedes independently in embryonic and extraembryonic tissues. Tissue-specific gene regulatory sequences are thereafter demethylated in the cells in which the gene is expressed. In these cases, the demethylation seems to take place in the absence of DNA replication probably in response to specific proteins.

The housekeeping genes are constitutively expressed, and the CG-rich islands are therefore thought to be protected from de novo methylation by binding of ubiquitous transcriptional factors [Antequera et al., 1989]. In cell culture a large proportion of nonessential genes are turned off, and CpG islands become de novo methylated [Antequera et al., 1989].

C. Allele-Dependant Methylation of DNA

Using probes that detect polymorphism in humans, Chandler et al. [1987] first showed that the two alleles of c-Ha-*ras*-1 were differentially methylated. The same analysis performed with several unidentified probes by Silva and White [1988] also demonstrated that a percentage of DNA sequences were differentially methylated on the two alleles of an individual and that this pattern was stably transmitted irrespective of parental origin. This suggests that differential methylation of alleles may be a common mechanism to regulate gene expression independently of the parent of origin. Based on these observations and on data with transgenic mice, McGowan et al., [1989] proposed a model based on mosaicism. The cross-talk hypothesis proposed by Monk [1990] explains how chromosomes could communicate and adapt their expression pattern.

D. Methylation and Imprinting

Methylation has been proposed as the most likely mechanism of imprinting, the inactive allele being de novo methylated at some point in the gamete or at the first stage of development. A de novo methylation process occurs between the spermatogonial stage and the first meiotic prophase, excluding regions that may play a role in the activation of the paternal genome during embryogenesis [Groudine and Conkin, 1985]. Genes may be imprinted during germ cell maturation by interaction of specific factors with regulatory sequences thus protecting the genes from de novo methylation in the early embryo. In both cases it suggests that some genes expressed during development rely for their expression on germ cell–specific factors.

1. Transgenes as probes for sex-specific methylation. Several examples of differential methylation of a transgene depending on the sex of origin and of parental effect on expression of the transgene have been reported. Swain et al. [1987] described a mouse strain in which an (RSV)LTR-*c-myc* fusion gene was specifically expressed in the heart only when the transgene was inherited from the father. The transmission by a female resulted in hypermethylation. A second transgenic mouse strain made with the same transgene shows the same parental pattern of expression and methylation [Chaillet et al., 1991]. Sapienza et al. [1987] reported that four of five lines containing the quail *troponin I (TNI)* gene showed evidence of gamete of origin-dependant methylation of the transgene. In three lineages, the transgene was hypermethylated following passage through the female germline and one following passage through the male germ line. In these cases and also in the mice described by Engler et al. [1991] containing the pHRD transgene, the nature of the transgene and not the locus in which it is integrated seems to specify the parental methylation pattern. It may be related to the action of strain-specific modifiers on plasmid or viral sequences present in the transgene.

We described one transgenic mouse strain containing hepatitis B virus (HBV) sequences and expressing the viral surface antigen specifically in the liver only when the transgene was of paternal origin [Hadchouel et al., 1987]. A second strain containing the same transgene did not show this sex-specific methylation. Reik et al. [1987] produced seven independant strains of mice containing a CAT-IgH-E fusion gene and showed that in one of these strains a hypermethylation of the transgene was a consequence of female transmission. Sasaki et al. [1991] have also described one strain out of three containing an MT-1-TTR construct in which hypermethylation is a consequence of maternal transmission.

In these three cases the differential methylation of the transgene most probably reflects that of the chromosome domain in which it is integrated. Sasaki et al. [1991] have localized the insertion on the proximal part of chromosome 11, which is an imprinted region according to the map of Beechey and Cattanach (Fig. 1). In the HBV-transgenic lineage the transgene is localized on chromosome 13, which is not involved in the parental effect.

In four cases described, de novo methylation of the transgene occurred when the gene was maternally transmitted. The transgene was of a different origin, and other transgenic mice containing the same transgene did not show a parental modification of their methylation status. The fact that de novo methylation is more frequently maternal may be a consequence of the injection into the male pronucleus. Integration in domains active in the male gamete and inactive in the oocyte may prevent the paternal allele from de novo methylation in the early embryo. It is of importance now to study if these regions are imprinted in the sense discussed above or if they only participate in some gamete specific function.

2. Factors affecting the transgene methylation. In four cases, the testes of progeny from a female were the only organs in which the transgene was not hypermethylated, and in subsequent generations from males the transgene was hypomethylated. The de novo methylation probably occurred in the female germ line or very early during development since the same pattern of *Hpa*II digestion was found in different organs. Thus the absence of methylation in the testis may be due either to demethylation, which would show the reversibility of the phenomenon, or to protection against de novo methylation in the male germline cells during development. Sasaki et al. [1991] showed that the methylation pattern of the MT-1-TTR transgene is established during spermatogenesis. Demethylation takes place before early prophase I. Similarly, Chaillet et al. [1991] reported that the parental-specific methylation of the *RSVIgmycA* transgene is established during gametogenesis and that further changes occur during embryogenesis. In the report of Sapienza et al. [1989] a maternal effect on transgene de novo methylation is clearly shown and varies among mouse strains. When the quail *TNI* transgene was inherited from a male, some, but not all, offspring exhibited transgene hypomethylation in their somatic tissues. This implied that both *cis*-acting elements (the progenitor transgene modification or imprinting) and *trans*-acting elements (the strain background of the nontransgenic female) are involved in the establishment of the methylation imprint [Sapienza et al., 1989]. It demonstrates that the pattern of methylation is imposed on the transgene in the fertilized eggs and depends on maternal factors acting in *trans*. Along the same lines, Engler et al. [1991] report the existence of a strain-specific modifier on chromosome 4 controlling the transgene methylation. Similar results were obtained with a transgenic mouse strain containing a mouse hsp68–*E. coli lacZ* construct. The *lacZ* transgene is expressed during days 10–14 of development. However, in some fetuses that contain a hypermethylated transgene no expression can be detected. Intermediates states of methylation are found as in the *TNI* line. This is due to the simultaneous presence of both hyper- and hypomethylated cells within the embryo and reflect a mosaicism [McGowan et al., 1989]. The existence of factors supplied by the maternal genome after fertilization is consistent with the data of Sanford et al. [1987] on postfertilization changes in the methylation status of genes.

3. Relevance to the parental effect. Although it is clear that transgenes can in some cases be differentially methylated depending on the gamete of origin, it remains to be demonstrated that this is due to the effect of neighboring imprinted mouse sequences. In the HBV-transgenic lineage, the transgene was methylated in every organ, including the testis, and transmission by males did not lead to demethylation. We think that the irreversibility of the transgene methylation is related to its nature and not to the nature of the integration region. Indeed we have shown that partial methylation of HBV DNA is reversed at each generation during liver development, the site of HBV gene expression, whereas total methylation is not [Pourcel et al., 1990]. A gene specifically expressed in the testis was identified a few kilobases from the transgene. The insertion of the trans-

gene has caused the deletion of genomic DNA, but the gene is not inactivated. In situ hybridization and analysis of testis RNA at different stages of maturation show that transcription is postmeiotic. In testis and sperm the gene is hypermethylated, although it is the site of transcription; however, we have not yet identified the regulatory sequences that may be the only region differentially methylated. Hypomethylation is found in the placenta and to a lesser extent in the embryo.

Sasaki et al. [1991] were unable to detect differential methylation in the genomic DNA surrounding the MT-1-TTR transgene and suggested that the transgene methylation status does not necessarily reflect that of the insertion locus.

IV. DISCUSSION

To understand the basis of the parental effect in mammals it is important to understand the constraints of intrauterine development. It requires the synthesis of extraembryonic tissues and invasion of the uterus and the immunological tolerance by the mother of the embryo.

The nuclear transplantation experiments have shown that paternal chromosomes were necessary for the production of extraembryonic tissues whereas maternal chromosomes contribute mainly to the development of the embryo. Surani et al. [1989] proposed that maternal chromosomes are essential for embryonic cells that are totipotential and/or pluripotential, whereas paternal chromosomes are preferentially required to sustain the proliferation of progenitor cells of specific differentiated tissues in both the embryonic and extraembryonic lineages. The results of Mann et al. [1990] suggest that paternally imprinted genes may be involved in development of muscle and skeleton. Additional observations show that imprinted genes are important for size regulation, cell proliferation, cell interactions, and differentiation [reviewed by Surani et al., 1990a,b].

It has been proposed that imprinting has evolved in mammals to restrain the proliferative growth of the placenta [Moore and Haig, 1991]. The paternal genes would help in acquiring nutrient from the mother, whereas the maternal genes would reduce this effect to survive pregnancy [Hall, 1990]. If this hypothesis is correct, the imprinted genes are expected to be involved in placental growth as well as functions necessary for postnatal growth and nutrition [Moore and Haig, 1991].

Interestingly, opposite effects are seen following the duplication of certain chromosomal regions from the mother or the father. This is not always the case, and the different penetrance observed in certain human disease suggests rather a mosaicism of expression of imprinted loci. The fragile X syndrome is an example [Laird, 1987; Vincent et al., 1991; Bell et al., 1991].

There is still no clear answer concerning the mechanism of imprinting, although methylation certainly plays a major role. It is difficult to imagine the existence of specific methylases in the gametes that would recognize and inactivate the parental allele that is not expressed during development. The most probable mechanism is the binding of gamete-specific factors that will protect the gene from general de novo methylation. Other transactivating factors may later be necessary to turn on the gene at a certain time in development. It is interesting to note that many testis-specific genes contain a CG-rich region characteristic of ubiquitous genes but are nevertheless controled by testis-specific factors.

V. EXPERIMENTAL PROCEDURES

A. Genetic Studies

The map of the mouse-imprinted regions has been derived from a genetic method making use of heterozygotes for robertsonian translocation. They exhibit high levels of nondisjunction, thus allowing the production of animals that are disomic for a chromosome arm of one parent and nullisomic for the equivalent chromosome arm of the other parent. Such mice are chromosomally balanced, but in some instances normal complementation is not achieved, suggesting the existence of imprinted gene(s) in the region involved. The method has been described in previous reviews. Figure 1 shows the autosomal chromosome imprinting map of the mouse obtained by Cattanach and Beechey [1990b].

B. Manipulation of the Mouse Embryo

Most of the techniques using mouse embryos and embryonic cells are described fully in the CSH laboratory manual *Manipulating the Mouse Embryo* [Hogan et al., 1986].

1. Pronuclear transplantation. The production of a diploid egg with two sets of chromosomes from one parent can be obtained from parthenogenetic activation (parthenogenone) or by transfer of a pronucleus into an haploid zygote [Solter, 1988]. Gynogenones contain two female pronuclei and androgenones contain two male pronuclei. Nuclear transplantation is performed as described by McGrath and Solter [1983].

2. Chimerae. Chimerae are made by aggregating embryos (four to eight cells) from different origins or by injection of embryonic cells (ICM cells, cultured stem cells, or teratocarcinoma cells) into a preimplantation embryo. The contribution of the two types of cells is estimated in the animal by use of coat color markers.

The isolation of pluripotential stem cell lines was originally reported by Evans and Kaufman [1981]. Both parthenogenetic and androgenetic stem cells have been reported to contribute to the embryo in aggregation chimerae [Mann et al., 1990].

3. Transgenic mice. Transgenic mice have been used during the last decade mainly in an experimental assay to study the regulation of gene expression. This technique also allows investigation of the effects of a gene product on the animal.

The method that has been most extensively employed to produce transgenic mice is the microinjection of DNA directly into the pronuclei of fertilized mouse eggs [Gordon et al., 1980]. This leads to the stable integration of the foreign DNA in diverse chromosomal locations. It is believed that in as much as 10% of the mice the DNA inserts into a gene or next to it. As a consequence insertional mutations are produced, and the DNA can be used to tag the mutated gene. In addition, interactions between the foreign DNA and adjacent mouse regulatory sequences are commonly observed [reviewed in Palmiter and Brinster, 1986].

C. DNA Methylation Analysis

DNA extracted from a tissue sample is digested with restriction enzymes whose recognition sites contain a CG pair. Some of these enzymes can only cut the DNA when C is unmethylated, whereas others always cut independently of the methylation status. The most widely used enzymes are the isoschisomers *Hpa*II and *Msp*I. *Msp*I cuts the sites CCGG and C^{m}CGG, and *Hpa*II only cuts CCGG. Using probes corresponding to a genomic region of interest, the restriction pattern of a DNA sample digested with the two enzymes is compared on a Southern blot. Complete digestion is obtained with *Msp*I, and a degree of methylation is seen with the *Hpa*II restriction pattern. The methylation analysis of a unique sequence by the Southern blot technique necessitates the use of 5–20 μg of DNA (1×10^6 diploid cells = 10μg DNA) and cannot be performed on very small samples.

REFERENCES

Allen ND, Cran DG, Barton SC, Hettle S, Reik W, Surani MA (1988): Transgenes as probes for active chromosomal domains in mouse development. Nature 333:852–855.

Antequera F, Macleod D, Bird A (1989): Specific protection of methylated CpGs in mammalian nuclei. Cell 58:509–517.

Ariel M, McCarrey J, Cedar H (1991): Methylation patterns of testis-specific genes. Proc Natl Acad Sci USA 88:2317–2321.

Barlow DP, Stöger R, Herrmann BG, Saito K, Schweifer N (1991): The mouse insulin-like growth factor type-2 receptor is imprinted and closely linked to the *Tme* locus. Nature 349:84–87.

Bartolomei MS, Zemel S, Tilghman SM (1991): Parental imprinting of the mouse H19 gene. Nature 351:153–155.

Becker PB, Ruppert S, Schütz G (1987): Genomic footprinting reveals cell type-specific DNA binding of ubiquitous factors. Cell 51:435–443.

Bird AP (1986): CpG-rich islands and the function of DNA methylation. Nature 321:209–213.

Cattanach BM, Beechey CV (1990a): Chromosome imprinting phenomena in mice and indication in man. In Fredgal K, Kihlman BA, Bennett MD, (eds): Chromosomes Today. London: Unwin Hyman, pp 135–148.

Cattanach BM, Beechey CV (1990b): Autosomal and X-chromosome imprinting. Development (Suppl):63–72.

Cattanach BM, Kirk M (1985): Differential activity of maternally, and paternally derived chromosome regions in mice. Nature 315:496–98.

Cedar H (1988): DNA methylation and gene activity. Cell 53:3–4.

Chaillet JR, Vogt TF, Beier DR, Leder P (1991): Parental-specific methylation of an imprinted transgene is established during gametogenesis and progressively changes during embryogenesis. Cell 66:77–83.

Chandler LA, Ghazi H, Jones PA, Boukamp P, Fusenig NE (1987): Allele-specific methylation of the human c-Ha-ras-1 gene Cell 50:711–717.

Chandra H, Brown SW (1975): Chromosome imprinting and the mammalian X chromosome. Nature 253:165–168.

de Chiara TM, Efstratiadis A, Robertson EJ (1990): A growth-deficiency phenotype in heterozygous mice carrying an insulin-like growth factor II gene disrupted by targeting. Nature 345:78–80.

de Chiara TM, Robertson EJ, Efstratiadis A (1991): Parental imprinting of the mouse insulin-like growth factor II gene. Cell 64:849–859.

DeLoia JA, Solter D (1990): A transgene insertional

mutation at an imprinted locus in the mouse genome. Development (Suppl): pp 73–79.

Engler P, Haasch D, Pinkert CA, Doglio L, Glymour M, Brinster R, Storb U (1991): A strain-specific modifier on mouse chromosome 4 controls the methylation of independent transgene loci. Cell 65:939–947.

Evans MJ, Kaufman MH (1981): Establishment or culture of pluripotential cells from mouse embryos. Nature 292:154–156.

Gordon JW, Scangos GA, Plotkin DJ, Barbosa JA, Ruddle FH (1980): Genetic transformation of mouse embryos by microinjection of purified DNA. Proc Natl Acad Sci USA 77:7380–7384.

Groudine M, Conkin KF (1985): Chromatin structure and de novo methylation of sperm DNA: implications for activation of the paternal genome. Science 228:1061–1068.

Hadchouel M, Farza H, Simon D, Tiollais P, Pourcel C (1987): Maternal inhibition of hepatitis B surface antigen gene expression in transgenic mice correlates with de novo methylation. Nature 329:454–456.

Hall JG (1990): Genomic imprinting: review and relevance to human diseases. Am J Hum Genet 46:857–873.

Hogan B, Costantini F, Lacy E (1986): Manipulating the Mouse Embryo. Cold Spring Harbor, NY: Cold Spring Harbor Laboratory.

Holliday R (1987): The inheritance of epigenetic defects. Science 238:163–170.

Holliday R (1990): Genomic imprinting and allelic exclusion. Development (Suppl): 125–129.

Jeffreys AJ, Royle NJ, Wilson V, Wong Z (1988): Spontaneous mutation rates to new length alleles at tandem-repetitive hypervariable loci in human DNA. Nature 332:278–281.

Johnnson DR (1975): Further observations on the hairpintail (*Thp*) mutation in the mouse. Genet Res 18:71–79.

Kaslow DC, Migeon B (1987): DNA methylation stabilizes X chromosome inactivation in eutherians but not in marsupials: evidence for multistep maintenance of mammalian X dosage compensation. Proc Natl Acad Sci 84:6210–6214.

Klar AJS (1990): The development fate of fission yeast cells is determined by the pattern of inheritance of parental and grandparental DNA strands. EMBO 9:1407–1415.

Kothary R, Clapoff S, Brown A, Campbell R, Peterson A, Rossant J (1988): A transgene containing *lac*Z inserted into the dystonia locus is expressed in neural tube. Nature 335:435–437.

Mann JR, Gadi I, Harbison ML, Abbondanzo SJ, Stewart CL (1990): Androgenetic mouse embryonic stem cells are pluripotent and cause skeletal

defects in chimeras: implications for genomic imprinting. Cell 62:251–260.

Marshall-Graves JA (1987): The evolution of mammalian sex chromosomes and dosage compensation: Clues from marsupials and monotremes. Trends Genet 3:252–256.

McGowan R, Campbell R, Peterson A, Sapienza C (1989): Cellular mosaicism in the methylation and expression of hemizygous loci in the mouse. Genes Dev 3:1669–1676.

McGrath J, Solter D (1983): Nuclear transplantation in the mouse embryo by microsurgery and cell fusion. Science 220:1300–1302.

McGrath J, Solter D (1984): Completion of mouse embryogenesis requires both the maternal and paternal genomes. Cell 37:179–83.

Monk M (1986): Methylation and the X chromosome. BioEssays 4:204–208.

Monk M (1990): Variation in epigenetic inheritance. Trends Genet 6:110–114.

Moore T, Haig D (1991): Genomic imprinting in mammalian development: A parental tug-of-war. Trends Genet 7:45–49.

Naveh-Many T, Cedar H (1981): Active gene sequences are undermethylated. Proc Natl Acad Sci USA 78:4246–4250.

Paldi A, Nagy A, Markkula M, Barna I, Dezso L (1989): Postnatal development of parthenogenetic-fertilized mouse aggregation chimeras. Development 105:115–118.

Palmiter RD, Brinster RL (1986): Germline transformation of mice. Annu Rev Genet 20:465–499.

Paroush Z, Keshet I, Yisraeli J, Cedar H (1991): Dynamics of demethylation and activation of the α-actin gene in myoblasts. Cell 63:1229–1237.

Pourcel C, Tiollais P, Farza H (1990): Transcription of the S gene in transgenic mice is associated with hypomethylation at specific sites and DNase I sensitivity. J Virol 64:931–935.

Reik W (1989): Genomic imprinting and genetic disorders in man. Trends Genet 5:331–336.

Reik W, Collick A, Norris ML, Barton SC, Surani MA (1987): Genomic imprinting determines methylation of parental alleles in transgenic mice. Nature 328:248–251.

Reitman M, Lee E, Westphal H, Felsenfeld G (1990): Site-independent expression of the chicken βA-globin gene in transgenic mice. Nature 348:749–752.

Renard P, Babinet C (1986): Identification of a paternal genotype effect on the cytoplasm of one-cell-stage mouse embryos. Proc Natl Acad Sci USA 83:6883–6886.

Sanford JP, Clark HJ, Chapman VM, Rossant J (1987): Differences in DNA methylation during oogenesis and spermatogenesis and their persistence during early embryogenesis in the mouse. Genes Dev 1:1039–1046.

Sapienza C, Paquette J, Tran TH, Peterson A (1989): Epigenetic and genetic factors affect transgene methylation imprinting. Development 107:165–168.

Sapienza C, Peterson A, Rossant J, Balling R (1987): Degree of methylation of transgenes is dependent on gamete of origin. Nature 328:251–254.

Sasaki H, Hamada T, Ueda T, Seki R, Higashinkakgawa T, Sakaki Y (1991): Inherited type of allelic methylation variations in a mouse chromosome region where an integrated transgene shows methylation imprinting. Development 111:573–581.

Shire JGM (1989): Unequal parental contributions: genomic imprinting in mammals. New Biol 1:115–120.

Silva AJ, White R (1988): Inheritance of allelic blueprints for methylation patterns. Cell 54:145–152.

Solter D (1988): Differential imprinting and expression of maternal and paternal genomes. Annu Rev Genet 22:127–146.

Surani MA, (1986): Evidence and consequences of differences between maternal and paternal genomes during embryogenesis in the mouse. In Rossant J, Pedersen RA, (eds): Experimental Approaches to Mammalian Embryonic Development. Cambridge: Cambridge University Press, pp 401–435.

Surani MA, Allen ND, Barton SC, Fundele R, Howlett SK, Norris ML, Reik W (1990a): Developmental consequences of imprinting of parental chromosomes by DNA methylation. Philos Trans R Soc [Lond] B326: 313–327.

Surani MA, Barton SC, Norris ML (1984): Development of reconstituted mouse eggs suggests imprinting of the genome during gametogenesis. Nature 308:548–550.

Surani MA, Barton SC, Norris ML (1987): Influence of parental chromosomes on spatial specificity in androgenetic-parthenogenetic chimeras in the mouse. Nature 326:395–397.

Surani MA, Kothary B, Allen ND, Singh PB, Fundele R, Fergusson-Smith AC, Barton SC (1990b): Genome imprinting and development in the mouse. Development (Suppl): 89–98.

Swain JL, Stewart TA, Leder P (1987): Parental legacy determines methylation and expression of an autosomal transgene: A molecular mechanism for parental imprinting. Cell 50:719–727.

Szyf M, Schimmer BP, Seidman JG (1989): Nucleotide-sequence specific de novo methylation in a somatic murine cell line. Proc Natl Acad Sci 86:6853–6857.

Thomas BJ, Rothstein R (1991): Sex, maps and imprinting. Cell 64:1–3.

Vergnaud G, Mariat D, Apiou F, Aurias A, Lathrop M, Lauthier V (1991): The use of synthetic tandem repeats to isolate new VNTR loci: Cloning of a human hypermutable sequence. Genomics 11:135–144.

Vincent A, Heitz D, Petit C, Kretz C, Oberlé I, Mandel J-L (1991): Abnormal pattern detected in fragile-X patients by pulsed-field gel electrophoresis. Nature 349:624–626.

Weintraub H, Groudine M (1976): Chromosomal subunits in active genes have an altered conformation. Science 193:848–846.

ABOUT THE AUTHOR

CHRISTINE POURCEL is *Chef de Laboratoire* at the Pasteur Institute, Paris. After receiving her *Maîtrise es Sciences* (B.S.) from Angers University in 1974, she received her Ph.D. under Pierre Tiollais at Paris VII University, where she concentrated in the cloning of the hepatitis B virus (HBV) genome and the production of a recombinant vaccine against HBV in mammalian cell culture. Dr. Pourcel preformed postdoctoral research in the laboratory of Jesse Summers at the Fox Chase Cancer Center, Philadelphia, where she studied the replication of the duck hepatitis B virus in dedifferentiating primary hepatocytes. At the Pasteur Institute in 1984, in collaboration with Charles Babinet, she developed an animal model for HBV infection using transgenic mice. Dr. Pourcel is currently studying the role of methylation in the control of gene expression and, in particular, in the establishment of imprinting. Her research papers have appeared in such journals as *Nature,* the *Proceedings of the National Academy of Sciences,* the *Journal of Virology, Science,* and *Cell.* Dr. Pourcel is on the editorial board for the journal *Transgenic Research.*

Genes in Mammalian Reproduction: 185–205
© 1993 Wiley-Liss, Inc.

Müllerian-Inhibiting Substance

Richard L. Cate and Cheryl A. Wilson

I. INTRODUCTION

Müllerian-inhibiting substance (MIS), also known as antimüllerian hormone (AMH), is a member of the transforming growth factor type β (TGF-β) family of growth and differentiation factors and is involved in sexual development. MIS was initially characterized as a testicular glycoprotein of 140 kD and purified by its ability to cause regression of the müllerian ducts in the male fetus. It is now recognized that MIS is also produced in the adult ovary, and may have an important role in the female. The discovery that MIS causes an apparent sex reversal of fetal ovaries has permitted the development of another bioassay and has suggested that MIS may have additional roles in the male. Consistent with this possibility, the expression of the human MIS gene in transgenic mice has produced a number of perturbations in the sexual development of both sexes.

The existence of MIS was first postulated by Alfred Jost. During the development of the male and female reproductive tracts, the müllerian ducts give rise to the uterus, fallopian tubes, and part of the vagina, while the wolffian ducts give rise to the epididymis, vas deferens, and seminal vesicles. The two ducts develop in the early embryonic stages of both sexes, and regression of one or the other duct occurs as a consequence of gonadal differentiation.

Testosterone is required for wolffian development in the male and for a long time also was thought to be responsible for müllerian duct regression. Jost [1947] showed that testosterone could not cause regression of the müllerian duct and suggested that the testis produces an additional factor. With the use of an organ culture assay developed by Picon [1969], Josso and colleagues in Paris [Picard and Josso, 1984] and Donahoe and colleagues in Boston [Budzik et al., 1983, 1985] were able to purify the protein to homogeneity. The genes have now been cloned for bovine [Cate et al., 1986; Picard et al., 1986a], human [Cate et al., 1986; Guerrier et al., 1990], mouse [Munsterberg and Lovell-Badge, 1991], and rat [Haqq et al., 1992] MIS.

This review focuses mostly on the reproductive roles of MIS in the male and female. The well-documented biological activities, as well as the supporting and negative evidence for the putative biological activities, are described. Although the structure of the MIS protein and gene is covered, readers should consult other reviews [Josso and Picard, 1986; Donahoe et al., 1987] for more information on the purification and characterization of the bovine protein. Also, since the focus is on the role of MIS in sexual development, the antiproliferative effects of MIS on tumor cells are not addressed, and readers are directed to reviews by Donahoe et al. [1987] and Cate et al. [1990].

II. STRUCTURE OF MIS

A. Protein Structure

1. Comparison of human, bovine, rat, and mouse MIS. cDNA or genomic clones have been isolated and sequenced for bovine [Cate et al., 1986; Picard et al., 1986a], human [Cate et al., 1986], rat [Haqq et al., 1992], and mouse [Munsterberg and Lovell-Badge, 1991] MIS. The bovine and human proteins are synthesized as 575 and 560 amino acid precursors, respectively, while the rat and mouse precursors are 553 and 555 amino acids (Fig. 1). The four MIS proteins share approximately 75% homology overall. The N termini show the least homology, with the bovine protein containing two insertions, while the C termini show a striking level of homology (approximately 95%). This is the region of MIS that shows similarity to the proteins of the TGF-β family [Barnard et al., 1990; Massague, 1990; Roberts and Sporn, 1990]. The region of MIS encoded by the second exon is also fairly well conserved, indicating that it may have some functional importance.

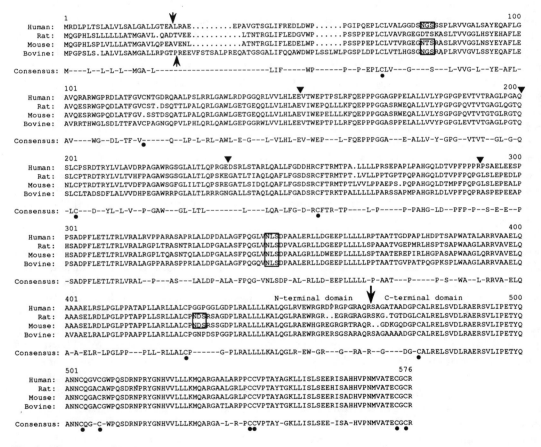

Fig. 1. *Alignment of the human, rat, mouse, and bovine MIS amino acid sequences. The mature N termini of the bovine and human proteins are indicated with small arrows, and the large arrow between Arg 427 and Ser 428 indicates the cleavage involved in proteolytic processing of human MIS. The potential N-glycosylation sites are marked with boxes, cysteines are denoted with dots, and the locations of the four introns are indicated by arrowheads.*

Bovine, human, and rat MIS contain 12 conserved cysteines, while mouse MIS contains only 11. Seven of the cysteines are in the highly conserved C-terminal domain; these cysteines are conserved in all members of the TGF-β family. Amino acids in the vicinity of the seven invariant cysteines also are conserved between family members. None of the MIS proteins contain the dibasic or tetrabasic proteolytic cleavage sites present in other members of the TGF-β family. The lack of such a site in MIS originally led to the suggestion that MIS, unlike the other related molecules, does not undergo proteolytic processing for biological activity. However, considerable evidence now exists that MIS does undergo processing (see below).

The human, bovine, and rat proteins contain two potential N-linked glycosylation sites, while the mouse protein has three. All four proteins preserve the site in exon 5. The level of glycosylation of the bovine MIS has been estimated at 8.3% [Budzik et al., 1980], 11% [Swann et al., 1979], and 13.5% [Picard et al., 1986b]. Glycosylation does not appear to be essential for biological activity of MIS, since tunicamycin treatment of cells producing the bovine protein [Josso et al., 1980] and endoglycosidase F treatment of the human protein (R Cate, unpublished data) does not affect the activity of MIS.

Like other secretory proteins, MIS undergoes signal sequence cleavage to generate the mature amino terminus. The mature N-termini of both the human (LRAEEP) and bovine (REEVFS) proteins have been determined by protein sequence analysis [Cate et al., 1986; Pepinsky et al., 1988], whereas the mature N-termini of the mouse (VENLAT) and rat (VEELTN) genes have been predicted by homology to the human and bovine proteins (Fig. 1). The mouse and bovine genes both contain proline residues within the last three amino acids of the leader sequence, a somewhat unusual feature for a signal peptide. Also, a von

Heijne [1986] analysis of the human and bovine proteins predicts cleavages within the leader sequences, six to eight amino acids upstream from the mature N termini. This suggests that either two processing steps are required for removal of the leader sequences or that the MIS proteins represent exceptions to the von Heijne rules.

2. Biosynthesis and proteolytic processing of MIS. Both human [Pepinsky et al., 1988] and bovine MIS [Picard and Josso, 1984; Budzik et al., 1985] are secreted as 140 kD homodimers. Secretion is very slow, as evidenced from pulse-chase experiments with human MIS in CHO cells (>10 hours; R Cate, unpublished data) and bovine MIS in Sertoli cells (48 hours [Vigier et al., 1985]). Immunohistochemistry [Tran and Josso, 1982; Hayashi et al., 1984; Tran et al., 1987] and endoglycosidase H digestion of intracellular protein (R Cate, unpublished data) demonstrate that the bulk of the MIS is sequestered in the endoplasmic reticulum (ER). The rate-limiting step for transit from the ER appears to be the proper folding of the C-terminal domain, since a truncated form of MIS containing only the N-terminal domain is secreted very rapidly from cells (R Cate, unpublished data).

Proteolytic processing is required for the generation of active TGF-β1 and involves two steps [Lawrence et al., 1985; Pircher et al., 1986]. First, the dimeric precursor is cleaved at a dibasic site 112 amino acids from the C terminus, and the two portions of the precursor remain noncovalently associated in a latent complex. In the second step, the complex undergoes dissociation, releasing the biologically active C-terminal domain, which can bind to the TGF-β receptor. Dissociation can be accomplished in vitro by acid, urea, or heat treatment [Lawrence et al., 1985; Sporn et al., 1987]. Currently, it is not known how dissociation occurs in vivo.

Until recently it was thought that proteolytic processing is not necessary for MIS activity, since the expected dibasic proteolytic cleavage site is missing and unprocessed MIS is active in organ culture assays. Since then it has been noted that approximately 5%–10% of human MIS secreted from CHO cells undergoes cleavage on both chains of the dimer at a site that is similar in location to the cleavage sites of the other members of the TGF-β family [Pepinsky et al., 1988]. This site between Arg 427 and Ser 428 is identical to the consensus sequence of monobasic cleavage sites described by Benoit et al., [1987]. It is conserved in the bovine, rat, and mouse proteins, although the mouse sequence has diverged somewhat on the C-terminal side of the cleavage (see Fig. 1). Like TGF-β, the N- and C-terminal dimers remain associated in a noncovalent complex.

Complete cleavage of the human protein at this site can be accomplished in vitro by plasmin treatment, and the resulting MIS is fully active in the müllerian duct regression assay [Pepinsky et al., 1988]. A mutant MIS molecule in which Arg 427 has been converted to a threonine cannot be cleaved by plasmin and is inactive in the müllerian duct regression assay [Cate et al., 1990] and the fetal ovary/aromatase assay (C Wilson et al., unpublished data), indicating that the cleavage is obligatory and must occur in both organ culture assays. By analogy with TGF-β, the noncovalent complex of the N- and C-terminal fragments may also be latent and have to undergo dissociation for biological activity.

There is strong evidence that the C-terminal domain of MIS is able to function alone in organ culture assays. Following plasmin cleavage and treatment with 1 Macetic acid, the C-terminus can be isolated and is active in the fetal ovary/aromatase assay and the müllerian duct regression assay (C Wilson et al., unpublished data). In addition, a monoclonal antibody (Mab 168) derived against bovine MIS was shown to recognize the C-terminal domain of MIS (C Wilson et al., unpublished data) and block MIS-induced

müllerian duct regression in vitro [Legeai et al., 1986]. Further evidence that the C terminus contains the epitope that binds to the MIS receptor is that antiidiotypic antibodies derived against Mab 168 exhibit MIS-like activity [Lefevre et al., 1989]. Thus MIS appears to be similar to other members of the TGF-β family, where the C terminus is the bioactive domain. Although the N-terminal domain has no biological activity itself, it can enhance the activity of the C-terminal fragment (C Wilson et al., unpublished data).

B. Gene Structure of MIS

1. Exon structure. The genes for human, rat, and mouse MIS have similar intron–exon structures (Fig 2A). All three genes are composed of five exons and separated by four similarly sized introns. The transcription initiation sites for the bovine [Cate et al., 1986] and human [Guerrier et al., 1990] genes have been mapped using primer extension and S1 nuclease digestion, while the rat initiation site [Haqq et al., 1992] has been predicted by comparison with the human and bovine sequences (Fig. 2B). The bovine, rat, and mouse genes have good TATA boxes at 28, 27, and 26 nucleotides, respectively, from their RNA initiation sites, while the human gene contains the sequence CTTAA 28 nucleotides upstream of its RNA start site. Although there is some heterogeneity in transcription initiation, 88% of the human transcripts initiate at the site indicated in Figure 2B [Guerrier et al., 1990]. All four MIS mRNAs contain very short 5'-untranslated regions, ranging from 10 nucleotides for bovine and human mRNAs to 8 nucleotides for the rat and mouse. The length of the 5'-untranslated regions appears to rule out any involvement in translational regulation of MIS expression.

The 3'-untranslated regions are also fairly short, ranging from 112 nucleotides for the human mRNA to 62 nucleotides for the rat mRNA. Recently it has been shown

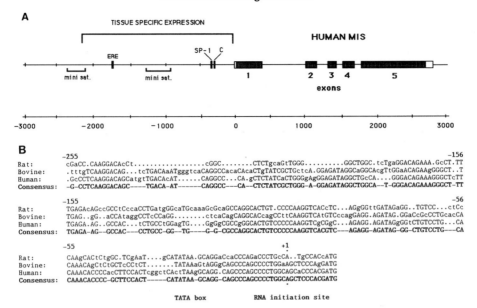

Fig. 2. *Structure of the human MIS gene and nucleotide sequences of the rat, bovine, and human promoters.* **A:** *Schematic diagram of the human MIS gene showing the relative location of potential regulatory sequences. The putative ERE, SP-1, and CRE (denoted C) sites and the minisatellite regions are discussed in the text. The region identified as being involved in the tissue-specific expression of MIS was deduced from the experiments of J. Peschon (unpublished data).* **B:** *Alignment of the rat, bovine, and human MIS nucleotide sequences immediately upstream of the transcription initiation sites. (Reproduced from Haqq et al., 1992, with permission of the publisher.)*

that the rat mRNA undergoes a change in the length of polyadenylation during fetal development [Lee et al., 1992]. At day 14 of fetal development, a single transcript is detected by Northern analysis in the testis. Between day 14 and birth at day 21, a second transcript can be detected with a shorter poly(A) tail. The longer transcript cannot be detected after birth. The significance of this developmental regulation is not clear, but may affect the translation or stability of the MIS mRNA during this period.

2. Chromosomal mapping. The human, bovine, and mouse MIS genes all have been mapped to autosomal chromosomes. Cohen-Haguenauer et al. [1987] demonstrated that the human MIS gene maps to the short arm of chromosome 19, band 19p13.3. Although the TGF-β1 gene also maps to chromosome 19 in humans, it is present on the long arm, subband q13.1–q13.3 [Fujii et al., 1986]. The bovine MIS gene maps to the same locus as osteonectin; the human homolog of osteonectin, SPARC, maps to human chromosome 5 [Rogers et al., 1991]. The mouse MIS gene has been mapped to the distal region of chromosome 10, between the phenyl hydroxylase and the zinc finger autosomal genes [King et al., 1991]. In addition, several polymorphic MIS-like sequences were mapped to other mouse chromosomes, suggesting the existence of pseudogenes in the mouse.

3. Promoter region. The nucleotide sequences immediately upstream of the transcription start sites of the human, bovine, and rat MIS genes are shown in Figure 2B.

Overall, these sequences show 74%–77% similarity, although gaps have been inserted to provide better alignment. In addition to the sequences shown, the nucleotide sequence of the human MIS gene has been determined from 3.6 kb upstream to the transcription initiation site [Guerrier et al., 1990]. A hybrid gene consisting of the human MIS promoter region (from 2.1 kb upstream to the start site) and the coding region of SV40 large T antigen causes tumors in the Sertoli cells of transgenic mice (J. Peschon, unpublished data). Thus most, if not all of the regulatory sequences that control the specific expression of MIS in Sertoli cells must reside within this region of the MIS promoter.

Within this region, there are some sequences that share homology with known regulatory elements [Jones et al., 1988]. There is an SP-1–binding site (CCGCCC) at −303 (and at −56 in the bovine promoter). SP-1 sequences are usually present in multiple copies or found together with AP-1 sites, which are not present in the MIS genes. In addition, the human MIS gene has such a high GC content (72% in the exons and 68% in the introns) that there are numerous CCGCCC sequences found throughout the gene, suggesting that the SP-1 sequences bear no function. A 13 bp palindrome similar to the estrogen response element (ERE) is present upstream of the human MIS gene at position −1772. [Guerrier et al., 1990]. This palindrome binds to the human estrogen receptor in vitro and activates transcription of a hybrid gene when several copies are inserted 5′ of a reporter gene. Whether the MIS gene is regulated by estrogen is not known, but the fact that MIS can downregulate aromatase and therefore estrogen levels in fetal ovary presents the interesting possibility of a feedback loop in granulosa cells. Two minisatellite-like regions that share a core TGGGGT element bracket the potential ERE at positions −2341 to −2150 and −1278 to −895. The rat and mouse promoters contain similar minisatellite regions upstream of −500 (R Cate, unpublished data).

Since gonadotropins, which use cyclic AMP for transducing signals, have been implicated in the regulation of the MIS gene [Voutilainen and Miller, 1987; Kuroda et al., 1990], it was possible that the cAMP response element (CRE) would be found in the promoter region. The consensus CRE (G/T) (A/T) CGTCA is not present, but a sequence with one mismatch is present at −262 in the human gene. Its significance is unclear since the sequence is not conserved in the other species and functional studies have not been done. A 9 bp sequence (GGAGATAGG) is found twice in the bovine and human promoters [Cate et al., 1990], but is not present in the rat and mouse promoter regions.

C. MIS Receptor

At present the MIS receptor has not been cloned, and very little is known about its expression. Clearly, the receptor is expressed in fetal müllerian ducts since they undergo MIS-induced regression in the male fetus. The müllerian ducts in female fetuses must also express the MIS receptor, since they undergo regression when exposed to MIS in vitro and in vivo in freemartins [Jost et al., 1972] and transgenic mice [Behringer et al., 1990]. In addition, MIS has been shown to suppress the induction of aromatase [Vigier et al., 1989] and induce cord formation in the fetal ovary [Vigier et al., 1987] indicating that cells within the fetal ovary must also possess the MIS receptor (see below). MIS has also been implicated in other biological processes such as germ cell maturation [Takahashi et al., 1986b; Munsterberg and Lovell-Badge, 1991], testicular development [Vigier et al., 1987; Behringer et al., 1990], testicular descent [Hutson and Donahoe, 1986], lung development [Catlin et al., 1988, 1990], and the growth regulation of tumor cells [Donahoe et al., 1987; Wallen et al., 1989; Chin et al., 1991]. If MIS is clearly involved in these processes, then the appropriate cell types must also bear the MIS receptor.

Both radiolabeling and immunological detection methods have failed to demonstrate specific binding of bovine or human MIS to tissue sections of müllerian ducts or fetal ovaries. Another approach has employed the anti-idiotypic antibodies raised against MAb 168, which blocks regression of the müllerian duct by bovine MIS [Lefevre et al., 1989]. MAb 168 has been shown to recognize the C terminus of MIS, the domain of MIS that has biological activity and presumably interacts with the MIS receptor (C Wilson et al., unpublished data). Since the anti-idiotypic antibodies exhibit MIS-like activity, they most likely interact with the receptor. However, these antibodies have not proved useful in detecting the receptor by immunohistochemistry.

Teng [1990] has detected binding of exogenous chick MIS by an indirect immunolabeling method to the mesenchymal cells that surround the chick müllerian duct, but controls without MIS were not performed. Using an antibody to chick MIS, Wang et al., [1990] have detected binding of endogenous chick MIS to the epithelial cells of the chick müllerian duct. If embryos were exposed to DES at day 5, a treatment believed to cause target cell insensitivity, binding of MIS was eliminated at day 7.5. Wang and Teng [1989] have reported the binding of chick MIS on rat müllerian ducts and human tumor cell lines. Whether the binding observed by the above authors reflects specific binding to the MIS receptor is not clear, since in some cases the binding occurs only on epithelial cells [Wang et al., 1990], and in other cases binding occurs on both epithelial cells and mesenchymal cells in the vicinity of the duct [Wang and Teng, 1989].

One of the problems in identifying the MIS receptor may be due to the requirement for proteolytic processing. A noncleavable mutant of MIS has no biological activity, indicating that the proteolytic cleavage is obligatory. Moreover, a large excess of the mutant cannot block the activity of wild-type MIS, indicating that it cannot bind to the MIS receptor [Cate et al., 1990]. Thus the full-length MIS molecule may not be a suitable probe for the receptor. In the future, attempts with the C-terminal fragment may provide a method for biochemical characterization. Also, molecular biology approaches may prove useful in isolating the gene for the receptor. Recently, the cDNA for the mouse activin receptor was isolated [Mathews and Vale, 1991], and, by homology to the nematode protein DAF, it was implied to code for a transmembrane serine-threonine kinase. If the MIS receptor is related to the activin receptor, a method for isolating the MIS receptor cDNA may now be available.

III. FUNCTION OF MIS

A. Overview

The expression of MIS is tightly regulated in both males and females from early development through adult life. In normal development, MIS is expressed in the gonads of both sexes, but with very distinct temporal patterns. The cell types that produce MIS, Sertoli cells in the testis [Blanchard and Josso, 1974] and granulosa cells in the ovary [Vigier et al., 1984a; Takahashi et al., 1986a], share many structural and functional features and are believed to be derived from a common progenitor cell [Fritz, 1982]. In both gonads, they form a diffusion barrier shielding germ cells from the somatic environment and release their secretion products into a fluid that bathes the germ cells. They share yet another similarity in that they both express MIS. However, expression of MIS in the Sertoli cell occurs early in fetal development of the male and after birth in the granulosa cell of the female.

Expression of MIS in the Sertoli cell may be regulated by the testis-determining factor on the Y chromosome, since MIS is the first known product of Sertoli cells. Recently a gene has been identified that appears to be the testis-determining gene in the mouse [Gubbay et al., 1990; Sinclair et al., 1990]. Expression of this putative DNA-binding protein (sry) precedes expression of MIS by 2 days, suggesting that regulation of MIS expression by sry would most likely be indirect [Munsterberg and Lovell-Badge, 1991]. The expression of MIS in the granulosa cell of the female gonad must involve a mechanism distinct from sry.

In Figure 3, the expression profile of MIS in the male and female is shown in relationship to its known and potential biological activities. The high level of expression of MIS in the male and the lack of expression in the female during fetal development is consistent with its role in regression of the müllerian duct in the male fetus, its best characterized activity. However, the continued expression of MIS in the male after the müllerian duct has regressed and the commencement of MIS expression in the female after birth have suggested that MIS may be involved in other biological processes. Also, abnormal expression of MIS leads to various perturbations in sexual development, as described in the following sections and outlined in Figure 3. That MIS might possess other biological activities is consistent with its position in the TGF-β family, whose members have many diverse activities. In the next two sections, the role of MIS in these processes will be examined and evaluated.

B. MIS in the Male

1. Ontogeny. MIS is one of the first proteins expressed during fetal development of the testis. Early in testis formation, the pre-Sertoli cells enlarge, aggregate with each other, and encompass the germ cells to form the seminiferous cords. In the mouse, MIS mRNA is detectable by day 12.5 postcoitum (pc) in the immature Sertoli cells within the cord region, while no MIS mRNA is detectable in the germ cells, in the interstitial and Leydig cells outside the cord region, or in any other fetal organ

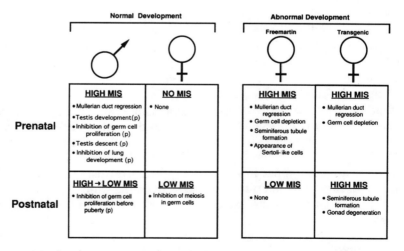

Fig. 3. *Expression profile and biological effects of MIS in normal and abnormal development. The well-documented and putative (p) effects of MIS are correlated with the expression levels of MIS at various times in development of the male and female.*

[Munsterberg and Lovell-Badge, 1991]. MIS activity is detectable at the time of seminiferous tubule formation in the testis of the pig at 25 days pc [Tran et al., 1977], of the calf at 43 days pc [Vigier et al., 1983], and of the rat at 13.5 days pc [Picon 1969, 1970; Tran et al., 1987]. Germ cells are not required for MIS expression, since mice homozygous for the W^e allele are devoid of germ cells and still express MIS [Munsterberg and Lovell-Badge, 1991]. In addition, seminiferous tubule formation is not a requirement for MIS production [Agelopoulou et al., 1984].

During fetal development, the level of MIS stays fairly constant in the mouse as judged by in situ hybridization [Munsterberg and Lovell-Badge, 1991] and in the rat as judged by bioactivity measurements [Picon, 1970; Donahoe et al., 1976; Tran et al., 1987]. In bovine fetuses, MIS levels plateau between 50 and 80 days pc, corresponding to the time when the müllerian duct regresses in the male calf [Vigier et al., 1983]. Vigier et al. [1982] have estimated the level of MIS in the serum of calf fetuses at 30–90 ng/ml.

Following birth, MIS levels progressively diminish until puberty and then sharply decrease. This decrease occurs after 2 weeks in the rat .[Picon, 1970; Donahoe et al., 1976; Kuroda et al., 1990], after 2 weeks in the mouse [Munsterberg and Lovell-Badge, 1991], and after 10–12 weeks in the calf [Donahoe et al., 1977c; Vigier et al., 1983]. The decrease in MIS production may coincide with the differentiation of the immature Sertoli cell to the mature Sertoli cell, but appears to precede the formation of the blood–testis barrier characterized by tight junctions between neighboring Sertoli cells[Josso and Picard, 1986]. These conclusions are supported by a study in the pig in which the decrease in MIS activity was found to occur after 60 days, prior to the formation of tight junctions and the blood–testis barrier, which occurs after 100 days [Tran et al., 1981]. Thus the decline in MIS

production may represent the first sign of functional maturation of the Sertoli cell [Josso and Picard, 1986]. Although the level of MIS usually falls below the level that permits bioactivity measurements, production of MIS by the mature Sertoli cell persists and has been detected in rete testis fluid in the boar [Josso et al., 1979] and bull [Vigier et al., 1983].

Human MIS levels show the same pattern after birth; serum levels decrease slowly from 40–150 ng/ml at 2 years to almost undetectable levels at the onset of puberty at 12 years [Baker et al., 1990; Hudson et al., 1990; Josso et al., 1990]. In these studies, Josso et al. [1990] observed significantly higher MIS levels in boys (12–18 years) with delayed puberty, while Baker et al. [1990] noted that MIS serum levels peaked in male infants aged 2–6 months, a time period that is coincidental with the transformation of the gonocytes to spermatogonia.

2. Müllerian duct regression. It is appropriate to begin the discussion of MIS function with its one unequivocal biological activity, the regression of the müllerian duct. Because the regression process involves multiple cell types, our understanding of how MIS binding to its receptor initiates the morphological changes that culminate in müllerian duct regression is still somewhat limited. Due to the lack of information on the MIS receptor, one must consider the possibility that either the epithelial cells, the mesenchymal cells, or both are the target cells for MIS binding. At present, most studies have focused on the cell biology of the regression process.

Programmed cell death has been proposed to play a role in regression, based on the appearance of lysosomes and the infiltration of the epithelium by macrophages [Price et al., 1977]. Dyche [1979] has challenged this conclusion, observing that phagocytic activity is also present in the persisting duct and may reflect normal remodeling. In addition, Hayashi et al. [1982] have shown that some of the cells in the müllerian epithelium remain viable, even after significant regression has occurred, while Trelstad et al. [1982] have argued that some epithelial cells may even reenter the mesenchymal cell population. Based on these observations and on others cited below, Trelstad et al. [1982] have favored an alternative hypothesis for müllerian duct regression in which müllerian duct morphogenesis is dependent on interactions between the müllerian duct epithelium and the surrounding mesenchyme, and MIS acts to disrupt these interactions.

Strong support for this hypothesis comes from the finding that one of the first histological changes initiated by MIS is the dissolution of the basement membrane [Trelstad et al., 1982]. Components of the basement membrane, including laminin, heparin sulfate proteoglycan, and type IV collagen, have been shown to disappear as the duct regresses [Ikawa et al., 1984b]. The loss of basement membrane could lead to the disassociation of the epithelial cells observed during regression. In addition, it has been shown that the mesenchymal matrix components, including fibronectin [Ikawa et al., 1984b] and hyaluronate [Hayashi et al., 1982], are lost from the mesenchyme near the duct. Loss of periductal fibronectin could contribute to epithelial cell disassociation and lead to condensation of the mesenchymal cells around the regressing duct. Thus MIS could interact with either the epithelial or mesenchymal cell to induce the synthesis of enzymes that degrade the matrix. Consistent with this possibility, regression of the chick müllerian duct can be blocked by physical manipulations that are thought to inactivate proteolytic enzymes [Lutz and Lutz-Ostertag, 1956; Salzgeber, 1961].

The müllerian duct is only sensitive to MIS for a defined period in development. Fetal rat ducts that are exposed to MIS after day 15 pc show no response to MIS, while those exposed to MIS up to 15 days pc will continue to regress, even if the MIS is removed [Picon, 1969; Donahoe et al., 1977a]. A similar "critical" period has been demonstrated for human müllerian ducts [Josso et al., 1977; Taguchi et al., 1984]. The sensitivity of the müllerian ducts to MIS action is also influenced by steroids and the second messenger cAMP. In vivo, DES prevents regression of the müllerian duct in the chick [Wolf, 1939] and the mouse [Newbold et al., 1984], while dibutyryl cAMP [Picon, 1976] and phosphodiesterase inhibitors [Ikawa et al., 1984a] inhibit MIS-induced regression in vitro. Since DES and dibutyryl cAMP do not suppress testicular production of MIS [Hutson et al., 1982; Ikawa et al., 1984a] it has been suggested that they alter the state of differentiation of the epithelial or mesenchymal cells, thereby affecting their sensitivity to the action of MIS.

The ability of MIS to cause regression of the müllerian duct can be measured in vitro with an organ culture assay developed by Picon [1969]. The urogenital ridge of a 14.5-day-old rat fetus is used as the target organ, and MIS-induced changes in the müllerian duct are assessed histologically after 3 days of exposure to MIS in organ culture. The main features of MIS-induced regression are basement membrane dissolution, epithelial cell disassociation and mesenchymal condensation (Fig. 4). Although the müllerian duct regression assay is quite specific for MIS and has been essential in the purification of the hormone, the assay is somewhat qualitative and subjective. Donahoe et al. [1977a] have proposed a grading system for the assay based on the level of müllerian duct regression, while Josso and colleagues used a statistical evaluation [Josso and Picard, 1986]. To achieve complete regression of the duct, a concentration of 2.5 µg/ml (17.5 nM) bovine or human MIS is required. More recently, an assay based on the ability of MIS to cause sex reversal of the fetal ovary has been developed [di Clement et al., 1992] which provides a more quantitative measure of

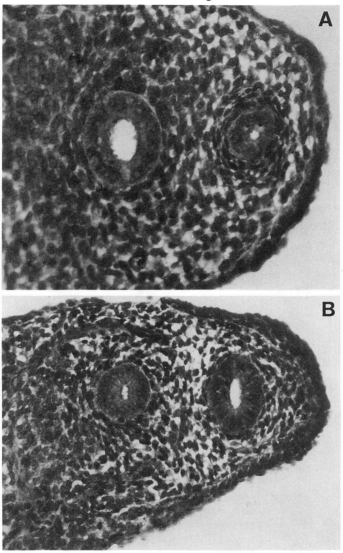

Fig. 4. *Regression of the rat müllerian duct by human recombinant MIS.*
A: *Partial regression of the müllerian duct produced by human MIS.* **B:**
The results of an organ culture assay performed in the absence of MIS.
The wolffian duct is on the left, and the müllerian duct is on the right.
(Reproduced from Cate et al., 1990, with permission of the publisher.)

MIS activity (see Effects of MIS in the Fetal Ovary, below).

3. Testis development. The hypothesis that MIS might play a role in testicular development is suggested essentially by two lines of evidence. The first concerns the tendency of XX/XY chimeras to develop as males even when the percentage of XY cells is low [McLaren, 1984] and the ability of ovaries to undergo sex reversal when

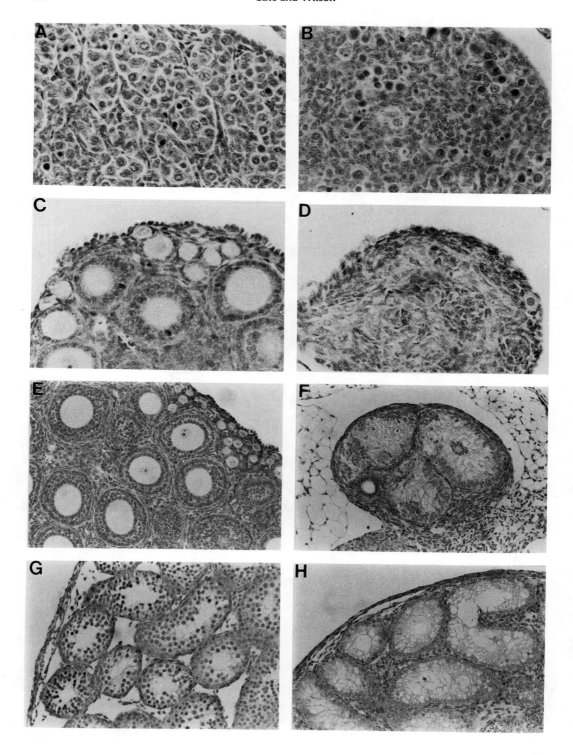

transplanted in males [Taketo et al., 1984; Maraud et al., 1990], from which it had been inferred that testes produce a secreted factor that diverts the ovary to a testicular pathway. The second concerns the ability of MIS to cause an apparent sex reversal of the fetal ovary. As discussed in the following section on the ovary, inappropriate exposure of fetal ovaries to MIS in vitro results in germ cell depletion, the formation of structures resembling seminiferous cords, and the appearance of cells that resemble Sertoli cells [Vigier et al., 1987]. In vivo, similar effects are observed in the gonads of freemartins, in which the female fetus is exposed to the MIS of her male twin [Jost et al., 1972], and in the gonads of transgenic female mice that chronically express MIS (Figs. 3, 5) [Behringer et al., 1990].

The above argument is circumstantial and obviously does not provide any degree of proof of MIS involvement in testicular development. Critics of this hypothesis have questioned whether the effects of MIS on the fetal ovaries actually represent a true sex reversal. Merchant-Larios and Taketo [1990] argue that the formation of seminiferous cords in MIS-treated ovaries is a consequence of oocyte depletion, because oocytes are not present to "split" the gonadal cords and establish the primordial follicles. At later stages, the retained gonadal cords become conspicuous and resemble seminiferous cords. This would explain cord forma-

Fig. 5. *Effects of human MIS in male and female transgenic mice. Hematoxylin and eosin stained histological sections of control (A,C,E,G) and MIS transgenic mice gonads (B,D,F,H). The transgenic female gonads show germ cell depletion at day 2 (A,B), a lack of follicle formation at day 9 (C,D), and the formation of seminiferous tubules by day 16 (E,F). The transgenic male gonad is from a mouse with undescended testes and is devoid of germ cells (G,H). (Reproduced from Behringer et al., 1990, with permission of the publisher.)*

tion in the XX gonads of W/W^v mutant mice or busulphan-treated rats, in which oogonia are absent or scarce [Merchant-Larios, 1976; Merchant-Larios and Centeno, 1981], and in fetal rat ovaries maintained 16 days in organ culture which also lose their germ cells [Prepin and Hida, 1989].

To assess true sex reversal, Merchant-Larios and Taketo [1990] suggest that ultrastructural studies must be performed to show that the epithelial cells within the cords have the characteristics of Sertoli cells. Such an analysis was not performed on the gonads from the MIS-expressing transgenic mice [Behringer et al., 1990], but Vigier et al. [1987] did show that the epithelial cells within the cords of MIS-treated ovaries had the characteristics of Sertoli cells. A final resolution of this debate may come when the MIS receptor status of the germ cells and granulosa cells is determined. Clearly, if only the germ cells possess the receptor, then all the MIS effects in the ovary must be mediated by them. If, on the other hand, the granulosa cell also has the receptor, then MIS may be responsible for its redifferentiation to a Sertoli-like cell.

Compelling evidence that MIS does not have an obligatory role in testicular development comes from a study of human males with persistent müllerian duct syndrome (PMDS), who have a defective gene for either MIS or the MIS receptor [Guerrier et al., 1989]. Typically, males with PMDS retain a uterus, but possess testes, which although not properly descended, appear to be normal. Knebelmann et al. [1991] have reported the isolation and sequence analysis of a mutant MIS gene from a PMDS patient. They showed that a mutation in the fifth exon changes the codon for Glu 358 to a stop codon, resulting in a truncated, unstable protein. The experiments of Tran et al. [1986] in which müllerian duct regression was blocked in rabbit fetuses by immunizing the mother with MIS, also argues against a role for MIS in testicular differentiation, since there was no effect on testis development.

4. Regulation of germ cell development. In normal development, XX germ cells in the ovary enter the prophase of meiosis well before birth, while XY germ cells in the testis enter a state of mitotic arrest. Mitotic proliferation and eventually meiosis begin after birth in the male. There is a close correlation between the arrest of germ cells at the prespermatogonia stage and the initiation of MIS production by adjacent somatic cells in the B6.Y^{DOM} ovotestis [Taketo et al., 1991]. MIS expression is switched off between 2 and 3 weeks after birth in the mouse, coincident with the first wave of spermatids undergoing meiosis [Munsterberg and Lovell-Badge, 1991]. Together, these results suggest that MIS may cause the mitotic arrest of male germ cells. Consistent with this hypothesis, the loss of oocytes in MIS-treated fetal ovaries is mainly due to inhibition of oogonial replication [Vigier et al., 1987]. Another function of MIS may be to eliminate any germ cells that enter meiosis prematurely [McLaren, 1990].

5. Descent of the testis. Descent of the testis can be separated into two phases: the first part consists of transabdominal movement of the testis from the posterior abdominal wall to the inguinal region, and the second phase occurs when the testis descends through the inguinal canal to the scrotum. The initial phase is independent of androgen and has been proposed by Hutson and Donahoe [1986] to be under the control of MIS. The hypothesis is based on the fact that MIS levels in infants with undescended testes are less than normal [Donahoe et al., 1977b], the fact that 70% of PMDS patients have completely undescended testes [Hutson and Donahoe, 1986], and the fact that fetal mice exposed to estrogen have undescended testes and retained müllerian ducts [Newbold et al., 1984]. Interestingly, transgenic male mice expressing high levels of human MIS some-

times have undescended testes that are devoid of germ cells (Fig. 5) [Behringer et al., 1990]. Although this appears to contradict the above hypothesis, it is possible that the high MIS levels desensitize the target tissue.

Against this model, Tran et al. [1986] did not see an effect on testicular descent in rabbits in which müllerian duct regression was blocked by immunization of the mother with MIS. Also, the high incidence of undescended testes in PMDS patients may be due to a mechanical impediment to descent [Josso et al., 1983].

6. Other potential roles in the male. Catlin et al. [1988] have investigated whether MIS might contribute to the deficiency of pulmonary surfactant observed in human male neonates who suffer from respiratory distress syndrome. They have reported that female lung fragments accumulated less of the major surfactant phospholipid when cultured together with testis or partially purified bovine and human recombinant MIS than do controls exposed to vehicle buffer. These effects have also been demonstrated in vivo [Catlin et al., 1990]. However, it should be noted that male transgenic mice expressing human MIS did not appear to have any perturbation in lung development [Behringer et al., 1990].

Interestingly, some of the male transgenic mice did show perturbations in reproductive development in addition to those described in the previous sections. Some males had mammary glands, feminized genitalia, and improper wolffian development [Behringer et al., 1990]. These changes suggest a problem with synthesis of androgen or androgen receptor and may reflect an effect of high MIS levels on Leydig cell function.

C. MIS in the Female

1. Ontogeny. Since exposure of the fetal ovary to MIS results in changes that are incompatible with reproductive function, it is important that the MIS gene be silenced during fetal development of the female. Thus it is not surprising that MIS is undetectable in the fetal ovaries of mouse [Munsterberg and Lovell-Badge, 1991], rat [Ueno et al, 1989a], and the ewe [Bezard et al., 1987]. A rather complex situation exists in the female chick embryo, where the right gonad becomes an ovotestis and produces MIS, causing the right müllerian duct to regress [Hutson et al., 1981]. The left müllerian duct does not regress and becomes the functional sex tract for the female chick. Unexpectedly, it was found that the left ovary also makes MIS, raising the question of why the left duct does not also undergo regression [Hutson et al., 1981]. Considerable evidence now suggests that estrogens protect the left duct from MIS action [MacLaughlin et al., 1983; Hutson et al., 1985; Stoll et al., 1990].

After birth, MIS is expressed in granulosa cells in a transient manner during follicular development. From ontogeny studies in the cow [Takahashi et al., 1986a], ewe [Bezard et al., 1987], rat [Ueno et al., 1989a,b], and mouse [Munsterberg and Lovell-Badge, 1991], the following picture has emerged. Low levels of MIS can be detected in developing follicles prior to puberty, but much higher levels are seen in the adult. As the follicle grows, the MIS levels of granulosa cells close to the basal lamina fade, while the level in the cells lining the antrum and in the cumulus oophorus increases. MIS production appears to be confined to the subpopulation of granulosa cells that is incapable of undergoing terminal differentiation under FSH stimulation [Erickson et al., 1985]. Just before ovulation, when the oocyte completes the first meiotic division, MIS becomes undetectable in cumulus cells [Ueno et al., 1989b; Munsterberg and Lovell-Badge 1991].

The level of MIS in the bovine adult ovary is low compared with newborn testis, as judged by Northern blot [Cate et al.,

1986; Picard et al., 1986a]. The level of MIS mRNA in the adult rat ovary is higher than that observed in the adult bovine ovary and may reflect the multiparous nature of the rat ovary [Haqq et al., 1992].

2. Effects of MIS in the fetal ovary. Although MIS is not produced during fetal development of the female, cells of the müllerian duct and the ovary clearly possess the receptor and will respond to MIS, if given the opportunity. The syndrome known as the *freemartin*, which has been recognized for centuries by cattle farmers, is the most vivid example of this. Müllerian duct regression occurs simultaneously in these male and female bovine twins due to the transfer of MIS [Vigier et al., 1984b] through their chorionic vascular anastomoses.

In addition to the müllerian effect, the ovaries of freemartins cease to grow, become depleted of germ cells, and eventually, in about one half of the cases, develop seminiferous tubules [Jost et al., 1972, 1973, 1975]. The possibility that MIS was also responsible for the ovarian effect was first considered by Jost et al. [1972] but could not be addressed without purified MIS. Vigier et al. [1987] proved definitively that MIS is indeed responsible, by exposing rat fetal ovaries to MIS in organ culture. MIS consistently produced a characteristic freemartin effect, characterized by a reduction of gonadal volume, germ cell depletion, and differentiation of epithelial cells to cells resembling rat fetal Sertoli cells. Chronic expression of human MIS in female transgenic mice produces a phenotype similar to the freemartin [Behringer et al., 1990]. Müllerian duct differentiation is inhibited, resulting in a blind vagina and no uterus or oviducts; germ cells are lost, and eventually the somatic cells organize as seminiferous tubules (Fig. 5). Unlike the freemartin condition in which the exposure to MIS is halted at birth and the abnormal ovary is maintained, the ovary in the transgenic female mouse continues to be exposed to

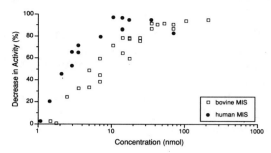

Fig. 6. *Concentration profiles for bovine and human MIS in the fetal ovary/aromatase assay. Data are from di Clemente et al. [1992], and show the percent decrease in aromatase activity produced by either bovine or human MIS relative to controls performed in the absence of MIS.*

lar development has suggested that it may have a physiological role in the adult ovary, perhaps in regulating oogenesis. Takahashi et al. [1986b] have reported that bovine MIS can inhibit the resumption of meiosis in both denuded and cumulus-enclosed rat oocytes in vitro, but Tsafriri et al. [1988] have not been able to repeat these results. Ueno et al. [1988] have shown that human recombinant MIS could inhibit germinal vesicle breakdown, but the effect required the presence of Nonidet P-40. Our understanding of MIS function in the adult ovary may be advanced when the MIS gene is "knocked out" in the mouse.

MIS and apparently degenerates by the adult stage (see Fig. 3).

In addition to causing a morphological sex reversal of the fetal ovary, MIS causes an endocrine sex reversal. Ovine fetal ovaries exposed to MIS were shown to release testosterone instead of estradiol due to the suppression of the enzyme aromatase [Vigier et al., 1989]. Aromatase activity of rabbit fetal ovaries was also decreased by MIS, as was the cAMP-stimulated aromatase activity of rat fetal ovaries. These results have formed the basis for developing another assay for MIS bioactivity: the fetal ovary/aromatase assay [di Clemente et al., 1992]. Rat fetal ovaries that are 16 days-old are explanted in organ culture for 3 days in the presence of dibutyryl cAMP, and aromatase activity is measured at the end of the culture period by the tritiated water technique [Ackerman et al., 1981]. As shown in Figure 6, a linear log/dose response to MIS treatment is observed between 1.5 and 30 nM bovine MIS and between 1 and 10 nM for human MIS. The assay is more sensitive than the müllerian duct regression assay and provides a level of quantitation that has not been possible until now.

3. Effects of MIS in the adult ovary. The transient expression of MIS during follicu-

IV. SUMMARY

Given the list of potential activities that have now been suggested for MIS, the original names of müllerian-inhibiting substance and antimüllerian hormone no longer seem appropriate. Over the next few years, it should be possible to clarify whether MIS is involved in testicular differentiation, germ cell development, testicular descent, and lung development. Clearly, the isolation of the MIS receptor will shed a great deal of light on these hypotheses. Also, further insight into MIS function in the adult ovary, as well as the testis, may come when mice are generated in which both MIS alleles have been rendered inactive.

REFERENCES

Ackerman GE, Smith ME, Mendelson CR, MacDonald PC, Simpson ER (1981): Aromatization of androstenedione by human adipose tissue stromal cells in monolayer culture. J Clin Endocrinol Metab 53:412–417.

Agelopoulou R, Magre S, Patsavoudi E, Jost A (1984): Initial phases of the rat testis differentiation in vitro. J Embryol Exp Morphol 83:15–31.

Baker ML, Metcalfe SA, Hutson JM (1990): Serum levels of müllerian inhibiting substance in boys from birth to 18 years, as determined by enzyme immunoassay. J Clin Endocrinol Metab 70:11–15.

Barnard JA, Lyons RM, Moses HL (1990): The cell biology of transforming growth factor beta. Biochim Biophys Acta 1032:79–87.

Behringer RR, Cate RL, Froelick GJ, Palmiter RD, Brinster RL (1990): Abnormal sexual development in transgenic mice chronically expressing müllerian inhibiting substance. Nature 345:167–170.

Benoit R, Ling N, Esch F (1987): A new pro-somatostatin-derived peptide reveals a pattern for prohormone cleavage at monobasic sites. Science 238:1126–1129.

Bezard J, Vigier B, Tran D, Mauleon P, Josso N (1987): Immunocytochemical study of anti-müllerian hormone in sheep ovarian follicles during fetal and post-natal development. J Reprod Fertil 80:509–516.

Blanchard MG, Josso N (1974): Source of the anti-müllerian hormone synthesized by the fetal testis: Müllerian-inhibiting activity of fetal bovine Sertoli cells in tissue culture. Pediatr Res 8:968–971.

Budzik GP, Donahoe PK, Hutson JM (1985): A possible purification of müllerian inhibiting substance and a model for its mechanism of action. In Lash JW, Saxens L (eds): Developmental Mechanisms: Normal and Abnormal. New York: Alan R. Liss, Inc., pp 207–223.

Budzik GP, Powell SM, Kamagata S, Donahoe PK (1983): Müllerian inhibiting substance fractionation by dye affinity chromatography. Cell 34:307–314.

Budzik GP, Swann DA, Hayashi A, Donahoe PK (1980): Enhanced purification of müllerian inhibiting substance by lectin affinity chromatography. Cell 21:909–915.

Cate RL, Donahoe PK, MacLaughlin DT (1990): Müllerian-inhibiting substance. In Sporn MB, Roberts AB (eds): Peptide Growth Factors and Their Receptors, II. Berlin: Springer-Verlag, pp 179–210.

Cate RL, Mattaliano RJ, Hession C, Tizard R, Farber NM, Cheung A, Ninfa EG, Frey AZ, Gash DJ, Chow EP, Fisher RA, Bertonis JM, Torres G, Wallner BP, Ramachandran KL, Ragin RC, Manganaro TF, MacLaughlin DT, Donahoe PK (1986): Isolation of the bovine and human genes for müllerian inhibiting substance and expression of the human gene in animal cells. Cell 45:685–698.

Catlin EA, Manganaro TF, Donahoe PK (1988): Müllerian inhibiting substance depresses accumulation in vitro of disaturated phosphatidylcholine in fetal rat lung. Am J Obstet Gynecol 159:1299–1303

Catlin EA, Powell SM, Manganaro TF, Hudson PL, Ragin RC, Epstein J, Donahoe PK (1990): Sex-specific fetal lung development and müllerian in-

hibiting substance. Am Rev Respir Dis 141:466–470.

Chin TW, Parry RL, Donahoe PK (1991): Human müllerian inhibiting substance inhibits tumor growth in vitro and in vivo. Cancer Res 51:2101–2106.

Cohen-Haguenauer HO, Picard JY, Mattei MG, Serero S, Nguyen VC, de Tand MF, Guerrier D, Hors-Cayla MC, Josso N, Frezal J (1987): Mapping of the gene for anti-müllerian hormone to the short arm of human chromosome 19. Cytogenet Cell Genet 44:2–6.

di Clemente N, Ghaffari S, Pepinsky RB, Pieau C, Josso N, Cate RL, Vigier B (1992): A quantitative and interspecific test for biological activity of anti-Müllerian hormone: The fetal ovary aromatase assay. Development 114:721–727.

Donahoe PK, Cate RL, MacLaughlin DT, Epstein J, Fuller AF, Takahashi M, Coughlin JP, Ninfa EG, Taylor LA (1987): Müllerian inhibiting substance: Gene structure and mechanism of action of a fetal regressor. Recent Prog Horm Res 43:431–467.

Donahoe PK, Ito Y, Hendren WH (1977a): A graded organ culture assay for the detection of müllerian inhibiting substance. J Surg Res 23:141–148.

Donahoe PK, Ito Y, Marfatia S, Hendren WH (1976): The production of müllerian inhibiting substance by the fetal, neonatal and adult rat. Biol Reprod 15:329–334.

Donahoe PK, Ito Y, Morikawa Y, Hendren WH (1977b): Müllerian inhibiting substance in human testes after birth. J Pediatr Surg 12:323–330.

Donahoe PK, Ito Y, Price JM, Hendren WH (1977c): Müllerian inhibiting substance activity in bovine fetal, newborn and prepubertal testes. Biol Reprod 16:238–243.

Dyche WJ (1979): A comparative study of the differentiation and involution of the müllerian duct and wolffian duct in the male and female fetal mouse. J Morphol 162:175–209.

Erickson GF, Hofeditz C, Unger M, Allen WR, Dulbecco R (1985): A monoclonal antibody to a mammary cell line recognizes two distinct subtypes of ovarian granulosa cells. Endocrinology 117:1490–1499.

Fritz IB (1982): Comparison of granulosa and Sertoli cells at various stages of maturation: similarities and differences. In Channing CP, Segals SJ (eds): Intraovarian Control Mechanisms. New York: Plenum, pp 357–384.

Fujii D, Brissenden JE, Derynck R, Francke U (1986): Transforming growth factor beta gene maps to human chromosome 19 long arm and to mouse chromosome 7. Somat Cell Mol Genet 12:281–288.

Gubbay J, Collignon J, Koopman P, Capel B, Economou A, Munsterberg A, Vivian N, Goodfellow P, Lovell BR (1990): A gene mapping to the sex-determining region of the mouse Y chromosome is a member of a novel family of embryonically expressed genes. Nature 346:245–250.

Guerrier D, Boussin L, Mader S, Josso N, Kahn A, Picard JY (1990): Expression of the gene for anti-müllerian hormone. J Reprod Fertil 88:695–706.

Guerrier D, Tran D, Vanderwinden JM, Hideux S, Van Outryve L, Legeai L, Bouchard M, Van Vliet G, DeLaet MH, Picard JY, Kahn A, Josso N (1989): The persistent müllerian duct syndrome: A molecular approach. J Clin Endocrinol Metab 68:46–52.

Haqq C, Lee MM, Tizard R, Wysk M, DeMarinis J, Donahoe PK, Cate RL (1992): Isolation of the rat gene for müllerian inhibiting substance. Genomics 12:665–669.

Hayashi A, Donahoe PK, Budzik GP, Trelstad RL (1982): Periductal and matrix glycosaminoglycans in rat müllerian duct development and regression. Dev Biol 92:16–26.

Hayashi M, Shima H, Hayashi K, Trelstad RL, Donahoe PK (1984): Immunocytochemical localization of Müllerian inhibiting substance in the rough endoplasmic reticulum and Golgi apparatus in Sertoli cells of the neonatal calf testis using a monoclonal antibody. J Histochem Cytochem 32:649–654.

Hudson PL, Douglas I, Donahoe PK, Cate RL, Epstein J, Pepinsky RB, MacLaughlin DT (1990): An immunoassay to detect human müllerian inhibiting substance in males and females during normal development. J Clin Endocrinol Metab 70:16–22.

Hutson JM, Donahoe PK (1986): The hormonal control of testicular descent. Endocr Rev 7:270–83.

Hutson JM, Donahoe PK, MacLaughlin DT (1985): Steroid modulation of müllerian duct regression in the chick embryo. Gen Comp Endocrinol 57:88–102.

Hutson J, Ikawa H, Donahoe PK (1981): The ontogeny of müllerian inhibiting substance in the gonads of the chicken. J Pediatr Surg 16:822–827.

Hutson JM, Ikawa H, Donahoe PK (1982): Estrogen inhibition of müllerian inhibiting substance in the chick embryo. J Pediatr Surg 17:953–959.

Ikawa H, Hutson JM, Budzik GP, Donahoe PK (1984a): Cyclic adenosine 3′, 5′-monophosphate modulation of müllerian duct regression. Endocrinology 114:1686–1691.

Ikawa H, Trelstad RL, Hutson JM, Manganaro TF, Donahoe PK (1984b): Changing patterns of fibronectin, laminin, type IV collagen, and a basement membrane proteoglycan during rat müllerian duct regression. Dev Biol 102:260–263.

Jones NC, Rigby PW, Ziff EB (1988): Trans-acting protein factors and the regulation of eukaryotic transcription: lessons from studies on DNA tumor viruses. Genes Dev 2:267–281.

Josso N, Fekete C, Cachin O, Nezelof C, Rappaport R (1983): Persistence of müllerian ducts in male pseudohermaphroditism, and its relationship to cryptorchidism. Clin Endocrinol 19:247–258.

Josso N, Legeai L, Forest MG, Chaussain JL, Brauner R (1990): An enzyme linked immunoassay for anti-müllerian hormone: A new tool for the evaluation of testicular function in infants and children. J Clin Endocrinol Metab 70:23–27.

Josso N, Picard JY (1986): Anti-müllerian hormone. Physiol Rev 66:1038–1090.

Josso N, Picard JY, Dacheux JL, Courot M (1979): Detection of anti-müllerian activity in boar rete testis fluid. J Reprod Fertil 57:397–400.

Josso N, Picard JY, Tran D (1977): The anti-müllerian hormone. Birth Defects 13:59–84.

Josso N, Picard JY, Tran D (1980): A new testicular glycoprotein: Anti-müllerian hormone. In Steinberger A, Steinbergers E (eds): Testicular Development, Structure and Function. New York: Raven, pp 21–31.

Jost A (1947): Recherches sur la differenciation sexuelle de l'embryon de lapin. Arch Anat Microsc Morphol Exp 36:271–315.

Jost A, Perchellet JP, Prepin J, Vigier B (1975): The prenatal development of bovine freemartins. In Reinborns R (eds): Symposium on Intersexuality. Berlin: Springer-Verlag, pp 349–406.

Jost A, Vigier B, Prepin J (1972): Freemartins in cattle: the first steps of sexual organogenesis. J Reprod Fertil 29:349–379.

Jost A, Vigier B, Prepin J, Perchellet JP (1973): Studies on sex differentiation in mammals. Recent Prog Horm Res 29:1–41.

King TR, Lee BK, Behringer RR, Eicher EM (1991): Mapping anti-müllerian hormone (Amh) and related sequences in the mouse: identification of a new region of homology between MMU10 and HSA19p. Genomics 11:273–283.

Knebelmann B, Boussin L, Guerrier D, Legeai L, Kahn A, Josso N, Picard JY (1991): Anti-müllerian hormone Bruxelles: A nonsense mutation associated with the persistent müllerian duct syndrome. Proc Natl Acad Sci U S A 88:3767–3771.

Kuroda T, Lee MM, Haqq CM, Powell DM, Manganaro TF, Donahoe PK (1990): Müllerian inhibiting substance ontogeny and its modulation by follicle-stimulating hormone in the rat testes. Endocrinology 127:1825–1832.

Lawrence DA, Pircher R, Jullien P (1985): Conversion of a high molecular weight latent beta-TGF from chicken embryo fibroblasts into a low molecular weight active beta-TGF under acidic conditions. Biochem Biophys Res Commun 133:1026–1034.

Lee MM, Cate RL, Donahoe PK, Waneck GL (1992): Developmentally regulated polyadenylation of two discrete mRNAs for müllerian inhibiting substance. Endocrinology 130:847–853.

Lefevre G, Tran D, Hoebeke J, Josso N (1989): Anti-idiotypic antibodies to a monoclonal antibody raised against anti-müllerian hormone exhibit anti-müllerian biological activity. Mol Cell Endocrinol 62:125–133.

Legeai L, Vigier B, Tran D, Picard JY, Josso N (1986): Monoclonal antibodies raised against bovine anti-müllerian hormone: Bovine, ovine, and caprine hormones share a set of identical epitopes. Biol Reprod 35:1217–1225.

Lutz H, Lutz-Ostertag Y (1956): Action des ultra-sons sur l'enzyme proteolytique des canaux de Müller de poulet male. CR Seances Acad Sci Ser III Sci 150:913.

MacLaughlin DT, Hutson JM, Donahoe PK (1983): Specific estradiol binding in embryonic Müllerian ducts: A potential modulator of regression in the male and female chick. Endocrinology 113:141–145.

Maraud R, Vergnaud O, Rashedi M (1990): New insights on the mechanism of testis differentiation from the morphogenesis of experimentally induced testes in genetically female chick embryos. Am J Anat 188:429–437.

Massague J (1990): The transforming growth factor-beta family. Annu Rev Cell Biol 6:597–641.

Mathews LS, Vale WW (1991): Expression cloning of an activin receptor, a predicted transmembrane serine kinase. Cell 65:973–982.

McLaren A (1984): Chimeras and sexual differentiation. In Le Douarin N. McLarens A (eds): Chimeras in Developmental Biology. London: Academic Press, pp 381–400.

McLaren A (1990): Sexual differentiation. Of MIS and the mouse. Nature 345:111.

Merchant-Larios H (1976): The role of germ cells in the morphogenesis and cytodifferentiation of the rat ovary. In Müller-Berats (ed): Progress in Differentiation Research. Amsterdam: North Holland, pp 453–462.

Merchant-Larios H, Centeno B (1981): Morphogenesis of the ovary from the sterile W/W^v mouse. In Vidrio A. Galinas C (eds): Advances in the Morphology of Cells and Tissues. New York: Alan R. Liss, pp 383–392.

Merchant-Larios H, Taketo T (1990): Testicular differentiation in mammals under normal and experimental conditions. J Electron Microsc Tech 19:158–171.

Munsterberg A, Lovell-Badge R (1991): Expression of the mouse anti-müllerian hormone gene suggests a role in both male and female sexual differentiation. Development 113:613–624.

Newbold RR, Suzuki Y, McLachlan JA (1984): Müllerian duct maintenance in heterotypic organ culture after in vivo exposure to diethylstilbestrol. Endocrinology 115:1863–1868.

Pepinsky RB, Sinclair LK, Chow EP, Mattaliano RJ, Manganaro TF, Donahoe PK, Cate RL (1988): Proteolytic processing of Müllerian inhibiting substance produces a transforming growth factor-beta-like fragment. J Biol Chem 263:18961–18964.

Picard JY, Benarous R, Guerrier D, Josso N, Kahn A (1986a): Cloning and expression of cDNA for anti-müllerian hormone. Proc Natl Acad Sci USA 83:5464–5468.

Picard JY, Goulut C, Bourrillon R, Josso N (1986b): Biochemical analysis of bovine testicular anti-müllerian hormone. FEBS Lett 195:73–76.

Picard JY, Josso N (1984): Purification of testicular anti-müllerian hormone allowing direct visualization of the pure glycoprotein and determination of yield and purification factor. Mol Cell Endocrinol 34:23–29.

Picon R (1969): Action du testicule foetal sur le developpement in vitro des canaux de Müller chez le rat. Arch Anat Microsc Morphol Exp 58:1–19.

Picon R (1970): Modifications, chez le rat, au cours du developpement du testicule, de son action inhibitrice sur les canaux de Müller in vitro. CR Acad Sci 271:2370–2372.

Picon R (1976): Testicular inhibition of fetal müllerian ducts in vitro: effect of dibutyryl cyclic AMP. Mol Cell Endocrinol 4:35–42.

Pircher R, Jullien P, Lawrence DA (1986): Beta-transforming growth factor is stored in human blood platelets as a latent high molecular weight complex. Biochem Biophys Res Commun 136:30–37.

Prepin J, Hida N (1989): Influence of age and medium on formation of epithelial cords in the rat fetal ovary in vitro. J Reprod Fertil 87:375–382.

Price JM, Donahoe PK, Ito Y, Hendren WH (1977): Programmed cell death in the müllerian duct induced by müllerian inhibiting substance. Am J Anat 149:353–75.

Roberts AB, Sporn MB (1990): The transforming growth factor-betas. In Sporn MB, Robertss AB (eds): Peptide Growth Factors and Their Receptors I. Heidelberg: Springer-Verlag, pp 419–472.

Rogers DS, Gallagher DS, Womack JE (1991): Somatic cell mapping of the genes for anti-müllerian hormone and osteonectin in cattle: identification of a new bovine syntenic group. Genomics 9:298–300.

Salzgeber B (1961): Evolution des gonades et des conduits genitaux de l'embryon de poulet soumis en culture a l'action des rayons x. Bull Biol Fr Belg 44:645.

Sinclair AH, Berta P, Palmer MS, Hawkins JR, Griffiths BL, Smith MJ, Foster JW, Frischauf AM, Lovell BR, Goodfellow PN (1990): A gene from the human sex-determining region encodes a protein with homology to a conserved DNA-binding motif. Nature 346:240–244.

Sporn MB, Roberts AB, Wakefield LM, de Crombrugghe B (1987): Some recent advances in the chemistry and biology of transforming growth factor-beta. J Cell Biol 105:1039–1045.

Stoll R, Faucounau N, Maraud R (1990): Action of estradiol on müllerian duct regression induced by treatment with norethindrone of female chick embryos. Gen Comp Endocrinol 80:101–106.

Swann DA, Donahoe PK, Ito Y, Morikawa Y, Hendren WH (1979): Extraction of müllerian inhibiting substance from newborn calf testis. Dev Biol 69:73–84.

Taguchi O, Cunha GR, Lawrence WD, Robboy SJ (1984): Timing and irreversibility of müllerian duct inhibition in the embryonic reproductive tract of the human male. Dev Biol 106:394–398.

Takahashi M, Hayashi M, Manganaro TF, Donahoe PK (1986a): The ontogeny of müllerian inhibiting substance in granulosa cells of the bovine ovarian follicle. Biol Reprod 35:447–453.

Takahashi M, Koide SS, Donahoe PK (1986b): Müllerian inhibiting substance as oocyte meiosis inhibitor. Mol Cell Endocrinol 47:225–234.

Taketo T, Merchant LH, Koide SS (1984): Induction of testicular differentiation in the fetal mouse ovary by transplantation into adult male mice. Proc Soc Exp Biol Med 176:148–153.

Taketo T, Saeed J, Nishioka Y, Donahoe PK (1991): Delay of testicular differentiation in the B6.YDOM ovotestis demonstrated by immunocytochemical staining for müllerian inhibiting substance. Dev Biol 146:386–395.

Teng CS (1990): Quantitative change in fibronectin in cultured müllerian mesenchymal cells in response to diethylstilbestrol and müllerian-inhibiting substance. Dev Biol 140:1–7.

Tran D, Josso N (1982): Localization of anti-müllerian hormone in the rough endoplasmic reticulum of the developing bovine sertoli cell using immunocytochemistry with a monoclonal antibody. Endocrinology 111:1562–1567.

Tran D, Muesy DN, Josso N (1977): Anti-müllerian hormone is a functional marker of foetal Sertoli cells. Nature 269:411–412.

Tran D, Meusy DN, Josso N (1981): Waning of anti-müllerian activity: An early sign of Sertoli cell maturation in the developing pig. Biol Reprod 24:923–931.

Tran D, Picard JY, Campargue J, Josso N (1987): Immunocytochemical detection of anti-müllerian hormone in Sertoli cells of various mammalian species including human. J Histochem Cytochem 35:733–743.

Tran D, Picard JY, Vigier B, Berger R, Josso N (1986): Persistence of müllerian ducts in male rabbits pas-

sively immunized against bovine anti-müllerian hormone during fetal life. Dev Biol 116:160–167.

Trelstad RL, Hayashi A, Hayashi K, Donahoe PK (1982): The epithelial-mesenchymal interface of the male rate müllerian duct: loss of basement membrane integrity and ductal regression. Dev Biol 92:27–40.

Tsafriri A, Picard JY, Josso N (1988): Immunopurified anti-müllerian hormone does not inhibit spontaneous resumption of meiosis in vitro of rat oocytes. Biol Reprod 38:481–485.

Ueno S, Manganaro TF, Donahoe PK (1988): Human recombinant müllerian inhibiting substance inhibition of rat oocyte meiosis is reversed by epidermal growth factor in vitro. Endocrinology 123:1652–1659.

Ueno S, Takahashi M, Manganaro TF, Ragin RC, Donahoe PK (1989a): Cellular localization of müllerian inhibiting substance in the developing rat ovary. Endocrinology 124:1000–1006.

Ueno S, Kuroda T, MacLaughlin DT, Ragin RC, Manganaro TF, Donahoe PK (1989b): Müllerian inhibiting substance in the adult rat ovary during various stages of the estrous cycle. Endocrinology 125:1060–1066.

Vigier B, Forest MG, Eychenne B, Bezard J, Garrigou O, Robel P, Josso N (1989): Anti-müllerian hormone produces endocrine sex reversal of fetal ovaries. Proc Natl Acad Sci USA 86:3684–3688.

Vigier B, Legeai L, Picard JY, Josso N (1982): A sensitive radioimmunoassay for bovine anti-müllerian hormone, allowing its detection in male and freemartin fetal serum. Endocrinology 111:1409–1411.

Vigier B, Picard JY, Campargue J, Forest MG, Heyman Y, Josso N (1985): Secretion of anti-Mullerian hormone by immature bovine Sertoli cells in primary culture, studied by a competition-type radioimmunoassay: Lack of modulation by either FSH or testosterone. Mol Cell Endocrinol 43:141–150.

Vigier B, Picard JY, Tran D, Legeai L, Josso N (1984a): Production of anti-müllerian hormone: Another homology between Sertoli and granulosa cells. Endocrinology 114:1315–1320.

Vigier B, Tran D, du Mesnil du Buisson F, Heyman Y, Josso N (1983): Use of monoclonal antibody techniques to study the ontogeny of bovine anti-müllerian hormone. J Reprod Fertil 69:207–214.

Vigier B, Tran D, Legeai L, Bezard J, Josso N (1984b): Origin of anti-müllerian hormone in bovine freemartin fetuses. J Reprod Fertil 70:473–479.

Vigier B, Watrin F, Magre S, Tran D, Josso N (1987): Purified bovine AMH induces a characteristic freemartin effect in fetal rat prospective ovaries exposed to it in vitro. Development 100:43–55.

von Heijne G (1986): A new method for predicting

signal sequence cleavage sites. Nucleic Acids Res 14:4683–4690.

Voutilainen R, Miller WL (1987): Human müllerian inhibitory factor messenger ribonucleic acid is hormonally regulated in the fetal testis and in adult granulosa cells. Mol Endocrinol 1:604–608.

Wallen JW, Cate RL, Kiefer DM, Riemen MW, Martinez D, Hoffman RM, Donahoe PK, Von Hoff D, Pepinsky B, Oliff A (1989): Minimal antiproliferative effect of recombinant müllerian inhibiting substance on gynecological tumor cell lines and tumor explants. Cancer Res 49:2005–2011.

Wang JJ, Teng CS (1989): Antibody against avian müllerian inhibiting substance (MIS) recognizes MIS on rat müllerian duct and human tumor cells. Proc Natl Sci Counc Repub China 13:267–275.

Wang JJ, Yin CS, Teng CS (1990): Lectin bindings and diethylstilbestrol effects on the recognition of müllerian inhibiting substance (MIS) on chick müllerian ducts by MIS-antiserum. Histochemistry 95:55–61.

Wolf E (1939): L'action due diethylstilboestrol sur les organes genitaux de l'embryon de Poulet. CR Seances Acad Si Ser III Sci 208:1532–1534.

ABOUT THE AUTHORS

RICHARD L. CATE is a senior scientist in the Department of Molecular Biology at Biogen, Inc., Cambridge, Massachusetts. He received his B.A. from Hofstra University in 1975, and his Ph.D. under Thomas Roche at Kansas State University in 1979; his Ph.D. thesis concerned the catalytic and regulatory reactions of mammalian pyruvate dehydrogenase complex. Dr. Cate pursued postdoctoral research at Harvard University in the laboratory of Dr. Walter Gilbert, focusing on the regulation of the rat insulin genes. Since joining Biogen in 1983, he has worked on Factor VIII, lipocortin, and mullerian inhibiting substance. In addition, he has authored publications on applications of oligonucleotide probes and chemiluminescence in molecular biology.

CHERYL A. WILSON is a postdoctoral fellow in the Department of Molecular Biology in the laboratory of Richard Cate at Biogen, Inc., Cambridge, Massachusetts. After receiving her B.S. in Medical Technology from the University of Missouri in 1978, she worked in Clinical Pathology laboratories for six years, specializing in and teaching immunology. She received her M.S. in Immunology from the University of Michigan in 1984, and her Ph.D. in Molecular Genetics and Microbiology under Trudy Morrison at the University of Massachusetts in 1989. Her graduate thesis work focused on the co-translational determination of membrane topology of the Type II glycoprotein, hemagglutinin-neuraminidase of the Newcastle disease virus. Dr. Wilson pursued postdoctoral research in the laboratory of Elliot Kieff at Harvard Medical School, where she studied the function of the latent membrane protein (LMP1) of the Epstein-Barr virus. She is currently developing procedures for isolating and studying the interactions and functions of the N- and C-terminal domains of müllerian inhibiting substance. Her research papers have appeared in journals such as the *Journal of Virology*, *Molecular and Cellular Biology*, and the *Journal of Cell Biology*.

Genes in Mammalian Reproduction: 207–228
© 1993 Wiley-Liss, Inc.

Molecular Biology of the Gonadal Proteins—Inhibin, Activin, and Follistatin

Shao-Yao Ying and Takuya Murata

I. INTRODUCTION

Recently, a great deal has been learned about inhibins, gonadal proteinaceous hormones, and their related proteins, activins, and follistatins. These gonadal proteins are responsible for a selective feedback on pituitary gonadotrophs, suggesting a modulating role of gonadal proteins on the secretion of FSH, a hormone that play a central role in steroidogenesis and gametogenesis in mammals.

The idea that proteins of gonadal origin selectively regulate the secretion of FSH by gonadotrophs has long been proposed [Mottram and Cramer, 1923; Martins and Rocha, 1931; McCullagh, 1932] and subsequently used to justify some optimism that an effective male contraceptive strategy might be developed [Gibbons, 1977]. However, the isolation and characterization of inhibin of testicular origin had been troublesome, and the inhibin molecule had evaded researchers for more than 50 years [Schwartz, 1991]. Within the past few years, marked advances have been made regarding the isolation, identification, and characterization of these gonadal proteins by the techniques of molecular biology. The rapid advances in this area of reproduction have also brought new concepts, challenging new problems, and new excitement to research in reproductive biology.

The aim of this review is to present re-cently obtained evidence of the structural and functional characteristics of these gonadal proteins, with emphasis on gene expression and on the control of gene expression. We discuss the structure and biological activities of inhibins, activins, and follistatins, the structures of the genes encoding these gonadal proteins, and the regulation of gene expression for these gonadal proteins. Because some newly discovered functions of the gonadal proteins might be indirectly related to mammalian reproduction, we will also include these observations in this review. We hope that elucidation of the diverse functions of the gonadal proteins will facilitate our understanding of their direct and indirect control over the wide spectrum of a complicated biological process involved with reproduction.

II. STRUCTURE OF THE GONADAL PROTEINS

A. Inhibin

Although original observations of inhibin activity were centered on the presence of this proteinaceous substance in testicular extract, it was subsequently demonstrated that ovarian follicular fluid contains an even higher inhibin activity [Grady et al., 1982]. Four laboratories obtained inhibin preparations, glycoprotein in nature [Jansen et al., 1981; de Jong et al., 1981; Godbout and Labrie, 1984], from ovarian follicular fluids which were homogeneous by HPLC and SDS/PAGE in the nonreduced condition [Ling et al., 1985; Miyamoto et al., 1985; Rivier et al., 1985; Robertson et al., 1985]. The biological activity was monitored based on the ability of inhibin to suppress the release of FSH but not that of LH, using cultured rat pituitary cells as a bioassay system. Inhibin from swine and cow has a molecular weight of 32 kD [Ling et al., 1985; Rivier et al., 1985; Fukuda et al., 1986; Robertson et al., 1986] while higher molecular weight forms (58 and 65 kD) from bovine follicular fluid were also reported [Robertson et al., 1985: Leversha et al., 1987]. The 32 kD form of bovine inhibin probably is a bioactive final product of proteolytic cleavage of an α-subunit precursor [McLachlan et al., 1986a,b] that contains pairs of basic amino acids or arginines serving as processing sites that could be used to form mature α-subunits of various molecular sizes [Mason et al., 1985].

There are two forms of 32 kD porcine inhibin, inhibins A and B, isolated based on their different time of elution in an HPLC system and on their similar but distinguishable amino acid sequences as determined by microsequencing of the N-terminal of the molecules [Ling et al., 1985]. Under reducing conditions of SDS/PAGE, 32 kD inhibin showed two bands; an 18 kD α-subunit and a 14 kD β-subunit, suggesting that inhibins are dimers of a common α-subunit (134 amino acids) and one of two similar but distinct β-subunits (βA of 116 amino acids, or βB, of 115) joined together by disulfide bonds. Similarly, 58 kD [Robertson et al., 1985] and 65 kD [Leversha et al., 1987] bovine inhibin can dissociate into 44 and 56 kD α-subunits, respectively, and into a 14 kD β-subunit. It has been postulated that an unprocessed β-subunit precursor attaches to various forms of inhibin to form molecules greater than 88 kD [Miyamoto et al., 1986]. The biosynthesis of inhibin probably involves the formation of a free larger precursor of α-subunit and processes at the proteolytic sites to yield an 18 kD α-subunit, which joins with a mature 14 kD β-subunit to form a 32 kD inhibin molecule [Bicsak et al., 1986].

Each inhibin subunit precursor possesses potential glycosylation sites in which N-linked carbohydrate chains could attach to specific asparagine residues [Mason et al., 1985]. The human α-subunit precursor (364 amino acids) contains three, one at the proregion and two within the α-subunit of mature 32 kD inhibin, whereas in other species studied there are only two potential N-linked carbohydrate chains (one at the proregion and one at the mature α-subunit). β-Subunit precursors (424 amino acids for βA or 407 for βB precursor) contain one at the proregion and none in the mature β-subunits [Mason et al., 1985, 1986]. Presumably, during translation of the α- and β-subunit precursors in the rough endoplasmic reticulum, signal peptides are cleaved and oligosaccharides are cotranslationally attached to the asparagine residues. The glycosylated precursors are then transported to the Golgi apparatus, where further modifications take place that may or may not include proteolytic cleavage to form 32 kD inhibins.

B. Inhibin/Activin β/TGFβ Family

During the isolation of inhibin it was observed that some side fractions were able to enhance FSH secretion, and isolation of these side fractions led to the isolation and characterization of molecules that are structurally similar to inhibins but have biologically opposite activity. These molecules were named *activins*. Activins are dimers of the β-subunit of inhibin with a molecular weight of 28 kD. Two types of activin, formed either by two identical β-subunits of inhibin A (βAβA) [Vale et al., 1986; Ling et al., 1986b] or by two similar but distinct β-subunits from inhibin A and inhibin B (βAβB) [Ling et al., 1986a], have been isolated and characterized. Conceivably, a third form of activin consisting of two β-subunits of inhibin B (βBβB) exists in follicular fluid [Ying, 1988].

Comparing the structure of inhibin and activin with that of transforming growth factor β1 (TGFβ1) [Derynck et al., 1985], it is clear that these gonadal proteins are members of a family of growth-related proteins, including TGFβ, that share many structural similarities [Mason et al., 1985]. Each member of this inhibin/activin/TGFβ family consists of the disulfide association of two subunits that possesses a unique 7 or 9 cysteine distribution pattern, suggesting highly conserved molecules in evolution. The β-subunits of activin and TGFβ share considerable amino acid homology, indicating a possibility that they are derived from a common precursor. There is also a high similarity in the amino acid sequence between α-and β-subunits of inhibin and β-subunits of TGFβ1, suggesting that these subunits share a common ancestral origin [Mason et al., 1985]. Both mature activin and TGFβ appear not to be glycosylated, although one glycosylated site in the precursors of β-subunits has been identified [Mason et al., 1985].

This inhibin/activin/TGFβ family also includes müllerian-inhibiting substance [Cate et al., 1986], the *Drosophila* decapentaplegic gene [Padgett et al., 1987], bone morphogenetic proteins [Wozney et al., 1988], and products of the Vg-related genes [Weeks and Melton, 1987]. Indeed, all members of this family are involved in cell differentiation during development. Furthermore, these gonadal proteins are widely distributed and involved in several other biological processes. For instance, activin-A was not only isolated from ovarian follicular fluid, but the same molecule also was isolated and characterized from a cell line of monocytes as erythroid differentiation factor [Eto et al., 1987], a cell line derived from eye tumor as nerve cell survival molecule [Schubert et al., 1990], and a

Xenopus differentiation cell line as embryonic development agent [Smith et al., 1990]. In addition, several cell lines either secrete activin or produce molecules similar to activin-A, including a human fibrous histocytoma cell line (as megakaryocyte differentiation activity, which has high sequence homology to the N-terminal sequence of activin A) [Fujimoto et al., 1991], human acute myeloid leukemia cells [Scher et al., 1990], chicken endoderm [Kokan-Moore et al., 1991], and a Leydig cell line, TM3 [Lee et al., 1989]. Inhibin is also produced by Leydig cells [Risbridger et al., 1989a; Drummond et al., 1989].

C. Monomer of α-Subunit of Inhibin

Recently, a monomeric α-subunit of 32 kD inhibin was isolated from follicular fluid and was found in the circulation [Sugino et al., 1989; Knight et al., 1989; Robertson et al., 1989]. Since there is a much higher mRNA level of the α-subunit in the ovary than of the β-subunit [Mason et al., 1985] and there is a high concentration of free α-subunit in the follicular fluid [Knight et al., 1989], it is possible that this inhibin α-subunit monomer is synthesized and secreted in excess of the 32 kD inhibin.

D. Follistatins as Activin-Binding Proteins

A third type of molecule, follistatin, which has no structural homology to inhibin, was isolated based on its ability to suppress the release of FSH by the pituitary [Ueno et al., 1987; Robertson et al., 1987]. Follistatins, glycoproteins composed of a single polypeptide chain, are 32, 35, or 37 kD, cysteine-rich molecules that probably are a result of different degrees of glycosylation and/or truncation of the carboxyl terminus of a single protein. There are two follistatin precursors, one 344 and the other 317 amino acids, with the latter being truncated at the carboxyl terminus. The precursors are organized into four domains; the first 29 amino acids comprise the signal peptide sequence (domain 1); the next 288 amino acids are cysteine rich (36 cysteines) and form three parallel and contiguous domains (domains 2, 3, and 4) that are highly similar in sequence to one another. The second domain of follistatin precursor is homologous to human pancreatic secretory trypsin inhibitor and epidermal growth factor, suggesting a common ancestral gene [Esch et al., 1987a; Shimasaki et al., 1988]. The larger precursor (344 amino acids) contains an additional 27 amino acids at the carboxyl end.

Follistatin has also been identified as an activin-binding protein. Sugino and coworkers [Nakamura et al., 1990; Sugino et al., 1990; Saito et al., 1991; Kogawa et al., 1991] have shown that a molecule binding to labeled activin isolated from the pituitary and the ovary contains an amino acid sequence identical to that of follistatin, strongly suggesting that follistatin itself is an activin-binding protein. Because inhibin contains a β-subunit derived from activin, it has been shown that activin has two binding sites for follistatin, whereas inhibin has only one binding site for follistatin, suggesting that follistatin binds to both activin and inhibin through the common β-subunit [Shimonaka et al., 1991].

E. Activin Receptor

Mathews and Vale [1991] recently cloned an activin receptor cDNA encoding for a protein of 513 amino acids that gives rise to the mature form of 494 amino acids. This mature activin receptor is a type I membrane protein of 56 kD, with a 116 amino acid N-terminal extracellular ligand-binding domain, which is cysteine rich, and a 352 amino acid intracellular kinase signaling domain with predicted serine/threonine specificity. There are two potential sites of N-linked glycosylation in the extracellular domain and a number of potential phosphorylation sites for protein kinase C and casein kinase II in the intracellular domain.

III. GENES ENCODING THE GONADAL PROTEINS

The structures of the inhibin α-subunit, inhibin/activin βA-subunit, and βB-subunit genes are very similar, whereas the inhibin α-subunit gene, although it undoubtedly evolved from a common ancestor to the β genes, has a slightly different composition among species studied. The cDNAs for the α-, βA-, and βB-subunits in the human [Mason et al., 1986; Mayo et al., 1986; Stewart et al., 1986], bovine [Forage et al., 1986], porcine [Mason et al., 1985], ovine [Keinan et al., 1987b; Feng et al., 1989a,b; Rodgers et al., 1989], and rat [Esch et al., 1987b; Woodruff et al., 1987] have been isolated and characterized, showing that all subunits are derived from precursor proteins that are eventually proteolytically processed to yield the mature proteins [Ling et al., 1985]. In each species, a single gene encoding the α-subunit of inhibin is composed of two exons, separated by an intron of 1.7–2.1 kb, and is localized on the distal portion of the long arm of chromosome 2. The positions of the introns and the size of the α-subunit gene are highly conserved in all five species. The first exon encoding the signal peptide and part of the amino-terminal end of the precursor sequence, but the major portion of the precursor sequence including the mature α-subunit is encoded by the second exon. Similarly, each gene of the β-subunits consists of two exons separated by introns of 10 and 2.8 kb for the βA and βB genes, respectively. The βA gene is on chromosome 7, whereas the βB-subunit gene is located near the centromere on the short arm of chromosome 2. The first exon of the βA or βB gene encodes its respective signal peptide and a part of the amino-terminal region of the βA or βB precursor, whereas the second exon encodes the respective bulk of the proregion of βA or βB as well as the mature βA- or βB-subunit. Therefore, the inhibins are encoded by separate genes localized on different chromosomes consisting of two exons separated by a variously sized intron [Stewart et al., 1986; Mason et al., 1985, 1986, 1989].

The α-subunit of inhibin shows a single species of mRNA by Northern hybridization to total Sertoli cell RNA of 1.6 kb [Toebosch et al., 1989] and in rat ovary [Davis et al., 1988]. The mRNAs of βA-subunit show at least three different cross-hybridizing bands by Northern analysis [Feng et al., 1989b]—the predominant species being 6.5 kb and two less abundant species being 3.0 and 1.8 kb, respectively. The βB mRNA is characterized by two distinct species, one 4.4 kb and one 3.3 kb in length; the 4.4 kb transcript is the predominant message present in total pooled testicular and ovarian RNAs, and the 3.3 kb mRNA is less abundant [Feng et al., 1989a].

Comparison of the nucleotide sequences of the coding regions of the three genes of different species reveals that there is a high degree of similarity with the greatest similarity in the βB-subunit. There are two Asn-linked glycosylation sites in the α-subunit precursor of all species studied except human, one at the proregion and the other the mature α-subunit, whereas there is only one glycosylation site at the preregion of the βA- or βB-subunit. There is an additional potential Asn-linked glycosylation site at amino acid 301 of the human α-subunit that makes mature human inhibin α-subunit containing 2 N-linked oligosaccharides. However, the importance of glycosylation on inhibin activity remains to be investigated. The human, porcine, bovine, ovine, and rat inhibin α-subunit show an approximate 80% similarity. There are only four amino acids different among the mature human, porcine, and rat βB-subunit, and the mature βB-subunits of all four species (human, bovine, porcine, and rat) are completely identical, but the βA-subunit of ovine testicular origin showed one amino acid difference from that of ovarian origin for all species studied. In addition, there is considerable similarity between the untranslated regions of the subunit genes of inhibins. Since there are potential glycosylation sites and multiple basic amino acid pairs for potential proteolytic cleavage sites present in these precursors, it is reasonable to postulate molecules of free α-subunits or partially processed dimer of α- and β-subunits.

As for follistatin, two species of cDNA encoding follistatin precursor were isolated, and follistatin cDNA and genes of rat, human, and porcine origin have been isolated and characterized [Esch et al., 1987a; Shimasaki et al., 1988, 1989; Michel et al., 1990]. These two species of cDNA, produced by alternative splicing, each encode the follistatin precursors of 344 or of 317 amino acids. Again, the follistatin gene, about 6 kb long and similar among the three species studied, consists of six exons separated by five introns. The first exon encodes the putative signal sequence; the following four exons encode the four domains of follistatin; and the last exon encodes an extract 27 amino acid domain only present in the carboxyl terminus of the bigger precursor (344 amino acids). The mRNA of follistatin is 1.5 kb.

The activin receptor cDNA was recently cloned [Mathews and Vale, 1991]; it encodes a protein of 513 amino acids with a 5′-untranslated region of 70 bp and a 3′-untranslated region of 951 bp. There are two hydrophobic regions; one is a 19 amino acid stretch at the N terminus, a putative signal peptide, and the other is a single putative 26 residue membrane-spanning region.

IV. BIOLOGICAL ACTIVITIES OF THE GONADAL PROTEINS

A. Endocrine Functions

Inhibins, activins, and follistatins, originally isolated and characterized based on their abilities to modulate the secretion of FSH by cultured pituitary cells, are highly potent, with EC_{50} values of ~4, ~25, and ~70 pM, respectively [Ying et al., 1987; Vale et al., 1986; Xiao et al., 1990; Wang et al., 1990]. The effect of the gonadal proteins on the secretion of FSH are of a delayed nature, usually 8–24 hours, which is different from that of hypothalamic releasing factors, which need only seconds to minutes to react. In addition, inhibin suppresses GnRH-induced secretion of both FSH and LH [Campen and Vale, 1988; Fukuda et al., 1987] and the upregulation of GnRH-binding sites [Sealfon et al., 1990; Gregg et al., 1991; Wang et al., 1989]. The gonadal proteins appear to affect the synthesis of the FSHβ subunit at the pretranslational level [Sealfon et al., 1990; Carroll et al., 1989; Mercer et al., 1987; Attardi et al., 1989; Attardi and Miklos, 1990]. The gonadal proteins modulate FSH secretion by regulating FSHβ mRNA but not LHβ mRNA at the level of the anterior pituitary gland. In cultured rat anterior pituitary cells, inhibin decreases FSHβ mRNA levels, whereas activin increases FSHβ mRNA 4 hours after the administration; follistatin at higher doses, to a lesser degree, also significantly reduces βFSH mRNA levels. None of these treatments produces changes in LHβ mRNA levels. This modulation of FSH secretion has been confirmed both in vitro and in vivo with recombinant gonadal proteins [Inouye et al., 1991; Huylebroeck et al., 1990; Schmelzer et al., 1990; Schwall et al., 1988, 1989; Rivier et al., 1991; Rivier and Vale, 1991; DePaolo et al., 1991; McLachlan et al., 1989]. Prolonged exposure of pituitary cells to activin stimulates both secretion of FSH and the storage of FSH [Vale et al., 1986] and increases the number of FSH-containing gonadotrophs [Katayama et al., 1990]. In addition, immunoneutralization of endogenous inhibin with a potent antirat inhibin serum significantly elevates FSHβ mRNA levels in immature rats, providing evidence that endogenous inhibin regulates FSHβ mRNA levels [Attardi et al., 1991].

Since inhibins, activins, and follistatins have been localized in the pituitary [Roberts et al., 1989] and there are known antagonistic actions of activin and inhibin in several biological systems [Ying et al., 1987a; DePaolo et al., 1991; Wang et al., 1990; Xiao et al., 1990], conceivably these gonadal proteins may act as autocrine agents to modulate the secretion of FSH by the anterior pituitary. Indeed, there are reports suggesting this type of function [Corrigan et al., 1991; Lucas et al., 1990], and follistatin is produced by the stellate-follicular cells of the pituitary [Gospodarowicz and Lau, 1989].

The action of activin appears not to be mediated through GnRH receptor; inhibin facilitates the binding of GnRH receptors while activin only inhibits GnRH receptors at very high concentration [Gregg et al., 1991; Braden et al., 1990]. Conversely, inhibin synthesis and secretion is controlled by FSH, LH, and gonadal steroids both in vivo and in vitro [Rivier et al., 1986; Rivier and Vale, 1989; Bicsak et al., 1987], whereas the control of activin synthesis and secretion remains to be determined.

Antisera against synthetic peptides corresponding to inhibin subunits [Rivier and Vale, 1986; Ying et al., 1987b, 1988b; Schanbacher, 1988; Shintani et al., 1991] or follistatin [Klein et al., 1991; Sugawara et al., 1990] conjugated to a protein have been developed using the approach documented by Lerner [1982]. In addition, antisera against the gonadal proteins using the native molecules or their subunits as immunogens were also reported [McLachlan et al., 1986a,b; Miyamoto et al., 1986]. With these antisera, the gonadal origins of inhibin have been established both in vitro and in vivo as primarily in the granulosa cells of ovary and Sertoli cells of the testis, but other gonadal cells such as the theca interna, the interstitial cells, and Leydig cells also are the source of inhibin [Cuevas et al., 1987; Bardin et al., 1989; Bicsak et al.,

1986]. To circumvent the possible presence of inactive forms of various inhibin α-subunits, a two-site enzyme-linked immunosorbent assay for inhibin, using monoclonal antibodies against the N-terminal region of the human inhibin α-subunit and a peptide sequence corresponding to the C-terminal of the human βA-subunit, appears to measure the active forms of inhibin [Betteridge and Craven, 1991]. Radioimmunoassay for activin-A and follistatin has been reported [Shintani et al., 1991; Sugawara et al., 1990].

On the other hand, extragonadal sources of the gonadal proteins have also been identified; inhibin is localized in the adrenal [Meunier et al., 1988; Crawford et al., 1987], the placental [Petraglia et al., 1987], and the bone marrow [Meunier et al., 1988], whereas activin is identified in kidney, the hypothalamus, and bone marrow [Meunier et al., 1988; Sawcheko et al., 1988] and follistatin in the kidney [Shimasaki et al., 1989]. In spite of such a wide anatomical distribution for sites of production of the gonadal proteins, the presumed immunoneutralization of endogenous inhibin (ovarian origin as its primary source) has increased serum FSH levels and increased the number of eggs in the oviducts in rats [Rivier and Vale, 1989], and sheep [Bindon et al., 1986; Cummins et al., 1986; Henderson et al., 1984; Forage et al., 1987; Findley et al., 1989; Mizumachi et al., 1990; Wrathall et al., 1990].

In addition to the modulation of FSH secretion, the gonadal proteins, particularly activin, also have other types of action on the anterior pituitary. These include inhibition of basal and GHRH-mediated secretion of growth hormone (GH), GH biosynthesis, and GHRH-stimulated proliferation of somatotroph cells [Vale et al., 1988; Corrigan et al., 1988; Billestrup et al., 1986; Kitaoka et al., 1988]. Activin also inhibits the basal secretion of ACTH by the pituitary [Vale et al., 1986; Corrigan et al., 1988]. Recently, inhibition by activin of prolactin

secretion in cultured rates pituitary cells also was reported [Kitaoka et al., 1988; Murata and Ying, 1991].

There are reports suggesting that activin has a direct effect on endocrine organs other than the anterior pituitary. These include the secretion of basal and glucose-stimulated insulin by rat pancreatic islets [Totsuka et al., 1988] and the inhibition of proliferation of human fetal adrenal cells [Spencer et al., 1990]. However, the importance of the gonadal proteins in regulating activity of these endocrine organs remains to be determined.

B. Paracrine/Autocrine Function

In addition to endocrine functions as their primary role in reproduction, the gonadal proteins have also been shown to have paracrine/autocrine activity within the gonads themselves. This type of action may indirectly modulate the hypothalamic–pituitary–ovarian axis through the alteration of steroidogenesis at a local level.

In cultured rat granulosa cells, TGFβ and activin enhance [Ying et al., 1986; Hutchinson et al., 1987; LaPolt et al., 1989], but inhibin inhibits [Ying et al., 1986; LaPolt et al., 1989] FSH-mediated aromatase activity. Activin and TGFβ also enhance the production of progesterone [LaPolt et al., 1989; Ying, 1988] whereas an inhibitory effect of activin on progesterone and no inhibitory effect of inhibin on aromatase activity were also reported [Hutchinson et al., 1987]. In gonadal cell lines, TGFβ and activin have similar paracrine function on steroidogenesis [Gonzales-Manchon and Vale, 1989].

In testicular Sertoli cells [Hsueh et al., 1987], an ovarian theca–interstitial cell culture system [LaPolt et al., 1989], and Leydig cells [Morris et al., 1988], this paracrine-type control of gonadotropin-mediated steroidogenesis was documented. In addition to paracrine action, an autocrine role of the gonadal proteins in the ovary has been implicated [Sugino et al., 1990]. Indeed, the identification of follistatin as a binding protein for activin in both ovaries and pituitaries [Saito et al., 1991; Kogawa et al., 1991; Sugino et al., 1990] and its ability to bind to both activin and inhibin [Shimonaka et al., 1991] strongly suggest an intragonadal autocrine regulation. By the same token, paracrine/autocrine actions of the gonadal proteins at the pituitary level have been reported [Corrigan et al., 1991; Kogawa et al., 1991].

Another example of a paracrine/autocrine role of the gonadal proteins is their ability to act on the oocyte of the follicle and on growth patterns during the embryogenesis (see below). Inhibin [O et al., 1989; Tsafriri et al., 1989] and activin [Itoh et al., 1990] have been found to modulate meiotic maturation of the oocyte, and the presence of immunoreactive inhibin in follicular fluid is directly related to the meiotic stage of the oocyte [Miller et al., 1991]. Incidentally, expression of mRNA for follistatin during early embryonic development of *Xenopus laevis* [Tashiro et al., 1991] suggests a role for the gonadal proteins in the fertilized oocyte.

C. Effects of Gonadal Proteins on the Neuronal Tissues

Using region-defined, inhibin/activin β-subunit-specific polyclonal antibodies, Sawcheko et al. [1988] have identified clusters of neurons showing positive staining for inhibin/activin β-subunit in the paraventricular, supraoptic, and dorsomedial nuclei of the hypothalamus. To assess the role of activin on hormone-containing neurons, Plotsky et al., [1988] were able to elevate plasma oxytocin and demonstrated a "activinergic" pathway that may be important in the regulation of milk ejection. Similarly, the regulation of neural oxytocin gene expression by gonadal steroids in pubertal rats has been reported [Chibbar et al., 1990]. Furthermore, Plotsky et al. [1991] have demon-

strated that the ventricular administration of activin also modulates the hypothalamic–pituitary–adrenal axis by stimulating corticotropin-releasing hormone (CRF) and adrenocorticotropin secretion. Indeed, the role of activin on the hypothalamic–pituitary–adrenal axis and oxytocin release has been further substantiated. Similar action of activin was observed in human adrenal tissues [Spencer et al., 1990; Rabinovici et al., 1990]. The regulation of gene expression for oxytocin [Chibbar et al., 1990] or proopiomelanocortin [Bilezikjian et al., 1991] is modulated by activin. Although the α-subunit of inhibin gene expression occurs in the adrenal cortex and is regulated by adrenocorticotropin [Crowford et al., 1987], the role of inhibin or the free α-subunit on the hypothalamic–pituitary–adrenal axis remains to be examined. Apart from this action on a hormone-containing neuron, activin-A also acts as an inhibitor of neural differentiation [Hashimoto et al., 1991], and activin-A has been isolated and characterized from a cell line derived from nickle-treated eye tumor. These findings may point out an important regulatory role of the gonadal proteins on neuronal tissues.

D. Cell Differentiation and Embryogenesis

The gonadal proteins also play an important role in cell proliferation, cell differentiation, and embryogenesis, which are closely related to growth and reproduction. The effects of the gonadal proteins on differentiation of 3T3 cells [Kojima and Ogata, 1989] and granulosa cell lines [Gonzalez-Manchon and Vale, 1989] have been reported. Recent observations that activin-A can be isolated from a mesodermal-inducing cell line in *Xenopus* [Smith et al., 1990] and that activin-A has mesodermal differentiation activity [Thomsen et al., 1990] have initiated a new series of studies on the biological activities of gonadal proteins on cell differentiation and embryogenesis. These findings suggest not only that activin is present in developing cells and tissues but also that it plays an important role in organogenesis [Asashima et al., 1991; Kokan-Moore et al., 1991; Roberts et al., 1991; Sokol and Melton, 1991; vander Eijndenvan Raaij et al., 1991]. In *Xenopus,* growth factors including TGFβ2, activin-A, and FGF induce isolated ectoderm explants to differentiate into mesodermal tissue types instead of epidermis; activin probably is responsible for induction of the dorsal mesoderm [Slack, 1991]. Indeed, activin was isolated in unfertilized eggs and blastulae of *Xenopus laevis,* and follistatin was antagonistic to the mesoderm-inducing activity on *Xenopus* animal-cap cells [Asashima et al., 1991]. These observations may shed light on how differentiated cells arise and form tissues and organs and how pattern is generated. To this end, there is evidence showing that activin and its related growth factors play a role in pattern formation during development [Melton, 1991; Sokol and Melton, 1991; Thomsen et al., 1990], and probably this type of protein is a part of the biochemical pathway leading to the generation of the body axis [Cho et al., 1991].

In chicken, activin-A was identified in the endoderm of cultured postgastulation explants [Kokan-Moore et al., 1991], and activin, bFGF, and TGFβ2 were active as inducers in a manner similar to amphibian blastula ectoderm, suggesting that a role of activin-A in pattern formation and/or generation of a body axis could extend to non-amphibian species as well [Cooke and Wang, 1991].

Both inhibin and activin may be involved in a hormonal regulation of the embryonic growth of several organs, including the heart, skin, hair, and bone in the rat, suggesting that the actions of these gonadal proteins may be coordinated with those of other members of the inhibin/activin β/TGFβ family during development [Roberts et al., 1991]. Furthermore, the gonadal proteins also exert a pivotal role in

folliculogenesis [Woodruff et al., 1991], spermatogonial proliferation [Mather et al., 1990], and proliferation of human luteinized preovulatory granulosa cells [Rabinovici et al., 1990] in the mammals. In addition, activin-A also exerts mitogenic activity on embryonal carcinoma cells [Schubert and Kimura, 1991; vander Eijndenvan Raaij et al., 1991].

Follistatin gene is highly expressed during early pregnancy [Ying et al., 1988b; Kaiser et al., 1990]. The mRNA is preferentially found in antimesometrial tissue and is confined to the endometrial-derived decidual cells [Kaiser et al., 1990]. Since activin is thought to act as a natural mesoderm-inducing factor [Smith et al., 1990], Tashiro et al. [1991] have isolated follistatin cDNA from *Xenopus* ovary cDNA library. There are two species of mRNAs for *Xenopus* follistatin being expressed during embryonic development, a 3.5 kb at the gastrula stage and a 2 kb mRNA from early neurula to tadpole. These findings suggest that follistatin may interact with activin in the induction of *Xenopus* embryogenesis.

E. Hematopoietic and Immunosuppressive Modulations

Eto et al. [1987] have isolated and characterized an activin-A molecule from the spent medium of a monocyte cell line, TPH 1, derived from a 1-year-old boy with leukemia. Subsequently, Yu et al. [1987, 1989, 1991] have demonstrated opposite effects of activin and inhibin on the induction of hemoglobin accumulation in the human erythroleukemic cell line K562. In addition, activin induces the differentiation of these cells, which is prevented by coincubation with inhibin. In human bone marrow cultures, activin potentiates the erythropoietin-mediated formation of precursors of proerythroblast, CFU-E, and BFU-E, whereas inhibin prevents the potentiation by activin, suggesting that the gonadal proteins play a role in selectively enhancing erythroid colony formation [Yu et al., 1989; Broxmeyer et al., 1988].

In addition to hematopoietic modulation, the gonadal proteins have the ability to regulate mitogen-mediated cell proliferation in thymocytes [Hedger et al., 1989] and splenocytes [Ying et al., 1989], indicating a potential role of the gonadal proteins in immunosuppression.

F. The Presence of Gonadal Proteins in Tumors—Including the Placenta

Inhibin is produced by trophoblasts of the placenta as determined by bioassay and radioimmunoassay [Hochberg et al., 1981; McLachlan et al., 1987; Petraglia et al., 1987; Qu et al., 1990], and its secretion is stimulated by human chorionic gonadotropin through adenylate cyclase. Reciprocally, activin stimulates the secretion of GnRH, hCG, and progesterone of trophoblast cultures, and this can be prevented by inhibin [Petraglia et al., 1989]. These findings suggest the presence of gonadal proteins in tumors or cultured cell lines. Indeed, immunoreactive inhibin has been detected in several tumors and cell lines, including granulosa cell tumors [Lappohn et al., 1989]. Several granulosa–theca cell as well as nongonadal tumors express inhibin subunit mRNA and gonadal proteins [Piquette et al., 1990; de Jong et al., 1990]. Immunoreactive serum inhibin levels are high in several tumors, including human juvenile granulosa cell tumor [Nishida et al., 1991], human virilizing ovarian tumors (Sertoli-Leydig tumors) [Ohashi et al., 1990], and dog Sertoli cell tumors [Grootenhuis et al., 1990]. It remains to be determined whether the gonadal proteins play a role in the development and/or regulation of the development of these tumors.

V. REGULATION OF GONADAL PROTEIN GENE EXPRESSION

The regulation of the secretion of gonadal proteins as determined by radioim-munoassay and immunohistochemistry goes hand in hand with the information derived from studies on gene regulation of the synthesis of the gonadal proteins. As discussed before, the gonadal proteins are widely distributed as determined by immunohistochemistry, Northern blot analysis, and in situ hybridization. However, it is noteworthy to indicate that measurements of mRNAs for subunits of inhibin may indicate the production of inhibins A and B, one of the three forms of activin. In addition, detection of follistatin mRNA at a cellular level may also indicate potential interreaction between inhibin/activin and follistatin, because the latter possesses inhibin-like activity and is an activin-binding protein.

Based on bioassay and information gathered with partially purified inhibin, FSH and inhibin have been postulated to have a reciprocal relationship. FSH stimulates the release of inhibin, which in turn suppresses further release of FSH by the pituitary. Since GnRH is the major hypothalamic releasing hormone of the reproductive cycle, the effect of GnRH on the gonadal proteins have also been investigated. Several experimental models and molecular biology techniques have been used, enabling investigators to study the regulation of steady-state mRNA levels and transcription of mRNAs of the gonadal proteins using cDNAs that encode the gonadal protein subunits. We will examine the effects of the following regulatory factors: gonadotropins, GnRH, hypophysectomy, steroids, and other factors.

A. The Regulation of Gonadal Proteins mRNA in the Female

In cultured rat granulosa cells, FSH stimulates secretion of immunoreactive inhibins of 32 and 50 kD; the latter probably is an unprocessed monomeric precursor of the inhibin α-subunit [Bicsak et al., 1986]. In addition, leutinizing hormone but not prolactin increases inhibin production in cells pretreated with FSH [Bicsak et al., 1986; Zhang et al., 1988]. IGF-1 and vasoactive intestinal peptide (VIP) also stimulates inhibin production, whereas GnRH or epidermal growth factor (EGF) inhibits FSH-stimulated inhibin production.

The gene expression of inhibin subunits has been demonstrated in cultured granulosa cells [LaPolt et al., 1989; Turner et al. 1989; LaPolt et al., 1990]. FSH-stimulated inhibin α and βA mRNA levels in a dose-dependent manner in cultured granulosa cells. Luteinizing hormone but not prolactin increases α and βA mRNAs levels in granulosa cells pretreated with FSH to induce functional luteinizing hormone and prolactin receptors. In a series of experiments, Turner et al. [1989] reported that FSH stimulated 60-, 70-, and 66-fold increases in levels of inhibin α-, βA-, and βB-subunit mRNAs, respectively, compared with those of untreated controls. Estrogen alone produces a four- and twofold increase in levels of inhibin α- and βB-subunit mRNA, respectively, but has no effect on the level of inhibin βA-subunit mRNA. Treatment with T or DHT alone has no consistent effect on the levels of any inhibin subunit mRNA. The stimulatory effects of FSH are not consistently altered by the presence of either androgen or estrogen [Turner et al., 1989]. The effects of FSH and luteinizing hormone on inhibin subunit mRNA levels are mimicked by forskolin. [LaPolt et al., 1990], suggesting the involvement of a second message pathway. Since activin is known to increase the secretion of 32 kD inhibin, activin, similar to FSH, stimulates inhibin α mRNA, and coincubation of FSH and activin augments the induction of α mRNA whereas treatment with FSH and/or activin increases expression of a predominant 6.8 kb βA-subunit mRNA. Incidentally, TGFβ also stimulates α mRNA levels.

In vivo, it has been shown that injection of FSH or PMSG into immature and mature rats produces a rapid increase in bio-

logical [Lee et al., 1981] and immunoreactive serum inhibin levels [Rivier et al., 1986; 1989]. Treatment with PMSG increases inhibin mRNA levels in the ovary [Davis et al., 1988; Meunier et al., 1988]. Administration of PMSG leads to a sharp increase, and injection of hCG following PMSG causes a decrease in the expression of all three mRNAs of inhibin subunits in large preovulatory follicles. Levels of inhibin α-subunit mRNA are low in small antral follicles immediately before ovulation, whereas inhibin βA- and βB-subunits mRNAs are present in large follicles.

In the ovaries of regularly cycling rats, the initial stage of follicular development stimulates α-subunit mRNA to a high peak a few hours after the preovulatory surge of gonadotropin (luteinizing hormone and FSH) late in the afternoon of proestrus, whereas the β-subunit mRNA decreases markedly [Meunier et al., 1989; Woodruff et al., 1988]. The βA mRNA levels also reach peak values at 1830 hours proestrus. These high mRNA levels of α- and βA-subunit decrease by 2400 hours proestrus, prior to the rupture of the preovulatory follicle, while other follicles in the ovary maintain an increased mRNA level. By the morning of estrus (0400 to 0700 hours), inhibin mRNA levels begin to increase. In addition, other cells in the ovary, including the corpus luteum, the theca cells, and interstitial cells, express α-subunit inhibin.

As for the localization of the mRNAs of inhibin subunits in different sizes of follicles, the α-subunit mRNA is in the granulosa layer of antral follicles >0.36 mm in diameter, while the α- and βA-subunit mRNA are both present in follicles of >0.8 mm. In these latter follicles, the thecal layer hybridizes with only the α-subunit mRNA. No hybridization of the α- or βA-subunit probe was found in the cells of the corpus luteum [Torney et al., 1989].

In primates, mRNA for the βB-subunit is expressed in granulosa cells of small antral follicles, whereas levels in dominant follicles are undetectable [Schwell et al., 1990]. Conversely, expression of α and βA mRNAs is detectable in granulosa cells of dominant follicles and in corpora lutea, but not in small antral follicles.

The endogenous FSH from the pituitary can be removed by hypophysectomy and administration of antagonists of GnRH; these treatments result in decreased secretion of serum immunoreactive inhibin [Keinan et al., 1987; Woodruff et al., 1989]. Hypophysectomy and administration of RnGH antagonist decreases the mRNA of inhibin α-subunit [Keinan et al. 1987; Woodruff et al., 1989]. From this information, the mRNAs of α- or β-subunit of inhibin are different at different stages of follicular development in regular cycling rats, suggesting different roles of these gonadal proteins in the fine tuning of the hypothalamic–pituitary–ovarian axis in the process of ovulation.

During pregnancy, levels of α and βB mRNAs remain fairly constant from day 7 after mating until parturition and then fall within 16 hours postpartum. In all ovaries, expression of inhibin gene is located in granulosa cells of healthy antral follicles. In general, the strongest signals for α and βA mRNAs are in large follicles, with weaker signals in smaller follicles; considerable α mRNA is detectable in some follicles in which βA is reduced or undetectable, despite strong signals for both α and βA in adjacent follicles [Penschow et al., 1990; Yohkaichiya et al., 1991; Woodruff et al., 1991].

In summary, FSH stimulates α-, βA-, and βB-subunit mRNA expression both in vitro and in vivo, whereas luteinizing hormone, activin, and TGFβ facilitate the α- and βA-subunit mRNA expression in vitro. After hypophysectomy and administration of GnRH antagonist, luteinizing hormone and FSH decrease the increment of α- and βB-subunit mRNA levels. Follistatin mRNA levels are also regulated by FSH.

B. The Regulation of Gonadal Protein mRNA in the Male

In cultured Sertoli cells, FSH stimulates inhibin production and inhibin secretion [Bicsak et al., 1987; Toebosch et al., 1989; Klaij et al., 1990], which is enhanced by phosphodiesterase inhibitor, forskolin, cholera toxin, and dibutyryl cAMP, suggesting that cAMP is the second messenger for FSH-stimulated inhibin production. FSH stimulates the α-subunit mRNA level, which reaches maximal levels within 1.5 hours. Testosterone has no effect on immunoreactive inhibin levels in either the presence of absence of FSH. Similarly, the expression of inhibin α-subunit mRNA is increased following FSH stimulation, whereas testosterone has no effect. The expression of inhibin βB-subunit mRNAs was not influenced by FSH or testosterone.

The stimulation of inhibin α-mRNA levels by FSH is mimicked by addition of dibutyryl cAMP, indicating that FSH action on the α-subunit gene is exerted via cAMP. Inhibition of translation by cycloheximide causes upregulation of the α-subunit mRNA and does not block the effect of FSH on the level of this mRNA. In FSH-stimulated cells, the half-life of the α-subunit mRNA is 6 hours, and this half-life is prolonged by inhibition of transcription using actinomycin D. On the other hand, the levels of two βB-subunit mRNAs (4.2 and 3.5 kb) in cultured rat Sertoli cells are not affected by FSH or dibutyryl cAMP. However, these mRNAs are also upregulated by cycloheximide treatment. Experiments using actinomycin D showed that the 4.2 kb mRNA is less stable than the 3.5 kb mRNA [Klaij et al., 1990].

Similarly, FSH stimulates the mRNA of inhibin subunits in Sertoli cells in vivo [Bardin et al., 1987; Rivier et al., 1988; Toebosch et al., 1988; Meunier et al., 1988]. In rat testis, two species of βB-subunit mRNA (4.4 and 3.3 kb) appear to be present in equal concentrations, as opposed to rat ovary, in which a predominant band of 4.4 kb and a minor band of 3.3 kb are observed. According to Feng et al. [1989], one major species of βA-subunit mRNA (6.5 kb) is identified in both testis and ovary. The concentration of βA-subunit mRNA in the testis is very low, representing only 0.5% of that in rat ovary. The accumulation of βB-subunit mRNA peaked at 20 days of age and declined thereafter in a pattern similar to that of the α-subunit gene.

Feng et al. [1989] reported that hypophysectomy causes a marked increase in the concentration as well as the total content of βB-subunit but no change in βA-subunit mRNA in rat testis. Contrary to the fact that FSH markedly increased α-subunit mRNA levels both in vivo and in vitro, neither FSH nor testosterone has any significant effect on the accumulation of βA- or βB-subunit mRNAs in hypophysectomized animals or Sertoli cell primary cultures. However, in immature animals, Krummen et al. [1990] observed that the testicular content of inhibin α- and βB-subunit mRNAs is decreased after hypophysectomy. In adult animals, hypophysectomy results in a lesser decrease in inhibin α-subunit mRNA but had no effect on βB-subunit mRNA. The level of inhibin α-subunit mRNA per testis is significantly lower in hypophysectomized rats than in intact controls at all time points after surgery. Replacement of FSH in hypophysectomized immature rats leads to a dose-dependent increase in α-subunit mRNA in testis. However, hypophysectomy and FSH replacement have no significant effect on βB-subunit mRNA [Krummen et al., 1989]. In adult rats, hypophysectomy significantly lowers and FSH replacement increases inhibin α-subunit mRNA levels. Replacement of testosterone in adult animals, either alone or in combination with FSH, has no effect on expression of inhibin α-subunit mRNA. βB mRNA levels in adult testis are not significantly altered by any of the treatments. βA-subunit mRNA levels were

below the detectable threshold of filter hybridization. On the other hand, replacement of luteinizing hormone to immature hypophysectomized animals does not alter levels on mRNA for either inhibin subunit. However, in adults luteinizing hormone restores testicular inhibin α-subunit mRNA content in testes of hypophysectomized animals to levels seen in intact, saline-treated control animals. Leutinizing hormone replacement also slightly but consistently decreases testicular βB-subunit mRNA content compared with levels seen in hypophysectomized saline-treated rats.

As for the localization of mRNAs of inhibin subunits, in 12-day-old rats immunostaining and mRNA signal for the α-subunit are found in Leydig cell clusters. The βA- and βB-subunit staining and βA-subunit message are detectable in isolated interstitial cells, but the clusters appear to lack these subunits. Positive immunostaining for each subunit is localized in a Sertoli cell-like pattern in seminiferous tubules, as is a positive mRNA signal for the α- and βB-subunit over regions containing these cell types. Treatment with hCG and PMSG greatly enhances the production of the α-subunit in the Leydig cell clusters, but not within the tubules, of these young rats. In adult rats, α- and βB-subunit staining and α-subunit mRNA signal are observed in the interstitial cells. As in the immature animals, all three subunits are localized in a Sertoli cell-like pattern in the tubules and positive mRNA signal for the α- and βB-subunits is found over these cells. There is, however, no obvious change in the expression of the subunits in the testis of adult rats after gonadotropin treatment [Roberts et al., 1989].

The α- and βB-subunit mRNAs also vary significantly in different stages of spermatogenesis, the highest levels of both α- and βB-subunit expressions are in stages XIII–I and the lowest in stages VII–VIII. The hybridization signals obtained with β-actin probe are not significantly different between different stages, indicating that the differences in the quantities of subunit mRNAs in different stages are not due to different amounts of RNA blotted. βA-subunit mRNA levels are below the detection limit of the filter hybridization methods [Bhasin et al., 1989].

The concentration of testicular inhibin α-subunit mRNA also varies depending on ages of the animal; it peaks between 20 and 25 days of age and gradually declines thereafter [Keinan et al., 1989].

In summary, FSH increased the expression of inhibin α-subunit mRNA in Sertoli cells in vitro. Hypophysectomy decreases the testicular content of both α and βB mRNA levels. Replacement of FSH to hypophysectomized rats restores the declined testicular concentration of inhibin α-subunit mRNA, but FSH had no apparent effect on the regulation of β-subunit mRNA. On the other hand, replacement of luteinizing hormone to hypophysectomized rats does not alter inhibin α-subunit mRNA levels in immature rats but restores it in adult rats.

VI. CONCLUSION

Inhibin, originally isolated and characterized as a gonadal protein hormone selectively suppressing the secretion of FSH by the pituitary gonadotrophs, has provided an avenue for exploring the structural and functional characteristics of an inhibin/activin β/TGFβ family, including activins and follistatins. Members of this family are widely distributed in numerous cells and tissues and exert diverse influences at different target cells. Among these multiple biological activities are functions of endocrine, paracrine/autocrine, neuronal, cell-proliferating, cell differentiation, embryo-developing, and tumor-relating in nature. All biological activities of the gonadal proteins studied thus far play an important role directly or indirectly in growth and reproduction. The studies needed to

elucidate this important gene family fully have just begun to be performed. Future investigations in this area will reveal further actions of gonadal proteins, which will be a major area of research in reproduction for the immediate future.

ACKNOWLEDGMENTS

The authors gratefully recognize the assistance of Dr. Richard L. Wood in the preparation of this manuscript. The research was supported in part by grants HD-22876 and HD-24648.

REFERENCES

Asashima M, Nakano H, Uchiyama H, Sugino H, Nakamura T, Eto Y, Ejima D, Nishimatsu S, Ueno N, Kinoshita K (1991): Presence of activin (erythroid differentiation factor) in unfertilized eggs and blastulae of *Xenopus laevis.* Proc Natl Acad Sci USA 88:6511–6514.

Attardi B, Keeping HS, Winter SJ, Kotsuji F, Maurer RA, Troen P (1989): Rapid and profound suppression of messenger ribonucleic acid encoding follicle-stimulating hormone β by inhibin from primate Sertoli cells. Mol Endocrinol 3:280–287.

Attardi B, Miklos J (1990): Rapid stimulatory effect of activin-A on messenger RNA encoding the follicle-stimulating hormone β-subunit in rat pituitary cell cultures. Mol Endocrinol 4:721–726.

Attardi B, Vaughan J, Vale W (1991): Regulation of FSHβ messenger ribonucleic acid levels in the rat by endogenous inhibin. Endocrinology 129:2802–2804.

Bardin CW, Morris PL, Shaha C, Feng ZM, Rossi V, Vaughan J, Vale WW, Voglmayr J, Chen CL (1989): Inhibin structure and function in the testis. Ann NY Acad Sci 564:10–23.

Betteridge A, Craven RP (1991): A two-site enzyme-linked immunosorbent assay for inhibin. Biol Reprod 45:748–754.

Bhasin S, Krummnen LA, Swerdloff RS, Morelos BS, Kim WH, DiZerega GS, Ling N, Esch F, Shimasaki S, Toppari J (1989): Stage dependent expression on inhibin α- and βB subunits during the cycle of the rat seminiferous epithelium. Endocrinology 124:987–991.

Bicsak TA, Tucker EM, Cappel S, Vaughan J, Rivier J, Vale W, Hsueh AJW (1986): Hormonal regulation of granulosa cell inhibin biosynthesis. Endocrinology 119:2711–2719.

Molecular Biology of the Gonadal Proteins   **221**

Bicsak TA, Vale W, Vaughan J, Tucker EM, Cappel S, Hsueh AJW (1987): Hormonal regulation of inhibin production by cultured Sertoli cells. Mol Cell Endocrinol 49:211–217.

Bilezikjian LM, Blount AL, Campen CA, Gonzalez-Manchon C, Vale W (1991): Activin-A inhibits proopiomelanocortin messenger RNA accumulation and adrenocorticotropin secretion of At20 cells. Mol Endocrinol 5:1389–1395.

Billestrup N, Gnzalez-Manchon C, Potter E, Vale W (1990): Inhibition of somatotroph growth and growth hormone biosynthesis by activin in vitro. Mol Endocrinol 4:356–362.

Bindon BM, Piper LR, Cahill LP, Driancourt MA, O'Shea T (1986): Genetic and hormonal factors affecting superovulation. Theriogenology 5:53–70.

Braden TD, Farnworth PG, Burger HG, Conn PM (1990): Regulation of the synthetic rate of gonadotropin-releasing hormone receptors in rat pituitary cell cultures by inhibin. Endocrinology 127:2387–2392.

Broxmeyer HE, Lu L, Cooper S, Schwall RH, Mason AJ, Nikolics K (1988): Selective and indirect modulation of human multipotential and erythroid hematopoietic progenitor cell proliferation by recombinant human activin and inhibin. Proc Natl Acad Sci USA 85:9052–9056.

Campen CA, Vale W (1988): Interaction between ovine inhibin and steroids on the release of gonadotropin from cultured rat pituitary cells. Endocrinology 123:1320–1328.

Corrigan AZ, Billestrup N, Bilezikjian LM (1988): FSH releasing protein inhibits growth hormone and adrenocorticotropic hormone production by rat anterior pituitary cells. 70th Annual Meeting of the Endocrine Society (abstract 163).

Carroll RS, Corrigan AZ, Charib SD, Vale W, Chin WW (1989): Inhibin, activin, and follistatin: regulation of follicle-stimulating hormone messenger ribonucleic acid levels. Mol Endocrinol 3:1969–1976.

Cate RL. Mattaliano RJ, Hession C, Tizard R, Farber NM, Cheung A, Ninfa EG, Frey AT, Gash DJ, Chow EP, Fisher RA, Bertonis JM, Torres G, Wallner BP, Ramachandran KL, Ragin RC, Manganaro TF, MacLaughlin DT, Donahoe PK (1986): Isolation of the bovine and human genes for müllerian inhibiting substance and expression of the human gene in animal cells. Cell 45:685–698.

Chibbar R, Toma JG, Mitchell BF, Miller FD (1990): Regulation of neural oxytocin gene expression by gonadal steroids in pubertal rats. Mol Endocrinol 12:2030–2038.

Cho KW, Moria EA, Wright CV, De Robertis EM (1991): Overexpression of a homeodomain protein confers axis-forming activity to uncommitted embryonic cells. Cell 65:55–64.

Cooke J, Wong A (1991): Growth-factor-related proteins that are inducers in early amphibian development may mediate similar steps in amniote (bird) embryogenesis. Development 111:197–212.

Corrigan AZ, Bilezikhjian LM, Carroll RS, Bald LN, Schmelzer CH, Fendly BM, Mason AJ, Chin WW, Schwall RH, Vale W (1991): Evidence for an autocrine role of activin B within rat anterior pituitary cultures. Endocrinology 128:1682–1684.

Crawford RJ, Hammond VE, Evans BA, Coghlan JP, Haralambidis J, Hudson B, Panschow JD, Richards RI, Tregear GW (1987):α-Inhibin gene expression occurs in the ovine adrenal cortex and is regulated by adrenocorticotropin. Mol Endocrinol 1:699–706.

Cuevas P, S-Y Ying, N Ling, N Ueno, F Esch, R Guillemin (1987): Immunochemical detection of inhibin in the gonad. Biochem Biophys Res Commun 141:23–30.

Cummins LJ, O'Shea T, Al-Obaidi SAR, Bindon BM, Findlay JK (1986): Increase in ovulation rate after immunization of Merino ewes with a fraction of bovine follicular fluid containing inhibin activity. J Reprod Fertil 77:365–372.

Davis SR, Burger HG, Robertson DM, Farnworth PG, Carson RS, Krozowski Z (1988): Pregnant mare's serum gonadotropin stimulates inhibin subunit gene expression in the immature rat ovary: dose response characteristics and relationships to serum gonadotropins, inhibin, and ovarian steroid content. Endocrinology 123:2399–2407.

Davis SR, Krozowski Z, McLachlan RI, Burger HG (1987): Inhibin gene expression in the human corpus luteum. J Endocrinol 115:R21–R23.

de Jong FH, Jansen EHJM, Hermans WP, Van der Molen HJ (1981): In Fijii T, Channing CP (eds): Non-Steroidal Regulators in Reproductive Biology and Medicine. Oxford: Pergamon Press, pp 73–84.

de Jong FH, Grootenhuis AJ, Klaij IA, Van Beurden WM (1990): Inhibin and related proteins: localization, regulation, and effects. Adv Exp Med Biol 274:271–293.

DePaolo LV, Shimonaka M, Schwell RH, Ling N (1991): In vivo comparison of the follicle-stimulating hormone surppressing activity of follistatin and inhibin in ovariectomized rats. Endocrinology 128:668–674.

Derynck R, Jarret JA, Chen EY, Eaton DH, Bell JR, Assoian RK, Roberts AB, Sporn MB, Goeddel DV (1985): Human transforming growth factor-β complementary DNA sequence and expression in normal and transformed cells. Nature 316:701–705.

Drummond AE, Risbridger GP, de Krester DM (1989): The involvement of Leydig cells in the regulation of inhibin secretion by the testis. Endocrinology 125:510–515.

Esch F, Shimasaki S, Mercado M, Cooksey K, Ling N, Ying S-Y, Ueno N, Guillemin R (1987a): Structural characterization of follistatin: A novel follicle-stimulating hormone release-inhibiting polypeptide from the gonad. Mol Endocrinol 1:849–855.

Esch FS, Shimasaki S, Cooksey K, Mercado M, Mason AJ, Ying S-Y, Ling N (1987b): Complementary deoxyribonucleic acid (cDNA) cloning and DNA sequence analysis of rat ovarian inhibins. Mol Endocrinol 1:388–396.

Eto Y, Tsiji T, Takegawa M, Takano S, Yakagawa Y, Shibai H (1987): Purification and characterization of erythroid differentiation factor (EDF) isolated from human leukemia cell line THP-1. Biochem Biophys Res Commun 42:1095–1103.

Feng Z-M, Bardin CW, Chen C-LC (1989a): Characterization and regulation of testicular inhibin β-subunit mRNA. Mol Endocrinol 3:939–948.

Feng Z-M, Li Y-P, Chen C-LC (1989b): Analysis of the 5′-flanking regions of rat inhibin α- and βB-subunit genes suggests two different regulatory mechanisms. Mol Endocrinol 3:1914–1925.

Findley JK, Doughton B, Robertson DM, Forage RG (1989): Effects of immunization against recombinant bovine inhibin alpha subunit on circulating concentrations of gonadotrophins in ewes. J Endocrinol 120:59–65.

Forage RG, Ring JM, Brown RW, McInermey BV, Cobon GS, Gregson RP, Robertson DM, Morgan FJ, Hearn MTW, Findlay JK, Wettenhall REH, Burger HG, deKretser DM (1986): Cloning and sequence analysis of cDNA species coding for the two subunits of inhibin from bovine follicular fluid. Proc Natl Acad Sci USA 83:3091–3095.

Forage RG, Brown RW, Oliver KJ, Atrache BT, Devine PL, Hudson GC, Goss NH, Bertram KC, Tolstoshev P, Robertson DM, deKretser DM, Doughton B, Burger HG, Findlay LK (1987): Immunization against an inhibin subunit produced by recombinant DNA techniques results in increased ovulation from bovine follicular fluid. Proc Natl Acad Sci USA 83:3091–3095.

Fujimoto K, Kawakita M, Kato K, Yonemura Y, Masuda T, Matsuzaki H, Hirose J, Isaji M, Sasaki H, Inoue T, et al. (1991): Purification of megakaryocyte differentiation activity from a human fibrous histiocytoma cell line: N-terminal sequence homology with activin A. Biochem Biophys Res Commun 174:1163–1168.

Fukuda M, Miyamoto K, Hasegawa Y, Nomura M, Igarashi M, Kanagawa K, Matsuo H (1986): Isolation of bovine follicular fluid inhibin of about 32 kDa. Mol Cell Endocrinol 44:55–60.

Fukuda M, Miyamoto K, Hasegawa Y, Ibuki Y, Ig-

arashi M (1987): Action mechanism of inhibin in vitro—cycloheximide mimics inhibin action on pituitary cells. Mol Cell Endocrinol 51:41–50.

Gibbons IR (1977): Sperm motility. In Greep RO, Koblinsky MA (eds): Frontiers in Reproduction and Fertility Control: A Review of the Reproductive Sciences and Contraceptive Development, Part II. Cambridge, MA: MIT, pp 452–457.

Godbout M, Labrie F (1984): Purification of inhibin from porcine and bovine follicular fluid. Biol Reprod 30[Suppl 1]:103.

Gonzalez-Manchon C, Vale W (1989): Activin-A, inhibin and transforming growth factorβ modulate growth of two gonadal cell lines. Endocrinology 125:1666–1672.

Gospodarowicz D, K Lau (1989): Pituitary follicular cells secrete both vascular endothelial growth factor and follistatin. Biochem Biophys Res Commun 165:292–298.

Grady R, Charlesworth MC, Schwartz NB (1982): Characterization of the FSH-suppressing activity in follicular fluid. Recent Prog Horm Res 38:409–456.

Gregg DW, Schwall RH, Nett TM (1991): Regulation of gonadotropin secretion and number of gonadotropin-releasing hormone receptors by inhibin, activin-A, and estradiol. Biol Reprod 44:725–732.

Grootenhuis AJ, van Sluijs FJ, Klaij IA, Steenbergen J, Timmerman MA, Bevers MM, Dieleman SJ, de Jong FH (1991): Inhibin, gonadotrophins and sex steroids in dogs with Sertoli cell tumours. J Endocrinol 127:235–242.

Hedger MP, Drummond AE, Robertson DM, Risbridger GP, de Kretser DM (1989): Inhibin and activin regulate [3H]thymidine uptake by rat thymocytes and 3T3 cells in vitro. Mol Cell Endocrinol 61:133–138.

Henderson KM, McNatty KP, Wards RL, Heath DA, Lun S (1991): Inhibin production in vitro by granulosa cells from Booroola ewes which were either homozygous or non-carriers of a fecundity gene influencing their ovulation rate. J Reprod Fertil 92:147–157.

Hishimoto M, Kondo S, Sakurai T, Etoh Y, Shibai H, Muramatsu M (1990): Activin/EDF as an inhibitor of neural differentiation. Biochem Biophys Res Commun 173:193–200.

Hochberg ZJ, Weiss J, Richman RA (1981): Inhibin-like activity in extracts of rabbit placentae. Placenta 2:259–264.

Hsueh AJW, Dahl KD, Vaughan J, Tucker E, Rivier J, Bardin CW, Vale W (1987): Heterodimers and homodimers of inhibin subunits have different paracrine action in the modulation of luteinizing hormone-stimulated androgen biosynthesis. Proc Natl Acad Sci USA 84:5082–5086.

Hutchinson LA, Findlay JR, de Vos FL, Robertson

DM (1987): Effects of bovine inhibin, transforming growth factor-β and bovine activin-A on granulosa cell differentiation. Biochem Biophys Res Commun 146:1405–1412.

Huylebroeck D Van Nimmen K, Waheed A von Figura K, Marmenout A, Fransen L, De Waele P, Jaspar JM, Franchimont P, Stunnenberg H, et al. (1990): Expression and processing of the activin-A/erythroid differentiation factor precursor: a member of the transforming growth factor-β superfamily. Mol Endocrinol 4:1153–1165.

Inouye S, Guo Y, DePaolo L, Shimonaka M, Ling N, Shimasaki S. (1991): Recombinant expression of human follistatin with 315 and 288 amino acids: Chemical and biological comparison with native porcine follistatin. Endocrinology 129:815–822.

Itoh M, Igarashi M, Yamada K, Hasegawa Y, Seki M, Eto Y, Shibai H (1990): Activin A stimulates meiotic maturation of the rat oocyte in vitro. Biochem Biophys Res Commun 166:1479–1484.

Jansen EHJM, Steenbergen J, de Jong FH, van der Molen HJ (1981): The use of affinity matrices in the purification of inhibin from bovine follicular fluid. Mol Cell Endocrinol 21:109–117.

Kaiser M, Gibori G, Mayo KE (1990): The rat follistatin gene is highly expressed in decidual tissue. Endocrinology 126:2768–2770.

Katayama T, Shiota K, Takahashi M (1990): Activin A increases the number of follicle-stimulating hormone cells in anterior pituitary cultures. Mol Cell Endocrinol 69:179–185.

Keinan D, Maadigan MB, Bardin CW, Chen C-LC (1989): Expression and regulation of testicular inhibin α-subunit gene in vivo and in vitro. Mol Endocrinol 3:29–35.

Kitaoka M, Kojima I, Ogata E (1988): Activin A: a modulator of multiple types of anterior pituitary cells. Biochem Biophys Res Commun 157:48–54.

Klaij IA, Toebosch AMW, Themmen APN, Shimasaki S, de Jong FH, Grootgoed JA (1990): Regulation of inhibin α- and βB-subunit mRNA levels in rat Sertoli cells. Mol Cell Endocrinol 68:45–52.

Klein R, Robertson DM, Shukovski L, Findlay JK, de Kretser DM (1991): The radioimmunoassay of follicle-stimulating hormone (FSH)-suppressing protein (FSP): stimulation of bovine granulosa cell FSP secretion by FSH. Endocrinology 128:1048–1056.

Knight PG, AJ Beard, JHM Wrathall, RRJ Castillo (1989): Evidence that the bovine ovary secretes large amounts of monomeric inhibin α subunit and its isolation from bovine follicular fluid. J Mol Endocrinol 2:189–200.

Kogawa K, Nakamura T, Sugino K, Takio K, Titani K, Sugino H (1991): Activin-binding protein is present in pituitary. Endocrinology 128:1434–1440.

Kojima I, Ogata E (1989): Dual effect of activin A on cell growth in Balb/c3T3 cells. Biochem Biophys Res Commun 159:1107–1113.

Kokan-Moore NP, Bolender DL, Lough J (1991): Secretion of inhibin βA by endoderm cultured from early embryonic chicken. Dev Biol 146:242–245.

Krummen LA, Morelos BS, Bhasin S (1990): The role of luteinizing hormone in regulation of testicular inhibin α and βB subunit messenger RNAs in immature and adult animals. Endocrinology 127:1097–1104.

Kruumen LA, Toppari J, Kim WH, Morelos BS, Ahmad N, Swerdloff RS, Ling N, Shimasaki S, Esch F, Bhasin S (1989): Regulation of testicular inhibin subunit messenger ribonucleic acid levels in vitro: effects of hypophysectomy and selective follicle-stimulating hormone replacement. Endocrinology 125:1630–1637.

LaPolt PS, Piquette GN, Soto D, Sincich C, Hsueh AJW (1990): Regulation of inhibin subunit messenger ribonucleic acid levels by gonadotropins, growth factors, and gonadotropin-releasing hormone in cultured rat granulosa cells. Endocrinology 127:823–831.

LaPolt PS, Soto D, Su J-G, Campen CA, Vaughan J, Vale W, Hsueh AJW (1989): Active stimulation on inhibin secretion and messenger RNA levels in cultured granulosa cells. Mol Endocrinol 3:1666–1673.

Lappohn R, Burger HG, Bouma J, Bangah M, Krans M, de Bruijn HWA (1989): Inhibin as a marker for granulosa-cell tumors. N Engl J Med 321:790–793.

Lee VWK, McMaster J, Quigg H, Findlay J, Leversha L (1981): Ovarian and peripheal blood inhibin concentrations increase with gonadotropin treatment in immature rats. Endcorinology 108:2403–2405.

Lee W, Mason AJ, Schwall R, Szonyi E, Mather JP (1989): Secretion of activin by interstitial cells in the testis. Science 243:396–398.

Lerner RA (1982): Tapping the immunological repertoire to produce antibodies of predetermined specificity. Nature 299:592–596.

Leversha LJ, Robertson DM, de Vos FL, Morgan FJ, Hearn MTW, Wettenhall REH, Findlay JK, Burger HG, de Kretser DM (1987): Isolation of inhibin from bovine follicular fluid. J Endocrinol 113:213–221.

Ling N, Ying S-Y, Ueno N, Esch F, Denoroy L, Guillemin R (1985): Isolation and partial characterization of a Mr 32,000 protein with inhibin activity from porcine follicular fluid. Proc Natl Acad Sci USA 82:7217–7221.

Ling N, Ying S-Y, Ueno N, Shimasaki S, Esch F, Hotta M, Guillemin R (1986a): Pituitary FSH is released by a heterodimer of the β-subunits from the two forms of inhibin. Nature 321:779–782.

Ling N, Ying S-Y, Ueno N, Shimasaka S, Esch F, Hotta M, Guillemin RA (1986b): Homodimer of the β-subunits of inhibin A stimulates the secretion of pituitary follicle stimulating hormone. Biochem Biophys Res Commun 138:1129–1137.

Lucas C, Bald LN, Fendly BM, Mora-Worms M, Figari IS, Patzer EJ, Palladino MA (1990): The autocrine production of transforming growth factor β1 during lymphocyte activation: A study with a monoclonal antibody-based ELISA. J Immunol 145:1415–1422.

Martins T, Rocha A (1931): Regulation of the hypophysis by the testicle and some problems of sexual dynamics. Endocrinology 15:421–434.

Mason AJ, Hayflick JS, Ling N, Esch F, Ueno N, Ying S-Y, Guillemin R, Niall H, Seeburg PH (1985): Complementary DNA sequences of ovarian follicular fluid inhibin show precursor structure and homology with transforming growth factor-β. Nature 318:659–663.

Mason AJ, Niall HD, Seeburg PH (1986): Structure of two human ovarian inhibins. Biochem Biophys Res Commun 135:957–964.

Mason AJ, Berkemeier LM, Schmeizer CH, Schwall RH (1989): Activin B: precursor sequences, genomic structure and in vitro activities. Mol Endocrinol 3:1352–1358.

Mather JP, Attie KM, Woodruff TK, Rice GC, Phillips DM (1990): Activin stimulates spermatogonial proliferation in germ–Sertoli cell cocultures from immature rat testis. Endocrinology 127:3206–3214.

Mathews LS, Vale WW (1991): Expression cloning of an activin receptor, a predicted transmembrane serine kinase. Cell 65:973–982.

Mayo KE, Cerelli GM, Spiess J, Rivier J, Rosenfeld MG, Evans RM, Vale W (1986): Inhibin A-subunit cDNAs from porcine ovary and human placenta. Proc Natl Acad Sci USA 83:5849–5853.

McLachlan RI, Robertson DM, Healy DL, de Kretser DM, Burger HG (1986a): Plasma inhibin levels during gonadotropin-induced ovarian hyperstimulation for IVF: A new index of follicular function? Lancet (8492):1233–1234.

McLachlan RI, Robertson DM, Burger HG, De Kretser DM (1986b): The radioimmunoassay of bovine and human follicular fluid and serum inhibin. Mol Cell Endocrinol 46:175–185.

McLachlan RI, Healy DL, Robertson DM, Burger HG, deKretser DM (1987): Circulating immunoreactive inhibin in the luteal phase and early gestation of women undergoing ovulation induction. Fertil Steril 48:1001–1005.

McLachlan RI, Dahl KD, Bremner WJ, Schwall R, Schmelzer CH, Mason AJ, Steiner RA (1989):

Recombinant human activin-A stimulates basal FSH and GnRH-stimulated FSH and LH release in the adult male macaque, *Macaca fascicularis*. Endocrinology 125:2787–2789.

McCullagh DR (1932): Dual endocrine activity of the testis. Science 76:19–20.

Melton DA (1991): Pattern formation during animal development. Science 252:234–241.

Mercer JE, Clements JA, Funder JW, Clarke IJ (1987): Rapid and specific lowering of pituitary FSHβ mRNA levels by inhibin. Mol Cell Endocrinol 53:251–254.

Meunier H, Rivier C, Evans RM, Vale W (1988a): Gonadal and extragonadal expression of inhibin alpha, beta A and beta B subunits in various tissues predicts diverse function. Proc Natl Acad Sci USA 85:247–251.

Meunier H, Cajander SB, Roberts VJ, River C, Sawchenko PE, Hsueh AJW, Vale W (1988b): Rapid changes in the expression of inhibin α-, βA-, and βB-subunits in ovarian cell types during the rat estrous cycle. Mol Endocrinol 2:1352–1363.

Meunier H, Roberts VJ, Sawchenko PE, Cajander SB, Hsueh AJW and Vale W (1989): Periovulatory changes in the expression of inhibin αA-, and βB-subunits in hormonally induced immature female rats. Mol Endocrinol 3:2062–2069.

Michael U, Albiston A, Findley JK (1990): Rat follistatin: Gonadal and extragondal expression and evidence for alternative splicing. Biochem Biophys Res Commun 173: 401–407.

Miller KF, Xie S, Pope WF (1991): Immunoreactive inhibin in follicular fluid is related to meiotic stage of the oocyte during final maturation of the porcine follicle. Mol Reprod Dev 28:35–39.

Miyamoto K, Hasegawa Y, Fukuda M, Nomura M, Igarashi M, Kangawa K, Matsuo H (1985): Isolation of porcine follicular fluid inhibin of 32 K daltons. Biochem Biophys Res Commun 129:396–403.

Miyamoto K, Hasegawa Y, Fukuda M, Igarashi M (1986): Demonstration of high molecular weight forms of inhibin in bovine follicular fluid (bFF) by using monoclonal antibodies to bFF 32 K inhibin. Biochem Biophys Res Commun 136:1103–1109.

Mizumachi M, Voglmayr JK, Washington DW, Chen CL, Bardin CW (1990) Superovulation of ewes immunized against the human recombinant inhibin alphasubunit associated with increased pre- and postovulatory follicle-stimulating hormone levels. Endocrinology 126:1058–1063.

Morris PL, Vale W, Cappel S, Bardin CW (1988): Inhibin production by primary Sertoli cell-enriched cultures: regulation by FSH, androgens and epidermal growth factor. Endocrinology 122:717–725.

Mottram JC, Cramer W (1923): Report on the general

effects of exposure to radium on metabolism and tumor growth in the rat and the special effects on testis and pituitary. QJ Exp Physiol 13:209–229.

Murata T, Ying SY (1991): Transforming growth factor-β and activin inhibit basal secretion of prolactin in a pituitary monolayer culture system. Proc Soc Exp Biol Med 198:599–605.

Nakamura T, Takio K, Eto Y, Shibai H, Titani K, Sugino H (1990): Activin-binding protein from rat ovary is follistatin. Science 247:836–838.

Nishida M, Jimi S, Haji M, Hayashi I, Kai T, Tasaka H (1991): Juvenile granulosa cell tumor in association with a high serum inhibin level. Gynecol Oncol 40:90–94.

O WS, Robertson DM, de Krester DM (1989): Inhibin as an oocyte meiotic inhibitor. Mol Cell Endocrinol 62:307–311.

Ohashi M, Hasegawa Y, Haji M, Igarashi M, Nawata H (1990): Production of immunoreactive inhibin by a virilizing ovarian tumor (Sertoli-Leydig tumour). Clin Endocrinol Oxf 33:613–618.

Padgett RW, St Johnson RD, Gelbart WM (1987): A transcript from a *Drosophila* pattern gene preducts a protein homologous to the transforming growth factor-β family. Nature 325:81–84.

Penschow JD, Aldred GP, Darling PA, Haralambidis J, Hammond VE, van Leeuwen BH, Mason AJ, Niall HD, Seeburg P and Coghlan JP (1990): Differential expression of inhibin α and βA subunit genes in rat and mouse ovarian follicles during pregnancy. J Mol Endocrinol 4:247–255.

Petraglia F, Sawchenko P, Lim AT, Rivier J, Vale W (1987): Localization, secretion, and action of inhibin in human placenta. Science 237:187–189.

Petraglia F, Vaughan J, Vale W (1989): Inhibin and activin modulate the release of genadotropin-releasing hormone, human choronic gonadotropin, and progesterone from cultured human placental cells. Proc Natl Acad Sci USA 86:5114–5117.

Piquette GN, Kenny RM, Sertich PL, Yamoto M, Hsueh AJ (1990): Equine granulosa-theca cell tumors express inhibin α- and β A-subunit messenger ribonucleic acids and proteins. Biol Reprod 43:1050–1057.

Plotsky PM, Kjaer A, Sutton SW, Sawchenko PE, Vale W (1991): Central activin administration modules corticotropin-releasing hormone and adrenocorticotropin secretion. Endocrinology 128:2520–2525.

Plotsky PM, Sawchenko PE, Vale W (1988): Evidence for inhibin β-Chain like-peptide mediation of suckling-induced oxytocin secretion. 18th Society for Neuroscience, Toronto, 1988, p 627.

Qu JP, Vankrieken L, Brulet C, Thomas K (1991): Circulating bioactive inhibin levels during human pregnancy. J Clin Endocrinol Metab 72:862–866.

Rabinovici J, Spencer SJ, Jaffe RB (1990): Recombinant human activin-A promotes proliferation of human leuteinized preovulatory granulosa cells in vitro. J Clin Endocrinol Metab 71:1396–1398.

Risbridger GP, Hancock A, Robertson DM, Hodgson YH, de Kretser DM (1989a): Follitropin (FSH) stimulation of inhibin biological and immunological activities by seminiferous tubule and Sertoli cell cultures from immature rats. Mol Cell Endocrinol 67:1–9.

Risbridger GP, Clements J, Robertson DM, Drummond AE, Muir J, Burger HG, de Krester DM (1989b): Immuno- and bioactive inhibin and inhibin alpha-subunit expression in rat Leydig cell cultures. Mol Cell Endocrinol 66:119–122.

Risbridger GP, Robertson DM, de Kretser DM (1990): Current perspectives of inhibin biology Acta Endocrinol (Copenh) 122:673–682.

Rivier J, Spiess J, McClintock R, Vaughan J, Vale W (1985): Purification and partial characterization of inhibin from porcine follicular fluid. Biochem Biophys Res Commun 133:120–127.

Rivier C, Rivier J, Vale W (1986): Inhibin-mediated feedback control of follicle-stimulating hormone secretion in the female rat. Science 234:205–208.

Rivier C, Vale W (1989): Immunoneutralization of endogenous inhibin modifies hormone secretion and ovulation rate in the rat. Endocrinology 125:152.

Rivier C, Schwall R, Mason A, Burton L, Vaughan J, Vale W (1991): Effect of recombinant inhibin on luteinizing hormone and follicle-stimulating hormone secretion in the rat. Endocrinology 128:1548–1554.

Rivier C, Vale W (1991): Effect of recombinant activin-A on gonadotropin secretion in the female rat. Endocrinology 129:2463–2465.

Robert DM, Foulds LM, Leversha L, Morgan FJ, Hearn MTW, Burger HG, Wettenhall REH, de Kretser DM (1985): Isolation of inhibin from bovine follicular fluid. Bichem Biophy Res Commun 126:220–226.

Roberts V, Meunier H, Sawchenko PE, Vale W (1989): Differential production and regulation of inhibin subunits in rat testicular cell types. Endocrinology 125:2350–2359.

Roberts VJ, Sawchenko PE, Vale W (1991): Expression of inhibin/activin subunit messenger ribonucleic acids during rat embryogenesis. Endocrinology 128:3122–3129.

Robertson DM, Foulds LM, Leversha L, Morgan FJ, Hearn MTW, Burger HG, Wettenhall REH, de Kretser DM (1985): Isolation of inhibin from bovine follicular fluid. Biochem Biophys Res Commun 126: 220–226.

Robertson DM, de Vos FL, Foulds LM, McLachlan RI, Burger HG, Morgan FJ, Hearn MTW, de Kretser DM (1986): Isolation of a 32 kDa form of inhibin from bovine follicular fluid. Mol Cell Endocrinol 44:271–277.

Robertson DM, Klein R, de Vos FL, McLachland RI, Wettenhall REH, Hearn MTW, Burger HG, de Kretser DM (1987): The isolation of polypeptides with FSH suppressing activity from bovine follicular fluid which are structurally different to inhibin. Biochem Biophys Res Commun 149:744–749.

Robertson DM, Giacometti M, Foulds LM, Lahnstein J, Goss NH, Hearn MTW, de Kretser DM (1989): Isolation of inhibin α-subunit precursor proteins from bovine follicular fluid. Endocrinology 125:2141.

Rodgers RJ, Stuchbery SJ, Findlay JK (1989): Inhibin mRNAs in bovine and bovine ovarian follicles and corpora lutea throughout the estrous cycle and gestation. Mol Cell Endocrinol 62:95–101.

Saito S, Nakamura T, Titani K, Sugino H (1991): Production of activin-binding protein by rat granulosa cells in vitro. Biochem Biophys Res Commun 176:413–422.

Sawcheko PE, Plotsky PM, Pfeoffer SW, Cunningham ETC, Vaughan J, Rivier J, Vale W (1988): Inhibin beta neural pathways involved in the control of oxytocin secretion. Nature 334:615–617.

Schanbacher BD (1991): Pituitary and testicular responses of beef bulls to active immunization against inhibin α. Int J Anim Sci 69:252–257.

Scher W, Eto Y, Ejima D, Den T, Svet-Moldavsky IA (1990): Phorbol ester–treated human acute myeloid leukemia cells secrete G-CSF, GM-CSF and erythroid differentiation factor into serum-free media in primary culture. Biochim Biophys Acta 1055:278–286.

Schmelzer CH, Burton LE, Tamony CM, Schwall RH, Mason AJ, Liegeois N (1990): Purification and characterization of recombinant human activin B. Biochim Biophys Acta 1039:135–141.

Schubert D, Kimura H, LaCorbiere M, Vaughan J, Karr D, Fischer WH (1990): Activin is a nerve cell survival molecule. Nature 344:868–870.

Schubert D, Kimura H (1991): Substratum-growth factor collaborations are required for the mitogenic activities of activin and FGF on embryonal carcinoma cells. J Cell Biol 114:841–846.

Schwall RH, Mason AJ, Wilcox JN, Bassett SG, Zeleznik AJ (1990): Localization of inhibin/activin subunit mRNA within the primate ovary. Mol Endocrinol 4:75–79.

Schwall RH, Nikolics K, Szonyi E, Gorman C, Mason AJ (1988): Recombinant expression and characterization of human activin A. Mol Endocrinol 2:1237–1242.

Schwall R, Schmelzer CH, Matsuyama E, Mason AJ (1989): Multiple actions of recombinant activin-A in vivo. Endocrinology 125:1420–1423.

Schwartz NB (1991): Why I was told not to study inhibin and what I did about it. Endocrinology 129:1690–1691.

Sealfon SC, Laws SC, Wu JC, Gillo B, Miller WL (1990): Hormonal regulation of gonadotropin-releasing hormone receptors and messenger RNA activity in ovine pituitary culture. Mol Endocrinol 4:1980–1987.

Shimasaki S, Koga M, Esch F, Cookway M, Mercado M, Koba A, Ueno N, Ying SY, Ling N, Guillemin R (1988): Primary structure of the human follistatin precursor and its genomic organization. Proc Natl Acad Sci USA 85: 4218–4213.

Shimasaki S, Koga M, Buscaglia ML, Simmons DM, Bicsak TA, Ling N (1989): Follistatin gene expression in the ovary and extragonadal tissues. Mol Endocrinol 3:651–659.

Shimonaka M, Inouye S, Shimasaki S, Ling N (1991): Follistatin binds to both activin and inhibin through the common beta-subunit. Endocrinology 128:3313–3315.

Shintani Y, Takada Y, Yamasaki R, Saito S (1991): Radioimmunoassay for activin A/EDF. Method and measurement of immunoreactive activin A/EDF levels in various biological materials. J Immunol Methods 137:267–274.

Slack J (1991): Embryology: Molecule of the moment (news). Nature 349:17–18.

Sokol S, Melton DA (1991): Pre-existent pattern in Xenopus animal pole cells revealed by induction with activin. Nature 6325:409–411.

Smith JC, Price BMJ, Van Nimmen K, Hylebroeck D (1990): Identification of a potent Xenopus mesoderm-inducing factor as a homologue of activin A. Nature 345:729–731.

Spencer SJ, Rabinovici J, Jaffe RB (1990): Human recombinant activin-A inhibits proliferation of human fetal adrenal cells in vitro. J Clin Endocrinol Metab 71:1678–1680.

Stewart AG, Milborrow HM, Ring JM, Crowther CE, Forage RG (1986): Human inhibin genes. Genomic characterisation and sequencing. FEBS Lett 206:329–334.

Sugawara M, DePaolo L, Nakatani A, DiMarzo SL, Ling N (1990): Radioimmunoassay of follistatin: Aplication for in vitro fertilization procedures. J Clin Endocrinol Metab 71:1672–1674.

Sugino K, T, Nakamur, K Takio, K Titani, K Miyamoto, Y Hasegawa, M Igarashi, H Sugino (1989): Inhibin alpha-subunit monomer is present in bovine follicular fluid. Biochem Biophys Res Commun 159:1323–1329.

Sugino H, Nakamura T, Kogawa K, Takio K, Titani K (1990): Activin-binding protein from bovine pituitary. Endocrinology 126:A219.

Tashiro K, Tamada R, Asano M, Hashimoto M, Muramatsu M, Shiokawa K (1991): Expression of mRNA for activin-binding protein (follistatin) during early embryonic development of *Xenopus laevis*. Bicohem Biophys Res Commun 174: 1022–1027.

Thomsen G, Woolf T, Whitman M, Sokol S, Vaughan J, Vale W, Melton DA (1990): Activins are expressed early in *Xenopus* embryogenesis and can induce axial mesoderm and anterior structures. Cell 63:485–493.

Toebosch AMW, Robertson DM, Trapman J, Klaaseen P, de Paus RA, de Jong FH Grootegoed JA (1988): Effects of FSH and IGF-1 on immature rat Sertoli cell inhibin α- and β-subunit mRNA levels and inhibin secretion. Mol Cell Endocrinol 55:101–105.

Toebosch AMW, Robertson DM, Klaij IA, de Jon FH, Grootegoed JA (1989): Effects of FSH and testosterone on highly purified rat Sertoli cells: Inhibin α-subunit mRNA expression and inhibin secretion are enhanced by FSH but not by testosterone. J Endocrinol 122:757–762.

Torney AH, Hodgson YM, Forage R, de Kretser DM (1989): Cellular localization of inhibin mRNA in the bovine ovary by in situ hybridization. J Reprod Fertil 86:391–399.

Totsuka Y, Tabuchi M, Kijima I, Shibai H, Ogata E (1988): A novel action of activin A: stimulation of insulin secretion in rat pancreatic islets. Biochem. Biophys. Res. Commun. 156:335–339.

Tsafriri A, Vale W, Hsueh AJ (1989): Effects of transforming growth factors and inhibin-related proteins on rat preovulatory graafian follicles in vitro. Endocrinology 125:1857–1862.

Turner IM, Saunders PTK, Shimasaki S, Hillier SG (1989): Regulation of inhibin subunit gene expression by FSH and estradiol in cultured granulosa cells. Endocrinology 125:2790–2792.

Ueno N, Ling N, Ying S-Y, Esch F, Shimasaki S, Guillemin R (1987): Isolation and partial characterization of follistatin, a novel Mr 35,000 monomeric protein which inhibits the release of follicle stimulating hormone. Proc Natl Acad Sci USA 84:8282–8286.

Vale W, Rivier J, Vaughan J, McClintock T, Corrigan A, Woo W, Karr D, Spiess J (1986): Purification and characterization of an FSH releasing protein from porcine ovarian follicular fluid. Nature 321:776–779.

Vale W, Rivier C, Hsueh A, Campen C, Meunier H, Bicsak T, Vaughan J, Corrigan A, Bardin W, Sawcheko P, Petraglia F, Plotsky P, Spiess J, Rivier C (1988): Chemical and biological characterization of the inhibin family of protein hormones. Recent Prog Horm Res, 44:1–34.

vanden Eijndenvan-Raaij AJ van Achterberg TA, vander Kruijssen CM, Piersma AH, Huylebroeck D, de Laat SW, Mummery CL (1991): Differenti-

ation of aggregated murine P19 embryonal carcinoma cells is induced by a novel visceral endoderm-specific FGF-like factor and inhibited by activin A. Mech Dev 33:157–165.

Wang QF, Franworth PG, Findlay JK, Burger HG (1990): Chronic inhibitory effect of follicle-stimulating hormone (FSH)-suppressing protein (FSP) or follistatin on activin- and gonadotropin-releasing hormone-stimulated FSH synthesis and secretion in cultured rat anterior pituitary cells. Endocrinology 127:1385–1393.

Wang QF, Farnworth PG, Findlay JK, Burger HG (1989): Inhibitory effect of pure 31-kilodalton bovine inhibin on gonadotropin-releasing hormone (GnRH)–induced up-regulation of GnRH binding sites in cultured rat anterior pituitary cells. Endocrinology 124:363–368.

Weeks DL, Melton DA (1987): A material mRNA localized to the vegetal hemisphere is *Xenopus* eggs coded for a growth factor related to TGFβ. Cell 51:861–867.

Woodruff TK, Ackland J, Rahal JO, Schwartz NB, Mayo KE (1991): Expression of ovarian inhibin during pregnancy in the rat. Endocrinology 128:1647–1654.

Woodruff TK, D'Agostino J, Schwartz NB, Mayo KE (1988): Dynamic changes in inhibin messenger RNAs in rat ovarian follicles during the reproductive cycle. Science 239:1296–1299.

Woodruff T, Meunier H, Jones PB, Hsueh AJW, Mayo KE (1987): Rat inhibn: molecular cloning of α- and β-subunit complementary deoxyribonucleic acids and expression in the ovary. Mol Endorcinol 1:561–568.

Wozney JM, Rosen V, Celeste AJ, Mitsock LM, Whitters MJ, Kriz RW, Hewick RM, Wange EA (1988): Novel regulators of bone formation: molecular clones and activities. Sci 242:1528–1534.

Wrathall JH, McLeod BJ, Glencross RG, Beard AJ, Knight PG (1990): Inhibin immunoneutralization by antibodies raised against synthetic peptide sequences of inhibin alpha subunit: effects in gonadotrophin concentrations and ovulation rate in sheep. J Endocrinol 124:167.

Xiao S, Findlay JK, Robertson DM (1990): The effect of bovine activin and follicle-stimulating hormone (FSH) suppressing protein/follistatin on FSH-induced differentiation of rat granulosa cells in vitro. Mol Cell Endocrinol 69:1–8.

Ying SY (1988): Inhibins, activins and follistatin: gonadal proteins modulating the secretion of follicle-stimulating hormone. Endocr Rev 9:267–293.

Ying SY, Becker A (1989): Follistan and activin are important immunomodulatory gonodal proteins. Endocrinology 124:A218.

Ying SY, Becker A, Baird A, Ling N, Ueno N, Esch F, Guillemin R (1986): Type beta transforming growth factor (TGFβ) is a potent stimulator of the basal secretion of follicle stimulating hormone (FSH) in a pituitary monolayer system. Biochem Biophys Res Commun 135:950–956.

Ying S-Y, Becker A, Swanson G, Tan P, Ling N, Esch F, Ueno N, Shimasaki S, Guillemin R (1987): Follistan specifically inhibits pituitary follicle stimulating hormone release in vitro. Bochem Biophys Res Commun 149:133–139.

Ying S-Y, Czvik J, Ling N, Ueno N, Guillemin R (1987): Secretion of follicle stimulating hormone and production of inhibin are reciprocally related. Proc Natl Acad Sci USA 84:4631–4635.

Ying S-Y, Swanson G, Becker A, Tan P, Wadleigh D, Ueno N, Shimasaki S, Chiang T-C, Hu R, Dong M-H (1988): Immunoneutralization and gene expression of follistatin isolated from porcine follicular fluid. Endocrinology 122:A566.

Yohkaichiya T, O'Connor A, de Kretser DM (1991): Circulating immunoreactive inhibin, gonadotro-pin, and prolactin levels during pregnancy, lactation, and postweaning estrous cycle in the rat. Biol Reprod 44:6–12.

Yu J, Maderazo L, Shao LE, Frigon NL, Vaughan J, Vale W, Yu A (1991): Specific roles of activin/inhibin in human erythropoiesis in vitro. Ann NY Acad Sci 628:199–211.

Yu J, Shao LE, Lemas V, Vaughan J, Rivier J, Vale W (1987): Importance of FSH-releasing protein and inhibin in erythroid differentiation. Nature 330: 765–767.

Yu J, Shao LE, Vaughan J, Vale W, Yu AL (1989): Characterization of the potentiative effect of activin on human erythroid colony formation in vitro. Blood 73:952–960.

Zang Z, Lee VWK, Carson RS, Burger HG (1988): Selective control of rat granulosa cell inhibin production by FSH and LH in vitro. Mol Cell Endocrinol 56:35–40.

ABOUT THE AUTHORS

SHAO-YAO YING is an Associate Professor in the Department of Anatomy and Cell Biology at the University of Southern California School of Medicine, where he teaches Microanatomy. After receiving his B.S. from the National Taiwan University, Taipei, in 1962, he received his M.S. and Ph.D. under Roland K. Meyer at the University of Wisconsin, Madison, where he concentrated on the steroidal control of a preovulatory surge of gonadotropin during ovulation. He pursued postdoctoral research with Roy O. Greep at the Laboratory for Human Reproduction and Reproductive Biology and Department of Anatomy, Harvard Medical School. From 1974 to 1978, Dr. Ying was Senior Lecturer at the Department of Anatomy and Histology, University of Adelaide, Adelaide, Australia. In 1978 he joined Roger Guillemin's group at the Laboratories of Neuroendocrinology, the Salk Institute, San Diego; there he worked on the isolation, characterization, and physiology of inhibin, activin, and follistatin. Dr. Ying's current research involves using probes and antibodies for growth factors to examine cell proliferation and cell differentiation in mammalian reproductive cells. His research papers have appeared in journals such as *Nature*, the *Proceedings of the National Academy of Sciences of the United States of America*, the *Journal of Molecular Biology*, and *Endocrinology*.

TAKUYA MURATA is a researcher with the Torii Nutrient-Stasis Project in Exploratory Research for Advanced Technology at the Research Development Corporation of Japan, where he is working on the mechanism of homeostatic control and its related compound (i.e., growth factors) under various states of nutrition. After receiving his B.A. in 1985 and M.A. in 1987 in reproductive physiology, he received his Ph.D. under Dr. Michio Takahashi of the University of Tokyo. His thesis work was on the effects of growth factors on reproductive function. Dr. Murata also earned a D.V.M. from The University of Tokyo in 1987. This was followed by postdoctoral research on the effects of growth factors on the pituitary function in culture in the laboratory of Dr. Shao-Yao Ying at the University of Southern California. Dr. Murata's research papers have appeared in such journals as the *Proceedings of the Society for Experimental Biology and Medicine* and *Life Sciences*.

Genes in Mammalian Reproduction: 229–245
© 1993 Wiley-Liss, Inc.

Advances in Gonadotropin-Releasing Hormone

Chris T. Bond and John P. Adelman

I. INTRODUCTION

Sexual reproduction accompanied the evolutionary appearance of eukaryotic organisms. Offering clear advantages for expanding the gene pool of any given species, this mode of reproduction constitutes the conduit of evolution. The decapeptide gonadotropin-releasing hormone (GnRH) regulates reproductive competence and represents a biological motif that has persisted across the evolutionary eons separating the human hypothalamus from unicellular algae. Since the first appearance of sexual reproduction, this remarkably uncomplex motif has regulated and coordinated multiple aspects of the reproductive process, performing this essential task in harmony with the tumultuous dynamics of ever-changing, ever more complex reproductive physiology. Much of our recent knowledge of these processes has been afforded by several rapidly evolving and powerful technologies, especially those of molecular biology. Thus in the last 7 years a great deal has been learned about the nature of GnRH-producing cells, the genes that encode GnRH, agents that influence the biosynthesis and regulate the secretion of GnRH, and the mechanisms by which they mediate their effects. This chapter presents an overview of these topics and a perspective on the exciting future of GnRH research.

A. Discovery of GnRH

The pituitary gland was long believed to be the "master gland"; pituitary extracts alter the functions of virtually every target tissue and physiological system. However, beginning in the mid-1930s, Harris performed a series of experiments that would forever change our view of the role of the pituitary. It had recently been demonstrated that application of an electrical shock to the brain of a rabbit could induce ovulation, a process previously believed to be controlled by the pituitary. When Harris lesioned the portal vasculature between the hypothalamus and the pituitary, electrical stimulus failed to induce ovulation. In concert, Harris demonstrated that transplantation of the pituitary to sites distal from the hypothalamus resulted in gonadal atrophy. Harris extended these initial studies by demonstrating that infusion of hypothalamic extracts onto pituitaries following transection of portal vessels is sufficient to induce ovulation [Harris, 1937, 1950; Green and Harris, 1947]. Appreciation of the hypothalamic contribution to reproductive function gave rise to the concept of the hypothalamic–pituitary–gonadal (HPG) axis and implied the existence of distinct endocrine chemical messengers from the brain that supercede pituitary hormones in the hierarchy of the reproductive physiological cascade, providing a link between the neural and endocrine systems. Thus was born the concept of neuroendocrinology. These pioneering studies initiated a monumental search to identify the hypothalamic factor(s) responsible for inducing ovulation, a search that would not be successfully completed for almost four decades.

The ability to assay for the elusive hypothalamic ovulatory factor was limited to cumbersome in vivo applications, which clearly were not sensitive or practical enough to be widely employed. However, during the ensuing decade the pituitary factors that regulate ovulation, the gonadotropins luteinizing hormone (LH) and follicle-stimulating hormone (FSH), were discovered, and it was quickly realized that the pituitary responded to application of hypothalamic extracts with release of LH and FSH [Harris, 1961]. Several rapid and sensitive techniques for measuring pituitary LH and FSH release were developed and applied as in vitro assays for the isolation and characterization of the hypothalamic gonadotropin-releasing factor. Ultimately, in 1971 two groups, one headed by Andrew Schally working with porcine hypothalamic extracts, and the other headed by Roger Guillemin using ovine material, purified and determined the primary amino acid sequence of a decapeptide with the ability to induce pituitary release of LH and FSH [Schally et al., 1971; Gurgus et al., 1972]. Guillemin and Schally received the Nobel Prize for the discovery of GnRH.

Today, we appreciate that GnRH is released in a pulsatile manner from terminals of a small set of hypothalamic neurons onto the pituitary, regulating pituitary gonadotropin release. The primary target tissues of the gonadotropins are the gonads. A complex and delicately balanced cascade of hormonal events occurs in the gonads in response to the gonadotropins. This cascade of events, which is the subject of other chapters in this book, results in gametogenesis and synthesis and release of factors such as the gonadal steroids. These gonadal

factors complete the HPG endocrine loop by influencing both hypothalamic and pituitary functions. Release, and possibly biosynthesis, of GnRH, pituitary synthesis and secretion of the gonadotropins, and pituitary sensitivity to GnRH all respond to circulating steroid levels. The role of GnRH cannot be limited to that of pituitary hormone-releasing factor, however. Much evidence, both new and old, suggests that GnRH has additional roles in reproduction and reproductive behavior, as well as in nonreproductive systems.

B. Molecular Biology of GnRH

1. **Mammalian coding sequences.** Following the determination of the sequence of the GnRH decapeptide, it was suggested that, due to its small size, GnRH might be enzymatically assembled in the cytoplasm without an mRNA template or the ribosomal translational apparatus. Alternatively, as had recently been shown for several other small endocrine peptides, bioactive moieties may be derived by site-specific proteolysis and postcleavage modifications from larger inactive precursor proteins, the precursor proteins being encoded by mRNAs and translated by the ribosome [Douglass et al., 1984]. The latter model turned out to be correct. Using degenerate radiolabeled oligonucleotide probes designed according to the possible nucleotide coding combinations for the decapeptide, a cDNA clone from human placenta that encodes a GnRH precursor protein was isolated and characterized [Seeburg and Adelman, 1984]. The predicted amino acid sequence of the GnRH precursor detailed a functionally tripartite molecule, which is consistent with the prototypic architecture of polyprotein precursors. The first 23 amino acids of the precursor are hydrophobic in nature and are believed to serve as a signal sequence, facilitating transport of the nascent polypeptide through the rough endoplasmic reticulum and on to the regulated secretory

pathway; the signal sequence is removed during this process. The GnRH decapeptide is in direct linkage with the signal sequence. Signal sequence cleavage exposes a glutamine residue that cyclizes, yielding the pyroglutamic acid found at the 1 position of the mature decapeptide. Following the GnRH moiety are three residues, Gly11-Lys12-Arg13, which are the sites for posttranslational proteolysis and tailoring that result in bioactive GnRH. Site-specific proteolytic enzymes with trypsin-like substrate specificity are believed to cleave the precursor on the C-terminal side of the paired basic residues, and a carboxypeptidase E-like enzyme then sequentially removes the arginine and lysine. Glycine 11 then donates its amide group to the glycine at position 10, a reaction catalyzed by peptidyl glycine α-amidating monooxygenase, resulting in an amide block at the C terminus of the mature decapeptide. Until recently, little was known about the specific molecules involved in this process. However, proteolytic processing molecules have now been identified and cloned [Fricker et al., 1989; Smeekens and Steiner, 1990]. The availability of these reagents will allow a precise examination of their roles in regulating the kinetics of GnRH biosynthesis. The remainder of the GnRH precursor is comprised of 56 amino acids and constitutes a previously unknown peptide, GnRH-associated peptide (GAP). Seeburg and Nikolics have demonstrated that this molecule is cosecreted with GnRH from hypothalamic terminals and has potent prolactin release–inhibiting activity [Nikolics et al., 1985; Phillips et al., 1985].

The cDNA encoding the GnRH precursor has also been characterized from rat, mouse, and human hypothalamus and predicts the same protein, with some minor sequence alterations between species. However, the 5'-untranslated region of the mRNA is different from that found in human placenta. Based on evolutionary considerations discussed below, it seemed possible that there may exist a family of related genes encoding GnRH. Examination of mouse, rat, and human genomic DNA by Southern blot and cloning, however, demonstrated that there is only a single homologous gene in those mammalian species, the difference in 5'-untranslated sequences resulting from differential splicing of the primary transcript [Adelman et al., 1986].

2. HPG mouse. One of the first applications of the cloned GnRH sequences was to provide direct proof that the GnRH gene encoded the hypothalamic releasing activity first suggested by Green and Harris. The homozygous hypogonadal (*hpg*) mouse does not reproduce, does not develop secondary sexual characteristics or demonstrate reproductive behaviors, and lacks detectable levels of GnRH. Seeburg and colleagues examined the structure of the GnRH gene in *hpg* mice and found a deletion in the genomic DNA, with one end point within the GnRH gene and the other approximately 33 kb away; this deletion removes most of GAP and all of the 3'-untranslated region of the GnRH mRNA, but leaves the promoter region intact as well as the 5'-untranslated region, the signal peptide, and the GnRH-encoding regions of the mRNA. Although low levels of a truncated GnRH mRNA can be detected in *hpg* mice, no mature GnRH decapeptide is produced, possibly due to mRNA or precursor protein instability and degradation. Next, in a dramatic demonstration of gene therapy, these researchers inserted a normal copy of the mouse GnRH gene into the germline of *hpg* mice. The resulting transgenic strain gave rise to animals that were cured of their reproductive deficiencies. Tissue-specific gene expression in accord with the pattern found in wild-type animals was demonstrated, and all of the endocrine functions missing in the parental *hpg* mice were reconstituted in their transgenic offspring [Mason et al., 1986a,b]. Thus GnRH accounts for all activities proposed by the earlier studies of Harris.

II. CURRENT STATUS

A. Evolution of the GnRH Motif

A GnRH-like peptide motif arose coincident with the appearance of sexual reproduction in eukaryotic organisms and plays a central role in regulating reproductive functions. In yeast, α mating factor is a tridecapeptide that controls the expression of a cascade of genes that determine mating type. α factor shares a striking homology with mammalian GnRH, being conserved at six positions, including those residues of mammalian GnRH shown to be most involved in gonadotropin release [Sherwood, 1987]. The conservation of structure and function between two evolutionarily distant molecules was dramatically illustrated when it was shown that application of the yeast peptide to dispersed mammalian pituitary cells resulted in low level release of LH and FSH [Loumaye et al., 1982]. Thus a relatively simple motif that has maintained the ability to regulate reproductive functions has been conserved with only subtle structural changes from ancient unicellular eukaryotes to the most recently emerged mammalian species.

During the period that separated yeast and mammals, life forms underwent a period of radiation as nature experimented with already successful functional patterns. As vertebrate organisms began to appear, expressing ever more complex structural systems, multiple GnRH-like peptides evolved. Characterization of these peptides from a number of vertebrate species, including the ancient hagfish and the modern chicken, has revealed that all members of this family are 10 amino acids in length and identical at the 1, 2, 4, 9 and 10 positions. The bioactive peptides contain a β turn between residues 5 and 6 that promotes interactions between the conserved N and

C termini. The conserved amino acid residues are thought to be important for stabilizing the bioactive structure as well as protecting the peptide from enzymatic degradation [Sherwood, 1986; King and Millar, 1990].

Antisera generated against alternate GnRH-like peptides used in combination with HPLC profiles of brain extracts from a variety of vertebrate species have indicated the existence of other novel forms. Indeed, more than one member of the GnRH family is expressed in most vertebrate species examined to date. The primary sequences of both forms found in the chicken have been established [Miyamoto et al., 1984; King and Millar, 1982].

B. Characterization of a Teleost GnRH Coding Sequence

Until very recently, the coding sequences and the deduced amino acid sequences of the GnRH preprohormone were known only for mouse, rat, and human, three species that express the same decapeptide. As immunological, chromatographic, and finally amino acid sequence data emerged, revealing the existence of multiple decapeptide sequences, a hypothesis was formulated that these multiple forms arose as a result of gene duplication events which may have produced either multiple decapeptide exons or multiple precursors that then drifted in sequence.

In the African cichlid fish *Haplochromis burtoni*, at least three decapeptides have been characterized by HPLC. Recently, the cDNA sequence encoding one of these decapeptide forms was cloned. The prohormone precursor sequence deduced from *H. burtoni* cDNA clones revealed an architecture identical to that of the mammalian precursors: a signal sequence followed by the decapeptide, processing site, and associated peptide. Alternate decapeptide sequences were not found on this precursor and further attempts to demonstrate related coding sequences by cDNA screening or genomic Southern analysis failed to yield evidence for the existence of homologous sequences [Bond et al., 1991a]. Does this mean that multiple, related decapeptides arose independently through convergent evolution? Alternatively, gene duplication events may have occurred very long ago, with sequence drift resulting in coding sequences of very low homology. Ultimately, these questions await the cloning of coding sequences for alternate GnRHs from a single species. In addition, future studies must clarify whether alternate GnRH-like decapeptides are expressed in mammals and what physiological roles they might play.

C. Ontogeny and Distribution of GnRH Neurons

Immunohistochemistry and in situ hybridization have been employed to determine the distribution of GnRH-expressing neurons in the central nervous system (CNS). A relatively small number of neurons express the GnRH gene, a few hundred to several thousand depending on the species examined; differing reports may result in part from varying sensitivities of the techniques employed. The distribution of GnRH immunoreactive fiber tracts and perikarya has been extensively documented in the adult rat. GnRH neurons are diffusely dispersed throughout the forebrain, primarily across the anterior hypothalamus, spanning neuronal areas that arise from different regions of the neuroepithelium, extending from the nervus terminalis through the medial basal hypothalmus and arcuate nucleus. Although GnRH fibers project to diverse regions of the CNS, two major terminal fields have been identified in the rat, the organum vasculosum lamina terminalis (OVLT) and the median eminence [Wray and Hoffman, 1986a].

Wray and colleagues and Pfaff and colleagues have studied the ontogeny of GnRH neurons and GnRH gene expres-

sion. Their results reveal that GnRH neurons express the decapeptide very early in development. On embryonic day 11.5 (E11.5) in the mouse and E4 in the chick, GnRH mRNA-expressing cells that also contain GnRH immunoreactivity are detected in the olfactory epithelium. At the earliest visualization, these cells are ovoid and lack processes. On subsequent days of embryonic development, the GnRH neurons grow processes, developing into at least two distinct subtypes (see below), and migrate as "tracks" of GnRH-expressing cells in extensive contact with each other, through the telencephalon and into the rostral diencephalon where they disperse to the typical adult distribution. This unique migration occurs along the pathway formed by the vomeronasal/nervus terminalis nerves from the olfactory placode to the rostral forebrain. Distribution of GnRH-expressing neurons similar to that seen in the adult in the preoptic area, anterior and medial basal hypothalamus is essentially established prenatally and does not greatly change throughout the life of the organism [Hoffman and Finch, 1986; Wray et al., 1989; Norgren and Lehman, 1991; Schwanzel-Fukuda and Pfaff, 1989b].

In both males and females, the levels of GnRH mRNA, determined by RNAse protection assays, and decapeptide, measured by radioimmunoassay (RIA), rise steadily from birth to sexual maturity. In mammals, the rise in mRNA is interrupted briefly during the postnatal suckling period, which may reflect an autoregulatory function of GnRH mediated by decapeptide in the mother's milk [Minaguchi and Meites, 1967; Smith White and Ojeda, 1984; Jakubowski et al., 1991].

D. Kallmann Syndrome

It is not surprising to find a relationship between GnRH neurons and olfaction. Although not all of the physiological functions of the nervus terminalis are thoroughly understood, the cells of this nucleus are generally thought to be involved in the regulation of reproductive behaviors, perhaps in mediating responses to pheromone stimuli. This is most clearly demonstrated in lower vertebrates, as well as in many mammalian species in which reproductive timing between females and males (ovulation and spermatogenesis) and a range of reproductively oriented behaviors are coordinated via olfactory pheromones [Lombardi and Vandenbergh, 1976; Demski and Northcutt, 1983; Demski, 1984]. In humans, Kallmann syndrome is a genetic disorder of reproductive and olfactory functions and is characterized by hypogonadotropic hypogonadism, with consequent infertility and anosmia. Although the severity of the symptoms varies substantially, all afflicted individuals lack normal levels of GnRH [Schwankhaus et al., 1989]. Following the demonstration that the reproductive deficiencies of the *hpg* mouse were due to a substantial deletion within the GnRH gene, genomic DNA from many Kallmann syndrome patients was examined by genomic Southern blot and at the nucleotide sequence level. In all cases the GnRH gene appeared normal. Examination of the brain of a fetus affected by Kallmann syndrome revealed the absence of olfactory bulbs and tracts and no migration of GnRH neurons from the region that normally constitutes the olfactory placode [Schwanzel-Fukuda et al., 1989a]. In contrast, the *hpg* mouse has a normal distribution of "would be" GnRH neurons in the adult hypothalamus as demonstrated by in situ hybridizations with undeleted GnRH gene sequences as probes. The results demonstrated that GnRH and GAP peptides are not necessary for proper migration of the GnRH neurons from the olfactory placode and indicated that Kallmann syndrome and the deficiencies in the *hpg* mouse do not result from a common molecular defect.

Recently, linkage analysis and chromosomal deletion mapping among families with a history of Kallmann syndrome identified a locus on the X chromosome associated with Kallmann syndrome. Further work resulted in the isolation and characterization of a gene that is probably responsible for at least some forms of Kallmann syndrome. The predicted protein product encoded at this locus shares striking homology with other molecules that are involved in cell adhesion and axonal pathfinding [Ballabio et al., 1989; Franco et al., 1991]. These results strongly suggest that a defect in this gene, *KALIG-1,* is responsible for the defect in neuronal migration that underlies Kallmann syndrome. Together, these two abnormal states, the *hpg* mouse and human Kallmann syndrome, indicate that both appropriate location and connectivity of GnRH neurons and expression of intact GnRH coding sequences are necessary for normal reproductive development and function.

E. Heterogeneity of GnRH Neurons

Although GnRH neurons in the mammalian CNS arise within a highly specific region of the developing brain, several lines of evidence suggest that by the time these cells have taken up residence in their adult locations they comprise a heterogeneous population. GnRH neurons can be divided into at least two distinct classes based on their morphology. One class has smooth aspinous processes and frequently extends bipolar projections. The other class possesses irregular spiny processes, with multiple projections. Furthermore, it has been shown in the rat that, while the total number of GnRH neurons does not vary between males and females at any time during development, the percent of spiny GnRH neurons in both sexes increases significantly during puberty. It is likely that these "new" spiny neurons are converted from smooth GnRH neurons and do not represent newly recruited GnRH neurons [Wray and Hoffman, 1986c].

As mentioned above, GnRH fibers are

widely dispersed throughout the CNS, and their axonal projections terminate on a variety of different target cell types; three types have been characterized. Projections responsible for gonadotropin release terminate on the portal vasculature in the median eminence. Long projections that leave the immediate vicinity of the cell soma terminate on other distal neurons. GnRH from these cells may act as a neurotransmitter or neuromodulator. A third pattern of projections overlaps with the first two: these cells send local projections, or collaterals, to neighboring neurons as well as to the median eminence or distal neurons. Indeed, some GnRH fibers within the preoptic hypothalamus synapse onto other GnRH neurons or projections [Bennett-Clark and Joseph, 1982; Shivers et al., 1983; King and Anthony, 1984; Leranth et al., 1985; Jennes, 1991]. Furthermore, using double-label immunocytochemistry, Wray and Hoffman have shown that although the number of catecholamine-innervated smooth GnRH neurons remains constant, the number of spiny GnRH neurons that receive catecholamine innervation increases as a function of age [Wray and Hoffman, 1986b]. Several different antigenic markers have been colocalized with GnRH in subsets of the overall population of GnRH neurons [López et al., 1991; Pu et al., 1991].

Heterogeneity of GnRH neurons is further indicated by results of studies examining GnRH mRNA levels. Zoeller and colleagues, using quantitative in situ hybridization, found that a subset of preoptic GnRH neurons in the rat contained levels of GnRH mRNA that differed as a function of the ovulatory cycle, possibly indicating heterogeneity in the response of these cells to changes in levels of gonadal steroids [Zoeller and Young, 1988]. In the fish *Haplochromis burtoni,* a species in which reproductive status is socially controlled, two distinct clusters of GnRH-immunoreactive cells exist in the diencephalon and telencephalon. These two populations of cells respond differently to transitions in reproductive status. In contrast to neurons in the telencephalon, diencephalon GnRH neurons undergo soma enlargement and dendrite arborization in reproductively active animals (Fernald et al. unpublished data).

F. Regulation of GnRH Release

1. Pulsatile release. Quantitative analysis of GnRH levels in the mammalian hypophysiotropic vasculature demonstrated that from prepubertal ages through sexual maturity GnRH is released from terminals in the median eminence with a discrete pulsatile frequency. The interpulse interval ranges from tens of minutes to an hour, depending on the species. This pulsatile form of release is essential to the development of reproductive capability; tonic release of GnRH is not sufficient to induce or maintain sexual fertility. Indeed, gonadotrope responses to GnRH reflect the GnRH pulse amplitude and interpulse period, a coordination essential for normal progression of the ovulatory cycle. Initial attempts to use GnRH therapeutically for several forms of infertility were unsuccessful, until the advent of pulsatile administration regimens [Simon et al., 1990; Southworth et al., 1991]. The exact nature and anatomical location of the GnRH pulse generator and whether pulsatility is an intrinsic capability of GnRH neurons or is established through distinct synaptic input is not known. Furthermore, it is not known whether GnRH release from neurons that project to regions other than the median eminence displays a pulsatile pattern. Lesion studies have implicated the arcuate nucleus as the site of the GnRH pulse generator. There is a limited population of GnRH neurons within the arcuate nucleus. Perhaps pulsatility is conferred through a network of GnRH–GnRH cell contacts, and the functions of specifically these arcuate GnRH neurons are disrupted by lesions that abolish pulsatility of release from the median eminence.

2. Puberty. Although the exact mechanisms underlying the sexual development that occurs at puberty are not fully characterized, an increase in the GnRH pulse frequency may play a pivotal role. Gradual increases in GnRH decapeptide levels during maturation, combined with increased pulse frequency, may trigger changes in gonadotropin levels, resulting ultimately in the LH surge pattern that stimulates gonadal maturation and gamete production [Grumbach, 1980; Bourguignon and Francimont, 1984; Delemarre van de Wall et al., 1989; Rodriguez and Wise, 1989; Bourguignon et al., 1990].

3. Neurotransmitters. A number of neurotransmitters are implicated in regulating pulsatile GnRH release. Blockade of NMDA receptors inhibits pulsatile release of GnRH. In addition, release is stimulated by kainate and NMDA in hypothalamic explants. These two excitatory amino acid neurotransmitters probably act through separate receptor types, since NMDA action is blocked by the NMDA receptor antagonist AP5 while kainate action is not. The action of NMDA is modulated by glycine, but at this time it is not known whether glycine acts on NMDA receptors or if glycine modulates whole-cell function through its own receptor [Bourguignon et al., 1989a,b].

A number of other neurotransmitters and neurohormones may be involved in regulation or coordination of GnRH synthesis and release. Endogenous opioid peptides may inhibit levels of GnRH or GnRH release [Ferin, 1987; Orstead and Spies, 1987; Thind and Goldsmith, 1988]. Dopamine is also implicated in control of the release of GnRH. Dopamine-containing neurons have been shown to synapse on axons of GnRH neurons in the median eminence, the predominant site of release of GnRH in its role as hormone-releasing-factor [Sarkar and Fink, 1981; Kuljis and

Advis, 1989]. Throughout the development of GnRH neurons, the extent of innervation by catecholiminergic fibers, at both the soma and axonic terminals, increases, although the significance of these inputs is not clear [Negro-Vilar et al., 1979].

The pineal hormone melatonin serves as the chemical transducer of photoperiod information. Melatonin has been shown to have profound effects on reproduction in seasonal breeders. Most dramatically, exogenously administered melatonin affects normal, seasonal gonadal regression and recrudescence in male hamsters. Melatonin probably serves as the coordinator of seasonal reproduction [Rusak, 1980; Tamarkin et al., 1985]. Although direct evidence for melatonin regulation of GnRH release has not been shown, the median eminence of the rat contains large numbers of melatonin-binding sites thought to be melatonin receptors. Future studies will reveal if these are melatonin receptors, if they are expressed on GnRH neurons, and what role in regulating GnRH release they may play.

G. Regulation of GnRH Gene Expression

1. Transcription. The essential role GnRH plays in reproduction has engendered interest in understanding the regulation of GnRH mRNA biosynthesis. Do changes in the rate of GnRH gene transcription or levels of GnRH mRNA participate in regulating reproductive status? In the female rat, some GnRH neurons have been shown by quantitative in situ hybridization to vary as much as twofold in their GnRH mRNA content relative to the estrous cycle, with a rise in mRNA levels following a bolus release of decapeptide on the afternoon of proestrous. It has been hypothesized from these results that GnRH synthesis may be coupled to GnRH release. Alternatively, the change in GnRH mRNA might be mediated by changing levels of gonadal steroids, which are known to affect levels of both gonadotropins and GnRH. Elucidation of the mechanism of steroid hormone action via receptors that act as transcription factors stimulated hypotheses of direct effect of gonadal steroids on GnRH gene expression. However, studies employing cloned GnRH promoter sequences in reporter gene constructs have shown that the GnRH gene does not respond to transcriptional regulation by the steroid hormone receptors. In addition, in situ and immunocytochemical (ICC) studies have failed to demonstrate the presence of estrogen receptors on GnRH neurons. Therefore, it appears that the effects gonadal steroids exert on GnRH levels are not direct transcriptional effects. Rather, these effects may be indirect, mediated by steroid-responsive neurons that interact with GnRH neurons [Bond et al., 1991b]. An intriguing twist is the demonstration by combined in situ, ICC, and electron microscopy of GnRH-containing terminals synapsing on estrogen receptor expressing neurons. These results introduce the possibility that GnRH modulates the effects of estrogen on non-GnRH neurons, perhaps representing a multicell loop of GnRH autoregulation [Langrub et al., 1991]. Alternatively, steroids may influence GnRH synthesis and/or release through direct interaction with cell surface molecules, independent of the classical steroid-hormone receptors. Puia et al. [1990] have shown that neurosteroids directly influence GABA receptors and their intrinsic ion channels. Changes in ion channel function could affect diverse properties of the cell, including biosynthesis or regulated release of stored products.

In the teleost *H. burtoni* social station dictates reproductive status. In native populations of this species, two groups of males can clearly be distinguished: dominant, brightly colored, aggressively territorial and reproductively competent males and nonterritorial, nonreproductively competent males that are held in a developmentally repressed state by the presence of their dominant counterparts. The soma size of neurons in the diencephalon that express GnRH changes relative to social and reproductive status; GnRH neurons become enlarged in the dominant male relative to those of the subdominant counterpart. Furthermore, social status, reproductive state, and enlargement of GnRH neurons remain plastic. Placing a dominant male in the presence of a larger, more dominant conspecific results in shrinkage of its GnRH neurons, loss of bright coloration, and regression of testes to a nonproductive state [Fernald and Hirata, 1977a,b].

GnRH mRNA levels have been examined in the *H. burtoni* male in relation to its reproductive and social status. While steady-state GnRH mRNA levels are very similar between the dominant and reproductively capable male and his subdominant, nonreproducing counterpart, large changes in GnRH mRNA levels are seen during the transition from one reproductive/social state to the other.

2. Posttranscriptional regulation. In addition to regulation of GnRH gene transcription, modulation of reproductive function via the GnRH neuron may occur at many points in the biosynthesis of GnRH, including posttranscriptional processing of primary transcripts, posttranslational processing of the prohormone into bioactive decapeptide, and release of decapeptide at nerve terminals. The cloning of GnRH-encoding sequences has added powerful new tools with which to study GnRH biosynthesis, including nucleotide probes specific for primary transcripts versus mature mRNA and antisera specific for prohormone or bioactive decapeptide. Studies employing combined immunocytochemistry and in situ hybridization have demonstrated that all cells that contain detectable levels of GnRH mRNA also contain protein precursor and decapeptide [Ronnekleiv et al., 1989]. These results, combined with other studies that indicate that the population of neurons that express GnRH is constant, imply that regulation of GnRH biosynthesis is accomplished through subtle modulation of decapeptide production rather than all-or-none effects on gene expression.

Studies are underway in a number of laboratories to examine the role of these posttranscriptional control points during the development and maintenance of reproductive capability. Relative levels of primary transcript to mature mRNA during the course of the estrous cycle in females and during development are being measured using specific nucleotide probes. Results from corollary studies of prohormone and mature decapeptide in paradigms of gonadectomy and steroid hormone treatment indicate that steroid hormone milieu may be reflected in relative levels of posttranslationally processed GnRH [Wetsel and Negro-Vilar, 1989; Roselli et al. 1990]. Further study is required to understand the effectors and mechanisms of posttranscriptional regulation of GnRH and to determine if control points such as posttranslational processing contribute significantly to the reproductive development and status of an organism.

Major new tools for the direct study of GnRH biosynthesis have recently been developed. Two groups, Mellon and coworkers, using rat GnRH promoter sequences, and Radovich and coworkers, using human GnRH promoter sequences to direct expression of the SV40 large T antigen, succeeded in producing transgenic mice that developed GnRH-expressing tumors. Subsequent cloning of transformed cells from these tumors has yielded the first GnRH-expressing cell lines. These cells demonstrate many characteristics of GnRH neurons. Their morphology is frequently neuron-like, with extensive neurite outgrowth and synaptic connections, and they contain secretory granules and secrete high levels of GnRH and GAP [Mellon et al., 1990; Radovick et al., 1991]. These cells may not completely reflect adult GnRH neurons, as

they were transformed very early in development in response to early activation of GnRH gene expression, and their repertoire of surface markers and other proteins may be abnormal as with many transformed cells. Nevertheless, they allow controlled, in vitro analysis of GnRH biosynthesis. Transfection of these cells with reporter gene plasmids driven by GnRH promoter sequences has revealed the presence of a potent and essential enhancer element approximately 2 kb 5' of the transcription start site that may underlie the highly cell type–specific expression of GnRH (Whyte et al., unpublished data). Cloning of transacting proteins that bind to this enhancer element will help to elucidate the mechanism of induction and maintenance of GnRH gene expression.

In addition to transcriptional studies, these GnRH-expressing cell lines are powerful tools for investigating regulatory points in the biosynthesis of GnRH. The decapeptide is efficiently produced, indicating that all components necessary for translation and posttranslational processing are present.

H. Other Biological Roles for GnRH

There is evidence for other physiological roles for GnRH in addition to stimulating gonadotropin release. Many of the postulated accessory activities of GnRH occur within the reproductive system, where GnRH may be involved in the development and function of reproductive organs and may mediate some reproductive behaviors. Recent studies have also suggested that GnRH mediates physiological activities outside the reproductive system.

Two major lines of investigation have contributed to understanding how the basic structure of GnRH may have been recruited to perform diverse functions within the CNS: anatomical diversity of GnRH-immunoreactive cells and projections and physiological assays. As few as half of the central neurons that are immunoreactive for GnRH project to the median eminence, and hypothalamic GnRH neurons fall into several different categories based on morphology and coexpressed peptides. Additional roles implicated for GnRH include actions as a neurotransmitter or neuromodulator in processes and behaviors related to reproduction. GnRH cells arise early in development near the olfactory tract, and cells are visualized there in the adult, suggesting that GnRH may transduce olfactory pheromone information into reproductive behaviors. In lower vertebrates, including teleosts, GnRH-immunoreactive terminal nerve cells are believed to function in this way. GnRH also enhances sexual receptivity and induces lordotic behavior in the rat [Moss and McCann, 1973; Pfaff, 1973; Stell et al., 1984].

GnRH may have other, nonreproductive CNS functions such as release of other pituitary hormones. GnRH has been shown to stimulate release of growth hormone as well as gonadotropins in fish, and there is evidence that GnRH acts as a neurotransmitter in fish retina and sympathetic ganglia of the bullfrog [Jones et al., 1984; Peter et al., 1990]. GnRH immunoreactivity has been reported in the spinal cord, midbrain, and hindbrain, and GnRH receptors have been localized to the limbic system, further supporting the hypothesis that GnRH may function as a neurotransmitter [Leblanc et al., 1988; Badr et al., 1989].

In many vertebrate species, including marsupial mammals, multiple GnRH decapeptides have been found. In chicken, two GnRH decapeptide amino acid sequences with different gonadotropin-releasing potencies have been determined, allowing generation of antibodies specific for each. Immunocytochemistry employing these antisera on chicken brain has revealed a differential distribution of neurons that express the two peptides. Chicken I GnRH, which differs from mammalian GnRH by only one amino acid, is expressed in neurons of the anterior and medial basal hypothalamus, a distribution that parallels mammalian GnRH neurons. Chicken II GnRH, possibly the more ancient of the two forms, differs in sequence at 3 of the 10 amino acid residues. Neurons expressing this decapeptide in the chicken brain are found laterally and posteriorly in the hypothalamus, implying that the functions performed by this molecule differ from those of Chicken I GnRH [Katz et al., 1990]. Future work will reveal the physiological role of Chicken II GnRH, whether the neurons expressing the separate GnRH moieties have similar origins in the olfactory placode, what synaptic connections are established by each, and if the two decapeptides have any regulatory influence on each other.

The GnRH gene is also expressed in nonneuronal tissues. Recent data demonstrate expression of GnRH mRNA and peptides in testes and ovary, where the decapeptide may act in an autocrine or paracrine fashion. GnRH synthesis has also been found in placenta and mammary gland, two relatively newly evolved tissues, and GnRH has been found in cells of the immune system.

1. GnRH as an autocrine factor. Electron microscopic immunohistochemistry in a number of mammalian species has shown that GnRH neurons of the preoptic area make synaptic contact on other GnRH neurons. The contacts between GnRH neurons in the adult brain may reflect their high degree of interconnectivity during the prenatal migration from the olfactory placode. The presence of GnRH–GnRH synapses formally allows the possibility of an autofeedback mechanism of GnRH on its own synthesis or release.

Studies, both in vivo and in vitro, employing potent GnRH agonists and antagonists administered in a variety of paradigms, have yielded evidence that GnRH release is inhibited by GnRH [DePaolo et al., 1987; Valenca et al., 1987]. This inhibition may play an important role in establishment and maintenance of the pulsatile pattern of GnRH secretion. There is some additional evidence that the action of NMDA on GnRH release is minimally affected by inhibition by GnRH, indicating that NMDA and GnRH may coordinate to control pulsatile release.

2. Paracrine functions of GnRH. In addition to the neurons of the CNS that express GnRH, several peripheral sites of synthesis are indicated based originally on the presence of GnRH immunoreactivity and more recently confirmed by the very sensitive molecular techniques of in situ hybridization and polymerase chain reaction (PCR). These peripheral sites of expression include tissues involved in the reproductive system, including both male and female gonads, and in mammals, placental and mammary cells [Petraglia et al., 1989; Ban et al., 1990; Oikawa et al., 1990]. GnRH has also been found in nonreproductive tissues, the thymus and spleen lymphocytes [Azad et al., 1991]. Further evidence for a functional, paracrine role for GnRH at these peripheral sites comes from the detection of GnRH-binding sites thought to represent GnRH receptors in ovary, testis, placenta, breast carcinoma, and thymus [Marchetti et al., 1989b; Szende et al., 1991].

3. Reproductive tissue expression of GnRH. The GnRH gene is expressed in human breast carcinomas, and GnRH immunoreactivity has been measured in maternal milk. The functional role for GnRH in mammary tissue has not been defined, although it is possible that this GnRH has an important role in the neonate rather than in maternal tissue. GnRH is also expressed in trophoblasts of the human placenta. Here GnRH may serve a classic paracrine role in stimulating synthesis and release of hCG, a placental-specific gonadotropin [Belisle et al., 1984]. GnRH immunoreactivity and, recently, GnRH mRNA have been shown in the rat testes and in the granulosa cells of the rat ovary. Again, the functional significance of these peripheral sites of synthesis is not known.

However, because GnRH has a very short half-life in the circulation, these peripheral sites of synthesis probably serve a paracrine role in the regulation of gonadal function. This hypothesis is supported by evidence for the presence of GnRH receptors on gonadal tissues.

4. Thymic and lymphocyte expression of GnRH. Interactions between the endocrine and immune systems have long been appreciated. Perhaps the most well-known interaction is the immunosuppressive effects of gucocorticoids. Understanding of the true extent of interaction and overlap of endocrine, neuroendocrine, and immune systems is an exciting question receiving more and more attention. One of the most recently recognized "cross" players in this interaction is GnRH. GnRH expression in rat spleen lymphocytes has been demonstrated by several techniques. GnRH-like immunoreactive material isolated from rat spleen lymphocytes coelutes on HPLC with synthetic GnRH decapeptide, and PCR amplification of reverse transcribed mRNA from these lymphocytes shows that the GnRH gene is transcribed in these cells. Furthermore, cellular extracts from spleen lymphocytes elicit a specific and dose-dependent release of LH from dispersed pituitary cells, release being blocked by GnRH antagonist [Emanuele et al., 1990].

GnRH-binding sites of low affinity but high specificity have been characterized on thymic membranes. Preincubation of thymocytes in GnRH followed by treatment with the mitogen concanavalin A resulted in approximately 40-fold higher thymocyte mitogenic activity than in response to concanavalin A alone. In the absence of treatment with mitogens, thymocyte expression of interleukin-2 receptors increased in response to treatment with GnRH. This effect was specifically blocked by simultaneous treatment with GnRH antagonist. Perhaps the most compelling evidence of a role for GnRH in immune system function comes from whole animal studies wherein neonatal (postnatal days 1–5) female rats were treated with a potent GnRH antagonist. Thymus morphology and weight, thymocyte profiles and responsiveness to mitogens, and responses to antigenic challenge at age 3 months were compared between control, GnRH antagonist treated, and neonatally castrated animals. Results from these studies suggest that the blockade of central and peripheral GnRH receptors by GnRH antagonist during a critical period of development of both the HPG axis and the brain–thymus–lymphoid axis dramatically impairs immune system development [Marchetti et al., 1989a; Batticane et al., 1991; Morale et al., 1991].

III. FUTURE PROSPECTS

In the next several years, GnRH research will involve examination of the regulation of GnRH biosynthesis, from the GnRH gene to decapeptide secretion at nerve terminals. Such information will illuminate the mechanisms by which GnRH neurons respond to multiple inputs. Ultimately, the knowledge gained through these studies will serve to formulate mechanisms to manipulate the reproductive system, impacting on methods of birth control, treatment of reproductive disorders, such as Kallmann syndrome and precocious puberty, and lead to an increased understanding of the intregration of neural and hormonal systems.

The growing appreciation of alternate physiological roles subserved by GnRH and the possibility that other members of the GnRH family are retained, in some form, in mammalian organisms promises to be one of the most exciting frontiers in GnRH research. This highly successful motif may serve as a model for understanding the transduction and integration of information across multiple physiological systems.

REFERENCES

Adelman JP, Mason AJ, Hayflick JS, Seeburg PH (1986): Isolation of the gene and hypothalamic cDNA for the common precursor of gonadotropin-releasing hormone and prolactin release-inhibiting factor in human and rat. Proc Natl Acad Sci USA 83:179–183.

Azad N, Emanuele NV, Hallorain MM, Tentler J, Kelley MR (1991): Presence of luteinizing hormone-releasing hormone (LHRH) mRNA in rat spleen lymphoctyes. Endocrinology 128:1679–1681.

Badr M, Marchetti B, Pelletier G (1989): Changes in hippocampal LH-RH receptor density during maturation and aging in the rat. Dev Brain Res 45:179–184.

Ballabio A, et al (1989): Contiguous gene syndromes due to deletions in the distal short arm of the human X chromosome. Proc Natl Acad Sci USA 86:10001–10005.

Ban E, Crumeyrolle-Arias M, Latouche J, et al. (1990): GnRH receptors in rat brain, pituitary and testis; modulation following surgical and gonadotropin-releasing hormone agonist-induced castration. Mol Cell Endocrinol 70:99–107.

Batticane N, Morale MC, Gallo, et al. (1991): Luteinizing hormone-releasing hormone signaling at the lymphocyte involves stimulation of interleukin-2 receptor expression. Endocrinology 129:277–286.

Belisle S, Guevin J-F, Bellabarba D, Lehoux J-G (1984): Luteinizing hormone-releasing hormone binds to enriched human placental membranes and stimulates in vitro the synthesis of bioactive human chronic gonadotropin. J Clin Endocrinol Metab 59:119.

Bennett-Clark C, Joseph SA (1982): Immunocytochemical distribution of LHRH neurons and processes in the rat: hypothalamic and extra hypothalamic locations. Cell Tissue Res 221:493–504.

Bond CT, Francis RC, Fernald RD, Adelman JP (1991): Characterization of complementary DNA encoding the precursor for gonadotropin-releasing hormone and its associated peptide from a teleost fish. Mol Endocrinol 5:931–937.

Bond CT, Seal RS, Simerly R, Adelman JP (1991b): Molecular Aspects of GnRH: Part I: 5′ flanking sequence of rat GnRH gene does not respond to estrogen in vivo or in vitro. Serono Symposia, GnRH, Modes of Action and Analogs. Scottsdale, Arizona.

Bourguignon JP, Franchimont P (1984): Puberty-related increase in episodic LHRH release from rat hypothalamus in vitro. Endocrinology 114:1943–1945.

Bourguignon JP, Gérard A, Franchimont P (1989a):

Direct activation of gonadotropin-releasing hormone secretion through different receptors to neuroexcitatory amino acids. Neuroendocrinology 49:402–408.

Bourguignon JP, Gérard A, Mathieu J, Simons J, Franchimont P (1989b): Pulsatile release of gonadotropin-releasing hormone from hypothalamic explants is restrained by blockade of N-methyl-D,L-aspartate receptors. Endocrinology 125:1090–1096.

Bourguignon JP, Gérard, Franchimont P (1990): Maturation of the hypothalamic control of pulsatile gonadotropin-releasing hormone secretion at onset of puberty: II. Reduced potency of an inhibitory autofeedback. Endocrinology 127:2884–2890.

Delemarre van de Wall HA, Plant TM, vanRees GP, Schoemaker J (1989): Control of the Onset of Puberty III. New York: Elsevier.

Demski LS (1984): The evolution of neuroanatomical substrates of reproductive behavior: sex steroid and LHRH-specific pathways including the terminal nerve. Am Zool 24:809–830.

Demski LS, Northcutt RG (1983): The terminal nerve: A new chemosensory system in vertebrates? Science 220:435.

DePaolo LV, King RA, Carrillo AJ (1987): In vivo and in vitro examination of an autoregulatory mechanism for luteinizing hormone-releasing hormone. Endocrinology 120:272–279.

Douglass J, Civelli O, Herbert E (1984): Polyprotein gene expression: generation of diversity of neuroendocrine peptides. Annu Rev Biochem 53:665–7.

Emanuele NV, Emanuele MA, Tentler J, et al. (1990): Rat spleen lymphocytes contain an immunoreactive and bioactive luteinizing hormone-releasing hormone. Endocrinology 126:2482–2486.

Ferin M (1987): A role for the endogenous opioid peptides in the regulation of gonadotropin secretion in the primate. Horm Res 28:119.

Fernald RD, Hirata N (1977a): Field study of Haplochromis burtoni: habitats and cohabitants. Environ Biol Fish 2:299–308.

Fernald RD, Hirata N (1977b): Field study of Haplochromis burtoni: quantitative behavioral observations. Anim Behav 25:964–975.

Franco B, Guioli, S, Pragliola A, et al (1991): A gene deleted in Kallmann's syndrome shares homology with neural cell adhesion and axonal path-finding molecules. Nature 353:529–535.

Fricker LD, Adelman JD, Douglass J, et al (1989): Isolation and sequence analysis of cDNA for rat carboxypeptidase E: A neuropeptide processing enzyme. Mol Endocrinol 3:666.

Green JD, Harris GW (1947): Neurovascular link between neurohypophysis and adenohypophysis. J Endocrinol 5:136.

Grumbach MM (1980): The Neuroendocrinology of Puberty. Sunderland, MA: Sinauer, pp 249–258.

Gurgus R, Butcher M, Amoss M, et al. (1972): The structure of the ovine hypothalamic luteinizing hormone-releasing factor (LRF). Proc Natl Acad Sci USA 69:278.

Harris GW (1937): Induction of ovulation in the rabbit by electrical stimulation of the hypothalamo-hypophyseal mechanism. Proc R Soc (Lond) 122:374.

Harris GW (1950): Oestrous rhythm, pseudopregnancy and the pituitary stalk in the rat. J Physiol [Lond] 111:347.

Harris GW (1961): The pituitary stalk and ovulation. In Villee C (ed): Control of Ovulation. New York: Pergamon Press, pp 56–78.

Hoffman GE, Finch CE (1986): LHRH neurons in the female C57BL/6J mouse brain during reproductive aging: no loss up to middle age. Neurobiol Aging 7:45–48.

Jakubowski M, Blum M, Roberts JL (1991): Postnatal development of gonadotropin-releasing hormone and cyclophilin gene expression in the female and male rat brain. Endocrinology 128:2702–2708.

Jennes L (1991): Dual projections of gonadotropin releasing hormone containing neurons to the interpeduncular nucleus and to the vasculature in the female rat. Brain Res 545:329–333.

Jones SW, Adams PR, Brownstein MJ, Rivier JE (1984): Teleost luteinizing hormone-releasing hormone: Action on bullfrog sympathetic ganglia is consistent with the role as neurotransmitter. J Neurosci 4:420.

Katz IA, Millar RP, King JA (1990): Differential regional distribution and release of two forms of gonadotropin-releasing hormone in the chicken brain. Peptides 11:443–450.

King JC, Anthony ELP (1984): LHRH neurons and their projections in humans and other mammals: species comparisons. Peptides 1:195.

King JA, Millar RP (1982): Structure of chicken hypothalamic luteinizing hormone-releasing hormone. II. Isolation and characterization. J Biol Chem 257:10729–10732.

King JA, Millar RP (1990): Genealogy of the GnRH family. Pro Clin Biol 342:54–59.

Kuljis RO, Advis JP (1989): Immunocytochemical and physiological evidence of a synapse between dopamine- and luteinizing hormone-releasing hormone-containing neurons in the ewe median eminence. Endocrionology 124:1579–1581.

Langrub MC Jr, Maley BE, Watson RE Jr (1991): Ultrastructural evidence for luteinizing hormone-releasing hormone neuronal control of estrogen responsive neurons in the preoptic area. Endocrinology 128:27–36.

Leblanc P, Crumeyrolle M, Latouche J, et al. (1988): Characterization and distribution of receptors for gonadotropin-releasing hormone in rat hippocampus. Neuroendocrinology 48:482–488.

Leranth C, Segura LMG, Palkovits M, et al. (1985): The LH-RH-containing neuronal network in the preoptic area of the rat: demonstration of LH-RH-containing nerve terminals in synaptic contact with LH-RH neurons. Brain Res 345:332–336.

Lombardi JR, Vandenbergh JG (1976): Pheromonally induced sexual maturation in females: regulation by the social environment of the male. Science 196:545–546.

Loumaye E, Thorner J, Catt KJ (1982): Yeast mating pheromone activates mammalian gonadotrophs: evolutionary conservation of a reproductive hormone? Science 218:1323.

López FJ, Merchenthaler I, Ching M, Wisniewski MG, Negro-Velar A (1991): Galanin: A hypothalamic-hypophysiotropic hormone modulating reproductive functions. Proc Natl Acad Sci USA 88:4508–4512.

Marchetti B, Guarcello V, Morale MC, et al (1989a): Luteinizing hormone-releasing hormone (LHRH) agonist restoration of age-associated decline of thymus weight, thymic LHRH receptors, and thymocyte proliferative capacity. Endocrinology 125:1037–1045.

Marchetti B, Guarcello V, Morale MC, et al (1989b): Luteinizing hormone-releasing hormone-binding sites in the rat thymus: Characteristics and biological function. Endocrinology 125:1025–1035.

Mason AJ, Hayflick JS, Zoeller RT, Young III WS, Phillips HS, Nikolics K, Seeburg PH (1986a): A deletion truncating the gonadotropin-releasing hormone gene is responsible for hypogonadism in the *hpg* mouse. Science 234:1366–1371.

Mason AJ, Pitts SL, Nikolics K, et al (1986b): The hypogonadal mouse: reproductive functions restored by gene therapy. Science 234:1372–1377.

Mellon PL, Windle JJ, Goldsmith P, et al. (1990): Immortalization of hypothalamic GnRH neurons by genetically targeted tumorigenesis. Neuron 5:1–10.

Minaguchi H, Meites J (1967): Effects of suckling on hypothalamic LH-releasing factor and prolactin inhibiting factor, and on pituitary LH and prolactin. Endocrinology 80:603–607.

Miyamoto K, Hasegawa Y, Nomura M, Igarashi M, Kangawa K, Matsuo H (1984): Identification of the second gonadotropin-releasing hormone in chicken hypothalamus: evidence that gonadotropin secretion is probably controlled by two distinct gonadotropin-releasing hormones in avian species. Proc Natl Acad Sci USA 81:3874–3878.

Morale MC, Batticane N, Baroloni G, et al. (1991): Blockade of central and peripheral luteinizing

hormone-releasing hormone (LHRH) receptors in neonatal rats with a potent LHRH-antagonist inhibits the morphofunctional development of the thymus and maturation of the cell-mediated and humoral immune responses. Endocrinology 1073–1085.

Moss RL, McCann SM (1973): Induction of mating behavior in rats by luteinizing hormone releasing factor. Science 181:177.

Negro-Vilar A, Ojeda SR, McCann SM (1979): Catecholaminergic modulation of luteinizing hormone-releasing hormone release by median eminence terminals in vitro. Endocrinology 104:1749–1757.

Nikolics K, Mason AJ, Szonyi E, Ramachandran J, Seeburg PH (1985): A prolactin-inhibiting factor within the precursor for human gonadotropin-releasing hormone. Nature 316:511–517.

Norgren RB Jr, Lehman MN (1991): Neurons that migrate from the olfactory epithelium in the chick express luteinizing hormone-releasing hormone. Endocrinology 128:1676–1681.

Oikawa M, Dargon C, Ny T, Hsueh AJW (1990): Expression of gonadotropin releasing hormone and prothymosin-alpha messenger ribonucleic acid in the ovary. Endocrinology 127:2350–2356.

Orstead KM, Spies HG (1987): Inhibition of hypothalamic gonadotropin-releasing hormone release by endogenous opioid peptides in the female rabbit. Neuroendocrinology 46:14.

Peter RE, Habibi HR, Chang JP, Nahorniak CS, Yu KL, Huang YP, Marchant TA (1990): Actions of gonadotropin-releasing hormone (GnRH) in the goldfish. Prog Clin Biol Res 342:393–398.

Petraglia F, Vaughan J, Vale W (1989): Inhibin and activin modulate the release of gonadotropin-releasing hormone, human chorionic gonadotropin, and progesterone from cultured human placental cells. Proc Natl Acad Sci USA 86:5114–5117.

Pfaff DW (1973): Luteinizing hormone-releasing factor potentiates lordosis behavior in hypophysectomized ovariectomized female rats. Science 182:1148–1150.

Phillips HS, Nikolics K, Branton D, Seeburg PH (1985): Immunocytochemical localization in rat brain of a prolactin release-inhibiting sequence of gonadotropin-releasing hormone prohormone. Nature 316:542–545.

Pu LP, Charnay Y, Ledugque P, et al. (1991): Light and electron microscopic immunocytochemical evidence that delta sleep-inducing peptide and gonadotropin-releasing hormone are coexpressed in the same nerve structures in the guinea pig median eminence. Neuroendocrinol 53:332–338.

Puia G, Santi M, Vicini S, et al. (1990): Neurosteroids act on recombinant GABA$_A$ receptors. Neuron 4:759–765.

Radovick S, Wray S, Lee E, et al. (1991): Migratory arrest of gonadotropin-releasing hormone neurons in transgenic mice. Proc Natl Acad Sci USA 88:3402–3406.

Rodriguez RE, Wise ME (1989): Ontogeny of pulsatile secretion of gonadotropin-releasing hormone in the bull calf during infantile and pubertal development. Endocrinology 124:248–256.

Ronnekleiv OK, Naylor BR, Bond CT, Adelman JP (1989): Combined immunohistochemistry for gonadotropin-releasing hormone (GnRH) and Pro-GnRH and in situ hybridization for GnRH messenger ribonucleic acid in rat brain. Mol Endocrinol 3:363–371.

Roselli CE, Kelly MJ, Ronnekleiv OK (1990): Testosterone regulates progonadotropin-releasing hormone levels in the preoptic area and basal hypothalamus of the male rate. Endocrinology 126:1080–1086.

Rusak B (1980): Suprachiasmatic lesions prevent an antigonadal effect of melatonin. Biol Reprod 22:148–154.

Sarkar DK, Fink G (1981): Gonadotropin-releasing hormone surge: possible modulation through postsynaptic alpha-adrenoreceptors and two pharmacologically distinct dopamine receptors. Endocrinology 108:862–867.

Schally AV, et al. (1971): Isolation and properties of the FSH and LH-releasing hormone. Biochem Biopshys Res Commun 43:393.

Schwankhaus JD, Currie J, Jaffe MJ, Rose SR, Sherins JJ (1989): Neurological findings in men with hypogonadotropic hypogonadism. Neurology 39:223–236.

Schwanzel-Fukuda M, Bick D, Pfaff DW (1989a): LHRH-expressing cells do not migrate normally in an inherited hypogonadal (Kallmann) syndrome. Mol Brain Res 6:311–326.

Schwanzel-Fukuda M, Pfaff DW (1989b): Origin of luteinizing hormone-releasing hormone neurons. Nature 338:161–163.

Seeburg PH, Adelman JP (1984): Characterization of cDNA for precursor of human luteinizing hormone releasing hormone. Nature 311:666–668.

Sherwood N (1986): Gonadotropin-releasing hormone: Differentiation of structure and function during evolution. In Fink G, Harmar AJ, McKerns KW (eds): Neuroendocrine Molecular Biology. New York: Plenum, pp 67–83.

Sherwood N (1987): The GnRH family of peptides. TINS 10:129–132.

Shivers BD, Harland RE, Pfaff DW (1983): The central nervous system role of luteinizing hormone releasing hormone. In Krieger DT, Brownstein MJ, Martin JB (eds): Brian Peptides. New York: Wiley, pp. 389–412.

Simon A, Birkenfeld A, Schenker JG (1990): Gonad-otropin releasing hormone (GnRH): mode of action and clinical applications: A review. Int J Fertil 35:350–362.

Smeekens SP, Steiner DF (1990): Identification of a human insulinoma cDNA encoding on novel mammalian protein structurally related to the yeast dibasic processing protease KEX 2. J Biol Chem 265:2997–3000.

Smith White SS, Ojeda SR (1984): Maternal modulation of infantile ovarian development and available ovarian luteinizing hormone-releasing hormone (LHRH) receptors via milk LHRH. Endocrinology 115:1973–1983.

Southworth MB, Matsumoto AM, Gross KM, et al. (1991): The importance of signal pattern in the transmission of endocrine information: pituitary gonadotropin responses to continuous and pulsatile gonadotropin-releasing hormone. J of Clin Endocrinol Metab 72:1286–1289.

Stell WK, Walker SE, Chohan KS, Ball AK (1984): The goldfish nervus terminalis: A luteinizing hormone-releasing hormone and molluscan cardioexcitory peptide immunoreactive olfactoretinal pathway. Proc Natl Acad Sci USA 81:940–944.

Szende B, Srkalovic G, Timar J, et al. (1991): Localization of receptors for luteinizing hormone-releasing hormone in pancreatic and mammary cancer cells. Proc Natl Acad Sci USA 88:4153–4156.

Tamarkin L, Caird CJ, Almeida OFX (1985): Melatonin: A coordinating signal for mammalian reproduction? Science 227:714–720.

Thind KK, Goldsmith PC (1988): Infundibular gonadotropin-releasing hormone neurons are inhibited by directed opioid autoregulatory synapses in juvenile monkeys. Neuroendocrinology 47:203–216.

Valenca MM, Johnston CA, Ching M, Negro-Vilar A (1987): Evidence for a negative ultrashort loop feedback mechanism operating on the luteinizing hormone-releasing hormone neuronal system. Endocrinol 121:2256–2259.

Wetsel WC, Negro-Vilar A (1989): Testosterone selectively influences protein kinase C–coupled secretion of proluteinizing hormone-releasing hormone-derived peptides. Endocrinology 125:538–547.

Wray S, Hoffman G (1986a): A developmental study of the quantitative distribution of LHRH neurons within the central nervous system of postnatal male and female rats. Comp Neurol 252:522–531.

Wray S, Hoffman G (1986b): Catecholamine innervation of LH-RH neurons: A developmental study. Brain Res 399:321–331.

Wray S, Hoffman (1986c): Postnatal morphological changes in rat LHRH neurons correlated with sexual maturation. Neuroendocrinology 43:93–97.

Wray S, Nieburgs A, Elkabes S (1989): Spatiotemporal cell expression of luteinizing hormone-releasing hormone in the prenatal mouse: evidence for an embryonic origin in the olfactory placode. Dev Brain Res 46:309–318.

Zoeller RT, Young WS III (1988): Changes in cellular levels of messenger ribonucleic acid encoding gonadotropin-releasing hormone in the anterior hypothalamus of female rats during the estrous cycle. Endocrinology 123:1688–1689.

ABOUT THE AUTHORS

CHRIS T. BOND and **JOHN P. ADELMAN** work together at the Vollum Institute for Advanced Biomedical Research at the Oregon Health Sciences University in Portland. Dr. Adelman is a scientist at the Vollum Institute and holds a joint appointment as Assistant Professor in the Cell Biology and Anatomy Department of the Medical School, where he teaches molecular biology and genetics. After receiving his B.S. and M.S. in microbiology from the University of Connecticut in 1979, he worked with Peter Seeburg at Genentech Inc., where he concentrated on molecular endocrinology and neurobiology. During this time, along with Dr. Seeburg, he cloned the genes encoding GnRH. He also met Ed Herbert and after several successful collaborative efforts, moved to Dr. Herbert's lab where his Ph.D. work included the discovery of a family of antisense transcripts encoded within the GnRH locus of the rat. Dr. Adelman then joined the faculty of the Vollum Institute and the Department of Cell Biology and Anatomy. Ms. Bond is a senior research associate in Dr. Adelman's laboratory. She received a B.S. in biochemistry from the University of Oregon in 1984 and a M.S. from OHSU in 1989. For the past seven years she has worked closely with Dr. Adelman; their current research involves continued study of the molecular and cellular mechanisms which regulate the biosynthesis of GnRH, the molecular basis of circadian rhythms, and the structure–function of potassium channels. Their research papers have appeared in such journals as *Nature, Science,* the *Proceedings of the National Academy of Sciences,* and *Neuron.*

Genes in Mammalian Reproduction: 247–269
© 1993 Wiley-Liss, Inc.

Regulation of Pituitary Gonadotropins by Gonadotropin-Releasing Hormone, Estradiol, Progesterone, Inhibin, and Activin

William L. Miller

I. INTRODUCTION

Two pituitary gonadotropins control gonadal function in mammals: luteinizing hormone (LH) and follicle-stimulating hormone (FSH). This review focuses on regulation of LH and FSH primarily in the female mammal. It emphasizes regulation at three major control points (Fig. 1) that are critical for gonadotropin synthesis and/or secretion. Some of the newest and most powerful tools of cellular and molecular biology are being used to define the molecular bases for these controls.

Figure 1 depicts three gonadotropin genes, each on separate chromosomes, that encode α, LH-β, and FSH-β subunits. These genes are thought to be expressed in one pituitary cell type to make the LH heterodimer (α and LH-β) and its related heterodimer FSH (α and FSH-β). Gonadotropin expression is regulated directly at the pituitary level by gonadotropin-releasing hormone (GnRH), estradiol-17β (E), progesterone (P), inhibin (IN), and activin (ACT). Although the effects of E and P are considerable at control point 1, these effects are discussed only briefly; it is the primary purpose of this review to document the effects of all these hormones at

control points 2 and 3 shown in Figure 1. Regulation at points 2 and 3 are likely to occur through changes in transcription, mRNA stabilities, translation, glycosylation [Liu et al., 1976], subunit combination, and secretion. The effects on transcription and/or mRNA steady-state levels of LH, FSH, or GnRH-Rec mRNAs, however, are emphasized.

Integration of the individual actions of GnRH, E, P, IN, and ACT into a comprehensive regulatory framework is important. Such a framework is considered later in this chapter (see section IX). Furthermore, the molecular mechanisms used by each hormone are innately important and are discussed (see section X). Finally, several novel developments involving gonadotropin genes and gonadotropin receptor genes will be introduced as they have the potential for rapidly advancing practical techniques in analyzing defects in reproductive performance.

Suggested background reading for this chapter includes two reviews on gonadotropin regulation at the mRNA and gene levels [Gharib et al., 1990; Mercer, 1990]. A review of gonadotropin subunit structures is also recommended [Ward et al., 1991].

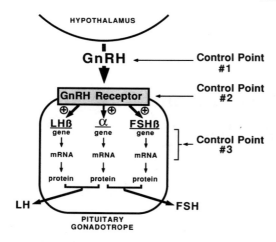

Fig. 1. *Triple-point control of gonadotropin synthesis and secretion. This diagram shows three critical control points that potentially regulate gonadotropin synthesis and secretion in vivo. Control point 1 involves GnRH output from the hypothalamus and is documented by Karsch et al. [1987] and Moenter et al. [1991]. Control point 2 involves GnRH-Rec number and/or sensitivity to GnRH and is documented by Laws et al. [1990a, b] and by Sealfon et al. [1990b]. Control point 3 involves regulation of gene expression as documented by Phillips et al. [1988], Miller et al. [1988], and Shupnik et al. [1989].*

II. TRIPLE-POINT CONTROL OF GONADOTROPIN PRODUCTION/SECRETION—A BRIEF OVERVIEW

Hypothalamic GnRH is the gonadotropin secretagog generally recognized as *the major stimulus* for both LH and FSH synthesis and release. Gonadotropins are not produced without GnRH. Pulsatile output of GnRH increases transcription of all the gonadotropin genes and the steady-state levels of their respective mRNAs; GnRH also releases stored LH and FSH.

The ovarian steroids E and P exert control at all three control points pictured in Figure 1. They can affect go-

nadotropin synthesis and secretion at control point 1 by regulating pulsatile output of GnRH from the hypothalamus (frequency and/or amplitude); they can change GnRH receptor number and/or efficiency of signal transduction at control point 2; they can directly alter gonadotropin gene transcription and/or modify gonadotropin subunit mRNA stabilities at control point 3. The peptide hormones IN and ACT alter gonadotropin synthesis and secretion at control points 2 and 3. Whenever a study on gonadotropin regulation is interpreted, all three control points, at a minimum, should be considered.

The *nature* and *importance* of each control point may vary with species. In sheep, there is good evidence for regulation at all three control points: GnRH pulse rate and amplitude are changed by 30- to 40-fold during the preovulatory surge [Karsch et al., 1987; Moenter et al., 1990]; 2) GnRH receptor (GnRH-REc) number and sensitivity to GnRH can be changed to a similar extent in ovine pituitary cultures [Laws et al., 1990b; Sealfon et al., 1990b]; and 3) E, P, and IN decrease transcription and/or mRNA stabilities of the gonadotropin subunits [Phillips et al., 1988] throughout the ovine reproductive cycle. By contrast, E and P stimulate FSH and LH production in rat pituitary cultures [Labrie et al., 1977]. Moreover, rat gonadotropes are reported to have an excess of GnRH receptors during the estrous cycle [Naor et al., 1980]. These data suggest that regulation at control point 2 may be relatively unimportant in rats and that regulation by E and P may have different effects at control point 3 in rats and sheep. There is less information about humans and other primates than for rats and sheep, but it appears that regulation at control point 2 is as important in humans and monkeys as it is in sheep [Knobil et al., 1980; Santoro et al., 1986]. Very limited data indicate that regulation at control point 3 in humans also seems to mimic sheep rather than rats [Miller and Wu, 1981]. Thus, triple-point control exists in all mammals, but the emphasis and character of these controls may vary by species.

III. MODELS FOR STUDYING GONADOTROPIN REGULATION AT CONTROL POINTS 2 AND 3

In vivo models have been cleverly devised to measure the direct effects of steroids and GnRH on the ovine pituitary [Clarke and Cummins, 1984, 1987]. Intricate surgery has permitted disconnection of the hypothalamus from the pituitary in sheep either by removing the hypophyseal stalk by aspiration or by severing the stalk and inserting an aluminum or Teflon plate (termed *hypothalamic disconnection* [HPD]). Exogenous GnRH can be given intravenously to these ewes at intervals of 1–2 hours to substitute for endogenous pulsatile GnRH. Often these ewes are ovariectomized (OVX) to remove gonadal influences. A ewe that has all the above procedures performed on it is an "HPD/ OVX/GnRH-pulsed ewe." Ovarian steroids or IN can be administered to these sheep to determine the direct effects of these hormones at the pituitary level. These experiments involve technically difficult surgery, however, and have been performed most convincingly in sheep. One can reasonably ask if the results obtained with sheep are valid for the rat or human. What if the species differences observed in pituitary cultures are, in fact, real? What if other differences exist? Moreover, in vivo studies are expensive, time consuming, and cannot reasonably be used for detailed time-course studies.

Primary dispersed cell pituitary cultures have been valuable for studying gene expression and secretion of LH and FSH. Gonadotropes in these cultures maintain functional levels of receptors for GnRH, E, IN and ACT. Some degree of FSH and LH synthesis is usually maintained. Gonadotropin synthesis is lower than that found in vivo, however, and responsiveness to P is decreased and lost after 1 week of culture [Batra and Miller, 1985a]. A major quandary facing studies with pituitary cultures is the fact that cultures from different species show important differences in gonadotrope regulation by GnRH, E, P, IN, and ACT (see below, sections VI, VII and VIII). One can reasonably ask if these differences are real or are caused by undefined factors peculiar to dispersed cell primary cultures.

Finally, immortal cell lines of gonadotropes from mice have recently been generated through the use of transgenic transformation technology [Windle et al., 1990]. Briefly, the tumor-promoting T antigen from SV40 virus was placed under control of 1.5 kb of the promoter region from the human α-subunit gene. Transgenic mice bearing this construct formed pituitary tumors that were propagated into several lines of immortal, α-subunit–producing gonadotropes. These cells represent an ideal model for studying GnRH signal transduction, since GnRH induces α expression in these cells. The entire array of GnRH-stimulated factors that promote transcription of the α gene may be studied in these cells. The cell lines have been of limited use for studying overall gonadotropin gene regulation, however, since none of the immortal gonadotrope lines express all gonadotropin subunits, and none appear to contain receptors for E, P, IN, or ACT.

No *one* model is suitable for studying all or even one aspect of gonadotropin regulation in its entirety. Presented below are significant examples from a number of well characterized models to illustrate specific points of gonadotropin regulation that occur directly at the pituitary level.

IV. OVARIAN HORMONES HAVE MAJOR EFFECTS ON SERUM GONADOTROPIN LEVELS

Before discussing GnRH and each gonadal hormone separately, the overall pattern of gonadotropin secretion throughout the estrous cycle will be reviewed. It is this pattern that must be explained by all of the molecular models discussed in this chapter.

It should be noted that in the absence of gonadal hormones (OVX mammals) the levels of LH and FSH are approximately 10-fold greater than midcycle levels of LH and FSH when ovaries are present [Goodman, 1988; Freeman, 1988]. Thus a major function of gonadal hormones is to lower the general level of gonadotropins by approximately 90%. This presumably permits the preovulatory surges of LH and FSH to be detected as massive signals against a relatively low background of gonadotropin secretion at ovulation. The lowest gonadotropin levels are associated with high serum P (midcycle), while the LH/FSH surge is coincident with a major decrease in P concentration, a substantial rise in E, and a doubling of IN (Fig. 2).

The rat estrous cycle is not presented here, but it is very similar to that of the sheep except that it occurs every 4 or 5 days instead of every 16. Another notable exception involves a double peak of P in the rat cycle. One peak occurs at midcycle, and another is synchronized with the LH surge. Progesterone can prematurely trigger the LH surge in rats; by contrast, administration of P to sheep totally blocks the LH

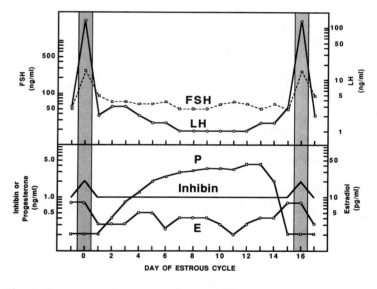

Fig. 2. *Serum gonadotropins and gonadal hormones throughout the ovine estrous cycle* [*Goodman, 1988*]*. The values for IN were taken from McNeilly et al.* [*1989*]*. Note that values for FSH are about 100-fold too high because an NIH–FSH–sheep standard was used that was only about 1% pure. The LH values are inflated by about twofold because an NIH–LH–sheep reference preparation was used that was only about 50% pure. Note the major inhibition of gonadotropin secretion when P is high. Note also the LH/FSH surges when P is low and E and IN are highest.*

surge. Another difference between rats and sheep is that the FSH surge has a marked double peak in rats, with the second peak clearly following the LH surge. A second FSH peak occurs in sheep also, but it is not as marked.

In the following pages, the individual effects of GnRH, E, P, IN, and ACT on sheep and rat gonadotropin regulation are discussed in detail. These effects, some of which are different between sheep and rats, are summarized in Table I for easy reference. How these effects can account for the ovine gonadotropin patterns (Fig. 2) is discussed immediately following Table I.

V. GnRH STIMULATES LH AND FSH PRODUCTION AT THE GENE/mRNA LEVELS

Although the search for a specific FSH-releasing factor continues, and although ACT is a candidate for being a selective stimulator of FSH production (see below), it is generally accepted that GnRH is the major tropic hormone for production of both LH and FSH. This is best illustrated in female mice that lack a functional gene for GnRH (hpg mice); these mice do not produce gonadotropins. When a normal GnRH gene is introduced into hpg mice by

transgenic technology, normal gonadotropin synthesis is restored [Mason et al., 1986a,b]. Similar conclusions have been drawn from studies with monkeys [Knobil, 1980], humans [Santoro et al., 1986], rats [Culler and Negro-Vilar, 1986, 1987], sheep [Hamernick and Nett, 1988], and pigs [Esbenshade et al., 1986]. Clearly, GnRH is a major positive regulator of pituitary gonadotropes in all species studied.

A. GnRH Increases Steady-State mRNA Levels of LH and FSH Subunits

1. **The case in sheep.** Clarke and Cummins [1984] used HPD/OVX ewes to directly study GnRH regulation at control point 3 in sheep. In one study using HPD/OVX ewes, α-subunit mRNA decreased 60%, 95% and 95% after 3, 7, and 30 days, respectively, following HPD. Decreases of 78%, 98% and 98% were recorded for LH-β mRNA during the same respective time periods [Hamernik et al., 1986]. A second study, limited to 7 days of HPD, showed decreases for α, LH-β, and FSH-β of 70%, 85% and 60%, respectively [Hamernik and Nett, 1988]. In the second study, one set of HPD ewes were pulsed with GnRH (250 ng every 2 hours) between days 3 and 7 after HPD. By day 7, α and FSH-β mRNA levels were normal while

TABLE I. Summary of the Effects of GnRH, E, P, IN, and ACT on Gonadotropin and GnRH-Rec Expression in Sheep and Rats

Effector hormone	Hormone Effect on Expression							
	FHSβ		LHβ		α		GnRH-Rec	
	Rat	Ewe	Rat	Ewe	Rat	Ewe	Rat	Ewe
*GnRH	++++	++++	++++	++++	++++	++++		
E	NC*	----	++	NC	N	----	ND	+++
*P	NC*	----	ND	----	ND	----	ND	---
IN	----	---	N	N	N	N	-	+++
ACT	+++	NC	N	ND	N	ND	ND	NC

+, Positive effect; −, negative effect; N, no effect; ND, no data available, NC, There are reports, but they are conflicting or inclusive. NC*, E and/or P increases FSH synthesis in rat pituitary cultures, but it is uncertain if this is accomplished by increasing FSH-β expression; *GnRH, different pulse rates have variable effects on gonadotropin subunit synthesis. *P, requires E either prior to or during treatment to maintain responsiveness to P.

LH-β mRNA had increased 3.7-fold over HPD levels. These studies indicate that hypothalamic GnRH is either the major tropic hormone that maintains normal gonadotropin subunit mRNAs in sheep or can substitute for it.

2. The case in rodents. As a nonsurgical variation on the HPD/OVX studies in sheep, several investigators have used either GnRH antagonists [Wierman et al., 1989; Rodin et al., 1989] or passive immunization to GnRH [Lalloz et al., 1988; Rodin et al., 1989] *specifically* to interrupt GnRH action in castrated female and male rats and/or mice. Results from all of these studies were similar to those in the sheep at the subunit mRNA levels.

In other studies, a male rat model (castrated, testosterone-replaced male rat) was used to show that the frequency of GnRH administration has a differential effect on gonadotropin expression [Dalkin et al., 1989]. No spontaneous pulses of GnRH or LH are detected in castrated rats that are given testosterone to lower gonadotropin levels; these data suggest that pulsatile GnRH secretion is very low or nonexistent in these animals. It was found that α and LH-β mRNAs increased substantially when supplemental GnRH was given at pulse frequencies of 1–4 per 30 minutes. By contrast, optimum frequencies for increasing FSH-β mRNA were 1–4 per 240 minutes. Similar results were observed at the level of secreted LH and FSH in HPD/OVX/GnRH-pulsed sheep [Clarke et al., 1987]. Overall, these results indicate that GnRH pulse frequency can *selectively* increase mRNA levels of specific gonadotropin subunits.

The hypogonadal (*hpg*) mouse provides another model exhibiting selective stimulation of gonadotropin synthesis by GnRH. The *hpg* mouse is completely and specifically devoid of GnRH because of a defective GnRH gene [Mason et al., 1986b]. Messenger RNA for the α-subunit can be increased in these mice 7- or 10-fold when GnRH is injected every 2 hours for 6 days (100 ng/pulse) [Stanley et al., 1988]. The α mRNA increases as early as 6 hours after GnRH injection and is increased maximally by 24 hours. These studies showed no increase in LH-β mRNA even when GnRH pulses were given every 30 minutes, as in the studies noted above with rats [Dalkin et al., 1989]. The authors could not explain their failure to increase LH-β mRNA, but their results reinforce the concepts that 1) GnRH increases steady-state levels of α mRNA and 2) GnRH does not necessarily enhance all gonadotropin expression equally.

3. The case in primary tissue culture. Primary rat pituitary cultures have been used to show that continuous GnRH stimulation (48 hours; 100 nM) can increase α mRNA up to sixfold (12-fold if α mRNA from thyrotrophs is assumed to contribute 50% of the α mRNA prior to stimulation) [Hubert et al., 1988]. In the same cultures LH-β mRNA levels did not increase, which is consistent with the studies on *hpg* mice (above). In this case there was no effort to pulse the cultures to optimize for LH-β mRNA.

4. The case in immortal gonadotropes. Several α-producing gonadotropin cell lines from mice show the stimulatory effect of GnRH on α mRNA levels. Three of five cell lines show three- to eightfold increases in steady-state levels of α mRNA when treated for 16 hours with 100 nM GnRH [Windle et al., 1990]. Longer treatments (48 hours) were even more effective (10-fold increases). Neither LH-β nor FSH-β subunit mRNAs are expressed in these cells.

5. Summary of mRNA studies. The data presented above were obtained using in vivo and in vitro models from sheep and rodents. They leave little doubt that GnRH is necessary to maintain normal steady-state levels of mRNA for all pituitary gonadotropin subunits in both female and male castrated animals.

Surprisingly, stimulation by GnRH is not

always equivalent for α, LH-β, and FSH-β mRNAs. Expression is greatly influenced by GnRH pulse rate and other factors that are not defined at present. The mechanism(s) responsible for pulse-specific regulation is totally unknown. Although not impossible, it would be incredibly novel if such specific differential regulation occurred at the level of mRNA stability. No precedent exists for such control. Regulation may be associated with differential control at a transcriptional level, however. The following section presents evidence that GnRH can alter gonadotropin gene transcription.

B. GnRH Stimulates Transcription of LH and FSH Subunit Genes

Quantitative analyses of mRNA depend on the ability of membrane-bound mRNA to bind radiolabeled cDNA probes. Transcription analysis (transcription run-on) involves the reverse process: Binding of radiolabeled nuclear-produced mRNAs to membrane-bound cDNAs. Haisenleder et al. [1991] used transcription run-on assays to measure transcription rates of the genes encoding the α, LH-β, and FSH-β subunits; the model system was the castrated, testosterone-replaced male rat (see above, The Case in Rodents). Pituitaries were taken from rats that had been given GnRH-pulse frequencies optimized for stimulating expression of selected gonadotropin subunits. Transcription increased three- to four-fold at gonadotropin subunit genes in the same patterns as mRNA levels increased after the same GnRH pulse rates. Although the kinetics of increased transcription and increased mRNA levels did not appear to follow classical cause-and-effect relationships, the authors concluded that GnRH does increase gonadotropin expression, at least in part, by increasing rates of gonadotropin subunit transcription.

Stimulation of gonadotropin subunit transcription by GnRH has also been demonstrated in primary rat pituitary cultures (hemipituitary slices) [Shupnik, 1990]. Pul-

satile administration of GnRH was not required to increase transcription of the α-subunit. Transcription of the α-subunit gene increased three- to four-fold with either continuous (1 nM GnRH; 2–6 hours) or pulsatile (1 pulse/hour for 4 hours) GnRH addition. Transcription of the LH-β gene increased two- to three-fold with pulsatile GnRH only, and transcription of the FSH-β gene increased only two-fold after one pulse of GnRH and then declined to control levels although GnRH pulses were continued.

Both in vivo and in vitro studies indicate that GnRH can increase transcription rates of all gonadotropin subunit genes. Importantly, these changes in transcription presage similar changes in gonadotropin subunit mRNA levels, which suggests that transcriptional regulation is the primary event.

VI. ESTRADIOL-17β AND TRIPLE-POINT CONTROL

With GnRH being the primary stimulus for maintaining gonadotropin transcription and mRNA levels, E must interfere with or augment GnRH effects when it acts at any or all three critical points of control (Fig. 1). Studies with sheep clearly show that E can act at control point 1 to increase hypothalamic GnRH secretion (20- to 40-fold) during the preovulatory LH surge [Moenter et al., 1990]. This type of regulation is discussed by J. Adelman (this volume).

E can also regulate how the gonadotrope receives the GnRH signal. It can change the number of GnRH receptors (GnRH-Rec) or alter steps in GnRH signal transduction. It can also act directly at the gene level to alter transcription rates of the gonadotropin subunit genes. Studies showing these actions of E at control points 2 and 3 are discussed below. Differences between the rat and sheep models become evident in this section.

254 Miller

A. Control by E in the Sheep Pituitary

1. Control by E at point 2. Receptors for GnRH (GnRH-Rec) increase sixfold in HPD/OVX/GnRH-pulsed ewes within 16 hours of E treatment [Clarke et al., 1988]. A similar study with HPD/OVX ewes (no GnRH given) showed a rise of 4.5-fold in GnRH-Rec 16 hours after treatment with E [Gregg and Nett, 1989]. These in vivo studies agree with in vitro data that show E increases GnRH-Rec number by four- to sixfold in ovine pituitary cultures [Laws et al., 1990b; Gregg et al., 1991]. The culture studies also measured GnRH-Rec mRNA "activity," which increased 10-fold. As noted below (see section VIII), inhibin plus E can boost GnRH-Rec mRNA activity by 20-fold in vitro. These changes in GnRH-Rec and GnRH-Rec mRNA activity are so large that they are likely to be physiologically important (see Fig. 3). The method used to measure GnRH-Rec mRNA "activity" is outlined below, since it is a novel method applicable to the estimation of mRNA for many membrane receptors.

The *Xenopus* oocyte assay was recently developed to measure the "bioactivity" of mRNA for hormone receptors in instances in which no sequence information exists for the receptor protein, mRNA, or gene that encodes it. In the case of the GnRH-Rec, total RNA was isolated from ovine pituitary cultures and injected into *Xenopus* oocytes. The oocyte, which has no endogenous GnRH-Rec, translated the mRNA to make ovine GnRH-Rec protein, which migrated to the oocyte membrane to interact with GnRH. Responsiveness to GnRH was

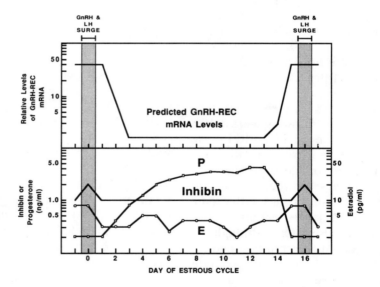

Fig. 3. *Predicted steady-state levels of GnRH-Rec mRNA during the ewe estrous cycle. The bottom panel shows reported values for levels of E and P [Goodman, 1988] and IN [McNeilly et al., 1989] throughout the ewe estrous cycle. The predicted pattern of GnRH-Rec mRNA is shown in the top panel. This pattern of GnRH-Rec mRNA is derived from in vitro data reported by Sealfon et al. [1990b see their Fig. 5, E+IN vs. IN+P]. The LH surge (30- to 100-fold increase in LH; see Fig. 2) and the GnRH surge (30- to 40-fold increase [Moenter et al., 1990]) are indicated by shaded bars.*

measured as a change in ion flux across the oocyte membrane in the presence of GnRH [Sealfon et al., 1990a]. The change in membrane potential is presumably proportional to the amount of GnRH-Rec protein and therefore the activity of GnRH-Rec mRNA. In all cases, measurement of GnRH-Rec mRNA activity from ovine pituitary cultures agreed with the in vitro binding studies except that mRNA activity changes were nearly double those observed for the GnRH-Rec [Laws et al., 1990b; Sealfon et al., 1990b].

Increasing the number of GnRH-Rec has biological consequences. It increases GnRH-stimulated LH secretion by threefold at high concentrations of GnRH (1–10 nM) in ovine pituitary cultures. Interestingly, this increased *responsiveness* is transient and lasts only 24–48 hours. By contrast, GnRH-Rec levels remain elevated as long as E is present. These results suggest that GnRH responsiveness (as measured by LH release) initially increases due to increased GnRH-Rec. Within 24–48 hours, however, E must also induce an inhibitory mechanism that reduces GnRH-stimulated LH secretion even while GnRH-Rec remains elevated. This E-induced inhibitory mechanism must be a postreceptor event or an event that makes it impossible for the GnRH-Rec to function properly. Moreover, E increases *sensitivity* to GnRH (ED$_{50}$ decreased by 90%–95%); this also seems to be a postreceptor phenomenon [Laws et al., 1990b]. Overall, the effects of E should create a major short-term boost to GnRH-stimulated LH secretion at the time of the preovulatory surge in sheep. They should also transiently increase GnRH-stimulated synthesis of, at least, α-subunit during the LH surge.

2. Control by E at point 3. Estradiol-17β inhibits FSH output directly at the pituitary level in HPD/OVX/GnRH-pulsed ewes [Clarke and Cummins, 1984]. This is observed in ovine pituitary cultures in which E (10 nM) inhibits FSH synthesis by 70%–

85% within 48 hours [Miller et al., 1977]. Synthesis of FSH in pituitary cultures from humans and pigs is similarly inhibited by E [Miller and Wu, 1981]. Thus in vitro and/or in vivo studies indicate that E blocks FSH output directly at the pituitary level in sheep, pigs, and humans.

Estradiol-17β also decreases mRNA levels for the FSH-β subunit by 70%–85% in ovine pituitary cultures [Alexander and Miller, 1982; Phillips et al., 1988]; E decreases α mRNA levels 70%–85% within 3 days [Hall and Miller, 1986; Phillips et al., 1988]. Transcriptional run-on studies have shown that E rapidly (2 hours) decreases transcription of both FSH-β and α-subunit genes by 85% and 70%, respectively. The effect of E seems to be directly at the α and FSH-β genes. The results indicate that E directly blocks transcription of both α and FSH-β genes in sheep (presumably in pigs and humans also).

Synthesis of LH appears unaffected by E in HPD/OVX/GnRH-pulsed ewes [Mercer et al., 1988]. These data are consistent with previous data from Clarke and Cummins [1984] showing that E acted directly at the pituitary to inhibit FSH but not LH output. More recently, however, Nett et al. [1990] used OVX/GnRH-pulsed sheep to show that E *plus* P act at the pituitary to decrease steady-state levels of both α and LH-β mRNAs by 80%–85%. Since GnRH receptors did not change during these experiments, the data suggest that E plus P act directly on gene transcription and/or mRNA stabilities. These dramatic results still do not indicate whether E or P is the active agent; it may be that the higher levels of E used in this latter study and longer treatment duration permitted the effect of E to be observed. It is also possible that E maintained responsiveness to P in these long-term castrated animals (see below, section VII) and that P actually decreased the LHβ and α-subunit mRNA levels.

3. Control by E in the sheep pituitary: Conclusions. At control point 2, E appears to act as a stimulator of GnRH action by increasing GnRH-Rec mRNA and GnRH-Rec and by activating postreceptor mechanisms to increase gonadotrope sensitivity to GnRH. The resulting stimulatory effect on GnRH-stimulated gonadotropin secretion is counteracted within 24 hours by the ability of E to lower maximal gonadotrope responsiveness to GnRH. Actually, this latter effect may help to limit the preovulatory LH surge in sheep (see section IX).

Presumably, E-induced increases in GnRH-Rec and sensitivity to GnRH stimulate gonadotropin subunit synthesis during the LH surge, especially since E also induces a simultaneous increase of GnRH during the surge. At control point 3, however, E strongly inhibits FSH synthesis by blocking transcription of both subunit genes (α and FSH-β). Levels of LH-β mRNA are unaffected by E alone, but are decreased dramatically by a combination of E plus P in vivo. The overall effect of E during the midluteal phase of the estrous cycle seems to be an inhibitory one. During the preovulatory surge the net effect of increasing E is probably minimal on gonadotropin production in sheep, since it generates both positive and negative forces toward synthesis.

B. Control by E in the Rat Pituitary (Note Species Differences)

1. Control by E at point 2. GnRH-Rec numbers change during the rat estrous cycle by two- to threefold [Clayton, 1989], but no evidence supports E action in this change. Actually, rat gonadotropes apparently have an excess number of GnRH-Rec at all times during the cycle [Naor et al., 1980]. A comparison of GnRH-Rec numbers in rat and sheep pituitary cultures reveals that the average rat gonadotrope in vitro contains about 15K receptors [Loumaye and Catt, 1982] and that the ovine gonadotrope contains between 0.5K

(P-treated; see section IX) to 20K GnRH-Rec when exposed to E and IN together (see section X). These data indicate a potential for major regulation of GnRH-Rec in the sheep. There is no evidence for such regulation in the rat.

It has been known since 1976 [Drouin et al., 1976] that E can decrease the ED$_{50}$ for GnRH in rat pituitary cultures by 30%–90%. This E-induced increase in sensitivity is like that found in ovine pituitary cultures. Therefore, it is possible that E can enhance the effect of GnRH on the rat gonadotrope, but its overall effect appears to be limited to this function alone.

2. Control by E at point 3. Estradiol-17β has either no effect or a slight positive effect on FSH synthesis in the rat pituitary. In one study, E has been reported to double FSH synthesis in rat pituitary cultures under conditions that inhibit synthesis of FSH by 80% in ovine pituitary cultures [Miller and Wu, 1981]. Moreover, E increases transcription of LH-β in primary rat pituitary cultures [Shupnik et al., 1989]; a classical E-response element (ERE) resides between -1159 and -1173 on the rat LH-β gene, and this functions as a positive regulator of gene expression in transient transfection assays when placed in front of a heterologous promoter [Shupnik and Rosenzweig, 1991]. Therefore, E appears to be either a neutral or positive regulator of gonadotropin synthesis at the pituitary level in the rat.

3. Control by E in the rat pituitary: Conclusions. At control point 2, gonadotrope sensitivity to GnRH increases in rats after E-treatment (ED$_{50}$ for GnRH decreases). This is similar to one of the three effects of E in sheep. At control point 3, E increases LH-β transcription severalfold, and FSH synthesis is usually reported to increase slightly after E treatment. These effects are clearly different from the inhibitory effects of E observed at the pituitary level in the sheep, pig, and human [Miller and Wu, 1981].

VII. PROGESTERONE AND TRIPLE-POINT CONTROL

Regulation by P has been difficult to study. Ovariectomy causes the pituitary to lose its responsiveness to P [Batra et al., 1986], but ovariectomy must be performed to remove ovarian P from the animal prior to studies with exogenous P. In the past, intervals between ovariectomy and P treatment have often been several weeks to several months. Such long-term ovariectomy easily causes loss of P sensitivity unless E is implanted in the animal at ovariectomy [Batra et al., 1986]. It is known that E permits P to act in many systems, presumably through preservation of the P receptor in some fashion [Chauchereau et al., 1992]. Pituitary cultures also lose sensitivity to P within 1 week [Batra and Miller, 1985a,b]. Nevertheless, several reports have shown that P can act at all three critical points of control shown in Figure 1.

Studies in sheep indicate that P can act at control point 1 to lower GnRH output [Karsch, 1987; plus new results from the Karsch laboratory indicating that P can block the E-induced surge of GnRH, Kasa-Vubu et al., 1992].

Data presented below show that P also regulates gonadotropin synthesis and secretion at critical points 2 and 3 (Fig. 1). Again, there are major species differences between sheep and rats.

A. Control by P in the Sheep Pituitary

1. Control by P at point 2. Midluteal levels of P decrease GnRH-Rec by 70% in ovine pituitary cultures [Laws et al., 1990a]. Similar decreases in GnRH-Rec mRNA activity also occur in these cultures [Sealfon et al., 1990b]. Even when cultures are pretreated with E, inhibin, or their combination (which increases GnRH-Rec mRNA activity by 20-fold), P decreases GnRH-Rec mRNA activity by 70%. These decreases in GnRH-Rec mRNA and GnRH-Rec also decrease GnRH-stimu-lated LH secretion by 70% [Batra and Miller, 1985a].

Experiments with HPD/OVX/GnRH-pulsed ewes show that neither E nor P alone alters GnRH-stimulated LH secretion; their combination, however, decreases GnRH-stimulated LH secretion by 70%–80% [Girmus and Wise, 1991]. Based on results from ovine pituitary cultures (see above), we interpret these in vivo data to mean that E maintains receptors for P in the OVX sheep, which permits P to inhibit GnRH-stimulated LH secretion.

2. Control by P at point 3. Synthesis of FSH is blocked 70% by physiological levels of P (10 nM) in ovine pituitary cultures [Batra and Miller, 1985b]. As with E, P decreases mRNA levels for both α and LH-β subunits [Phillips et al., 1988]. Finally, P is as effective as E in blocking FSH-β transcription (85% in 2 hours) and α transcription (70% in 2 hours). Therefore it appears that E and P are equally potent in inhibiting FSH synthesis directly at the gene level in sheep.

The only in vivo study showing that P has a direct effect on the pituitary was reported by Nett et al. [1990] after studying OVX/GnRH-pulsed ewes. In that study E and P were administered together to block LH synthesis and to decrease mRNA levels for LH-β and α-subunits. One interpretation of these data, regarding LH-β, is that E maintained P receptors in the OVX ewes and that P was the active hormone that decreased LH-β mRNA levels.

B. Control by P in the Rat Pituitary

1. Control by P at point 2. Both negative and positive regulation have been reported for P in rat pituitary cultures. In all cases, E must be present for 12–24 hours before P is added, presumably to maintain P receptors [Lagacé et al., 1980]. Acute treatment with P (<12 hours) increases sensitivity to GnRH as measured by secretion of LH (ED_{50} for GnRH is lowered);

longer treatment with P (48 hours) decreases GnRH-stimulated LH secretion [Massicotte et al., 1984; Hsueh et al., 1979]. There are no reports that either E or P changes rat pituitary GnRH-Rec numbers to cause these effects.

2. Control by P at point 3. As with all effects of P noted above, it is necessary to have E present to observe the effect of P on FSH synthesis in rat pituitary cultures. In contrast to ovine pituitary cultures, rat pituitary cultures increase FSH production up to 100% when treated with a combination of E plus P [Massicotte et al., 1984]. There are no reports on the effects of E plus P on mRNA levels for either FSH or LH in the rat.

3. Overall conclusions for P in rats and sheep. The effects of P depend heavily on the presence of E, a fact that has made it difficult to study the effects of P. Thus many studies that report "no effect" of P at the pituitary level may have overlooked a major effect because they used animals that were ovariectomized (which removes E as well as P) too long before being treated with P. The reliance of P action on E is further complicated by the fact that E and P often act in a similar manner. For example, they both inhibit FSH synthesis in sheep and both stimulate it in rats.

Nowhere is the observance of species differences between the rat and sheep greater than in the actions of P in pituitary culture. Before labeling these differences as possible artifacts of the model system, however, it should be remembered that the main peak of P and the LH surge are coincident in the rat; an early rise in P can actually advance the preovulatory LH surge in rats [Freeman, 1988]. Thus a stimulatory function for P is reasonable and to be expected in the rat. By contrast, P completely blocks the LH surge in sheep. It is not surprising that these physiological differences should derive from differences at the molecular level.

VIII. INHIBIN/ACTIVIN AND TWO-POINT CONTROL

IN and ACT are structurally related hormones that may be produced not only in the gonads, but in many tissues. Inhibin is an α/β heterodimer of general structure, $\alpha\beta_A$ or $\alpha\beta_B$. Activin is a related dimer consisting of $\beta_B\beta_B$, $\beta_A\beta_A$ or $\beta_A\beta_B$ [Ying, 1988]. The α- and β-subunits of IN and ACT are not related in any way to the gonadotropin subunits. IN and ACT were originally named for their respective abilities to inhibit or enhance FSH output. Since that time, it has been learned that inhibin also affects GnRH-Rec levels. At this site of action, inhibin actually increases GnRH-Rec in sheep gonadotropes, so the term *inhibin* may be somewhat misleading in certain instances [Laws et al., 1990a].

Physiological interactions between IN, ACT, and a third gonadal protein, follistatin, are not yet fully understood. It was recently reported that follistatin binds to and inactivates ACT [Kogawa et al., 1991]. Prior to this report, follistatin had been considered to be a hormone that inhibited FSH production directly at the pituitary level because it decreased FSH synthesis in rat pituitary cultures [Ying, 1988; Carroll et al., 1989]. Follistatin is clearly an important molecule that can regulate FSH synthesis, but, because it acts by binding to and incapacitating ACT and does not act directly on the pituitary, it is not discussed at length in this review.

The observation that follistatin *does* inhibit FSH production in rat pituitary cultures suggests, however, that ACT is made in these cultures and that follistatin inactivates pituitary-made ACT. This concept is further strengthened by the recent finding that inactivating antibodies to ACT can also inhibit FSH synthesis in rat pituitary cultures [Corrigan et al., 1991]. Thus the concept is emerging that bioactive ACT (and perhaps IN also) is made in the pituitary where it (they) may have autocrine functions. This concept, when fully explored, may explain the wide variation in levels of FSH that are produced in pituitary cultures from different species. It may change the current views on FSH regulation in vivo.

A. Control by Inhibin and Activin in the Sheep Pituitary

1. Control by IN/ACT at point 2. Surprisingly, porcine follicular IN increases GnRH-Rec mRNA activity by 10-fold in ovine pituitary cultures, as measured by the *Xenopus* oocyte assay [Sealfon et al., 1990b]. Importantly, this effect of IN is completely additive with the 10-fold increase in GnRH-Rec mRNA activity caused by E. IN also increases GnRH-Rec numbers (six-fold) and responsiveness to GnRH (two- to three-fold) [Laws et al., 1990a]. In contrast to E, IN increases responsiveness to GnRH as long as IN is present in the culture media (>72 hours). Thus responsiveness to GnRH and GnRH-Rec numbers reflect each other at all times after IN treatment.

That IN increases gonadotrope responsiveness to GnRH in sheep is supported by a single in vivo study [Wrathall et al., 1990]. Midcycle, intact ewes were passively immunized to IN and then challenged 24 hours later with a bolus of GnRH. Ewes immunized to IN gave only 50% the response of control ewes, a finding predicted by studies with ovine pituitary cultures.

Recombinant human ACT does not alter GnRH-Rec levels in ovine pituitary cultures [Gregg et al., 1991], but this may be due to synthesis of large amounts of activin in these cultures. This possibility is currently under investigation in several laboratories.

2. Control by IN/ACT at point 3. Inhibin selectively and rapidly (6 hours) decreases the level of FSH-β mRNA (80% decrease) in HPD/OVX/GnRH-pulsed ewes [Mercer et al., 1987]. Inhibin also selectively decreases FSH synthesis (50%–90% decrease

in 24 hours) and FSH-β mRNA levels (30%–70% decrease in 24 hours) in ovine pituitary cultures [Miller et al., 1988]. It is not known if inhibin acts at a transcriptional level. Since activin is reported to stabilize FSH-β mRNA selectively in rat pituitary cultures [Carroll et al., 1991], it is possible that inhibin works at a posttranscriptional site to destabilize FSH-β mRNA.

Recombinant human ACT shows very little effect on FSH synthesis in ovine pituitary cultures (38% increase only [Gregg et al., 1991]). ACT has not been tested for its effects in ewes.

3. Control by IN/ACT in the sheep pituitary: Conclusions. Inhibin appears to have opposing effects on the ovine gonadotrope; it inhibits FSH synthesis by as much as 80%, but it increases GnRH-Rec thereby increasing the ability of GnRH to stimulate synthesis of at least α gonadotropin subunit. These opposing effects may cancel each other at all times during the estrous cycle. Activin has little or no effect on either FSH synthesis (only 38% increase in FSH synthesis in ovine pituitary cultures) or GnRH-Rec (no effect).

B. Control by Inhibin and Activin in the Rat Pituitary

1. Control by IN/ACT at point 2. In contrast to the stimulatory effect of inhibin on ovine GnRH-Rec, inhibin decreases rat GnRH-Rec by 50% in rat pituitary cultures [Wang et al., 1988] and prevents GnRH-stimulated up regulation of GnRH-Rec in these same cultures [Wang et al., 1988, 1989]. The effect of ACT on rat GnRH-Rec is unknown.

2. Control by IN/ACT at point 3. Inhibin specifically decreases FSH-β mRNA levels by 90%–100% within 4 hours in rat pituitary cultures [Attardi et al., 1989; Carroll et al., 1989]. By comparison, ovine pituitary cultures show essentially no effect of IN treatment before 4 hours. Furthermore, the maximum level of FSH-β mRNA inhibition rarely reaches 80% in ovine pituitary cultures [Miller et al., 1988]. Synthesis of FSH-β in the rat pituitary seems exquisitely sensitive to inhibin. In vivo studies with rats immunized against inhibin showed that the level of FSH increased, indicating that inhibin plays a physiological role in lowering FSH serum levels in rats [Rivier and Vale, 1989].

Activin increases FSH-β mRNA levels by two- to three-fold within 24 hours in rat pituitary cultures [Carroll et al., 1989, 1991; Attardi and Miklos, 1990]. This action is accomplished, at least in part, by stabilizing FSH-β mRNA, although a mechanism for such an action is unknown.

3. Control by IN/ACT in the rat pituitary: Conclusions. Inhibin is a true inhibitor in all capacities (FSH synthesis and GnRH-Rec levels) in the rat. This has led some to suggest that IN is a general translational inhibitor in gonadotropes. However, since IN increases GnRH-Rec in ovine pituitary cultures, this assertion is unlikely to be true. The ability of ACT to stabilize FSH-β mRNAs in rat pituitary cultures is novel. It may be that IN destabilizes FSH-β mRNA, since IN and ACT are related hormones. Much more work needs to be done to understand the effects of IN and ACT in sheep, rats, and other mammals.

IX. PUTTING IT ALL TOGETHER—CREATION OF THE MAMMALIAN ESTROUS CYCLE

Enumeration of individual control mechanisms used by GnRH, E, P, IN, and ACT does not convey the integrated power of these mechanisms to regulate pituitary gonadotropes to create the mammalian reproductive cycle. Presented below is a framework designed to explain coordinate control of gonadotropin synthesis and secretion by hypothalamic and gonadal hormones (see also Table I). This framework is based on the results presented above and represents a series of hypotheses, each of

which can be experimentally tested. The overall scheme is designed to be instructive, but also to stimulate further experimentation.

Regulation of the *ovine* estrous cycle is discussed in detail because our laboratory has worked closely with the sheep model. Specific predictions for the ewe *may* apply to other species, but not necessarily, since the nature and importance of each triple-point control can vary by species (Fig. 1).

A. Regulation of Tonic Production of LH and FSH in the Ewe

A mammal lacking gonadal hormones (OVX) produces 5–10-fold more LH and FSH than does an intact animal [Clayton et al., 1985]. High production rates of LH and FSH in OVX animals reflect relatively high levels of subunit mRNAs [Nilson et al., 1983; Papavasiliou et al., 1986; Gharib et al., 1987]. One job that gonadal hormones perform, therefore, is to keep gonadotropin mRNA levels low. Their combined effects are inhibitory and keep gonadotropin production and secretion to about 10% of their potential during the estrous cycle.

Figure 3 (bottom panel) shows serum levels of E, P, and IN during the ovine estrous cycle. Presumably enough E is present in this cycle to maintain P responsiveness. Since P is at high levels during most of the cycle, the effects of P should dominate the cycle. Between days 4 and 14, levels of P are about 10 nM, which is 10 times the ED_{50} for P action in ovine pituitary cultures [Batra and Miller, 1985a]. Therefore, between days 4 and 14, P, by itself, should provide highly effective negative pressure at all three control points pictured in Figure 1. As noted previously in this chapter, P can lower GnRH stimulation by 1) decreasing GnRH output from the hypothalamus and 2) reducing GnRH-Rec levels by 70%. Additionally, P directly inhibits α and FSH-β gene transcription. Based on the recent work of Nett et al. [1990], the combination of E and P also decreases LH-β mRNA levels directly at the gene and/or mRNA levels.

Inhibin is reported by McNeilly et al. [1989] to be relatively abundant and relatively invariant during the entire ewe estrous cycle. The ED_{50} value for highly enriched IN is reported to be 0.5–1.0 ng/ml [Carroll et al., 1989], so the average level of IN during the cycle (1 ng/ml) is one to two times the ED_{50} for IN. During the LH surge IN is two to four times the ED_{50}, a concentration capable of keeping FSH-β mRNA levels very low (control point 3). Additionally, transcription of α and FSH-β subunit genes can be inhibited by E during the LH surge when E also increases to just above its ED_{50} value (about 1.5 ED_{50} as observed in ovine pituitary cultures [Miller et al., 1977]). Thus throughout the entire ovine estrous cycle there is highly effective negative control on gonadotropin *synthesis,* even during the several days surrounding the preovulatory LH surge.

As predicted above, the steady-state levels of all gonadotropin subunit mRNAs do, in fact, remain relatively low and stable throughout the entire ewe estrous cycle [Leung et al., 1988]. During the midcycle, mRNA levels for α, LH-β, and FSH-β are kept within a fairly narrow range and vary only 10%–40%. During the 4 days surrounding the preovulatory LH surge; 1) α mRNA doubles, 2) LH-β mRNA increases by 50%, and 3) FSH-β mRNA rapidly decreases by 80% and then increases to twice its midcycle level. None of these changes persists for more than 24 hours, and none appears dramatic enough to cause much change in overall gonadotropin synthesis. They certainly do not appear large enough to create a preovulatory surge of either LH or FSH.

In summary, E, P, and/or IN suppress mRNA levels for α, FSH-β, and LH-β by about 90% during most of the ewe cycle. When inhibition by P is released and GnRH stimulation is increased during the preovulatory LH surge, α, LH-β, and FSH-β mRNA levels do increase 50%–200%, which may help build LH and FSH reserves for the next cycle.

B. Creation of the Preovulatory LH Surge in the Ewe

The preovulatory LH surge does not appear to be driven by abrupt changes in LH or FSH synthesis, as discussed above. The changes that do occur are too small and too short lived to cause the 30- to 100-fold changes in LH secretion that occur during the surge. Rather, it appears that a well-coordinated effort to increase both GnRH and GnRH-Rec by at least 30- and 20-fold, respectively, seems to be involved.

Major increases in hypothalamic GnRH output (30- to 40-fold) have recently been measured during the ovine LH surge [Moenter et al., 1990]. This work is well documented, and the timing of this GnRH surge is marked by shaded bars in Figures 2 and 3. These bars also denote the position of the 30- to 100-fold LH surge, which lasts about 14 hours. The top panel of Figure 3 diagrams the pattern of the 20-fold increase in GnRH-Rec mRNA that is predicted to occur in vivo. These data on GnRH-Rec are derived from results in primary ovine pituitary cultures. Removal of the inhibitory effects of P and addition of stimulatory E and IN can cause GnRH-Rec mRNA activity to jump by 20-fold [Sealfon et al., 1990b]. Such large increases in GnRH-Rec number are not observed in vivo, but this may be due to the fact that increased GnRH output may rapidly cause internalization and degradation of newly made GnRH-Rec. If increased GnRH-Rec is synchronized with increased GnRH output, the *steady-state levels* of GnRH-Rec will remain unchanged. The predicted increase in GnRH-Rec mRNA levels is diagrammed in the upper panel of Figure 3.

In summary, E, P, and IN appear to coordinate a dramatic increase in both GnRH output and GnRH-Rec. Both increases appear to be about 20- to 40-fold in magnitude, and the combination appears quite capable of causing the preovulatory LH surge in sheep.

X. PUTTING IT ALL TOGETHER— REGULATION OF THE GENES ENCODING GONADOTROPIN SUBUNITS AND GnRH-REC

It is not known why the pituitary has the capacity to produce more LH and FSH than is ever used in intact animals. It is clear, however, that E, P, and/or IN keep LH and FSH synthesis low by blocking maximal production by 90%. In the sheep, much of this inhibition can occur at the transcriptional level for α and FSH-β subunits [Phillips et al., 1988]. The LH-β gene may also be inhibited by E *plus* P at transcription [Nett et al., 1990], and P may inhibit transcription of the gene encoding GnRH-Rec. By contrast, transcription of the same genes appears to be either unaffected or increased by E and/or P in the rat [Shupnik et al., 1989]. It is uncertain as to what makes the difference between positive and negative regulation. Nevertheless, all of the gonadotropin genes offer rare opportunities to study the molecular mechanisms responsible for negative and positive regulation by the superfamily of steroid/thyroid hormone receptors.

Inhibin action can also be studied in a unique way using the ovine FSH-β and GnRH-Rec genes. Inhibin blocks FSH-β expression in the first case, but it stimulates GnRH-Rec expression in the second instance. The mechanism of action of inhibin is completely unknown at present.

A. Transcriptional Regulation by E and/or P

While many examples of *positive* regulation exist for steroid hormones, few examples of *negative* regulation are characterized. The E-induced increase in transcription of the rat LH-β gene represents the classical form of transcriptional

enhancement observed with many genes [Shupnik and Rosenzweig, 1991]. Only one example of negative regulation at the transcriptional level has been reported, however, for P except those noted above in the sheep [Chauchereau et al., 1992]. None have been reported for E. Several examples of transcriptional regulation (negative and positive) are noted below that may have analogies in gonadotropin regulation.

1. **Negative regulation by glucocorticoids.** The best known examples of steroid-mediated negative regulation involve glucocorticoid action. In the few reported cases, glucocorticoid receptors (GR) appear to interfere with a nuclear transcription factor that normally stimulates transcription. In one study the GR-DNA–binding site (GRE) overlaps a CREB binding site (CREB is a transcriptional activator associated with cAMP action [Yamamoto et al., 1988]). The overlapping of these two proteins destroys both proteins' abilities to activate the α gene in choriocarcinoma cells [Akerblom et al., 1988]. Overlapping DNA-binding sites for GR and JUN/FOS heterodimers cause a similar inhibitory phenomenon [Diamond et al., 1990]; JUN/FOS is a transcriptional activator complex associated with protein kinase C action. Two reports indicate that GR can interact directly with JUN/FOS even in the absence of an overlapping GRE [Schule et al., 1990; Yang-Yen et al., 1990]. Thus the few known examples of glucocorticoid-mediated negative regulation suggest that steroids regulate transcription negatively by an interference type of mechanism.

2. **Negative regulation of gonadotropins and GnRH-Rec.** The examples of negative regulation noted above may be particularly insightful for understanding negative regulation of the genes encoding α, LH-β, FSH-β, and GnRH-Rec. It is believed that JUN/FOS (activated by protein kinase C) and CREB may be activated by GnRH [Horn et al., 1991; Naor et al., 1979]. Fur-

thermore, progesterone response elements (PRE) are identical to glucocorticoid response elements (GRE) and similar to ERE. Thus there could be a similarity between negative regulation by G,P, and E in terms of gonadotropin and GnRH-Rec regulation.

Our laboratory has focused most recently on the gene encoding the ovine FSH-β subunit and its regulation by P. There are at least six high-affinity binding sites for PR in the 5'-flanking region of the oFSH-β gene [Webster et al., 1991; Guzman et al., 1991]. At least four more PRE exist within the translated portion of the gene. Moreover, several of these PRE overlap putative JUN/FOS- or CREB-binding sites. It will be important to determine the interactions of PR with these highly potent transcriptional activators. It will also be important to determine if similar PRE occur in the α and LH-β ovine genes.

Thus far no ERE have been found on any of the gonadotropin genes [Keri et al., 1991; Hirai et al., 1990] except the rat LH-β gene, where positive regulation is observed [Shupnik et al., 1989]. This observation opens the possibility that ERE (or PRE) may not be necessary and that direct interactions between PR and/or ER and transcriptional activators may cause negative regulation as reported by Schule et al. [1990] and Yang-Yen et al. [1990]. Important basic questions can be answered about negative regulation if regulation of the gonadotropin genes can be understood.

Because the negative actions of E or P might be cell specifically regulated, it will be important to use the current immortal gonadotrope cell lines developed by Windle et al. [1990] and to generate more immortal cell lines to achieve expression of all the gonadotropin subunits. Neither has the GnRH-Rec gene nor its cDNA been isolated. It will be particularly informative to isolate this gene and to express it permanently or transiently in the immortal GnRH-Rec-expressing mouse gonado-

tropes [Windle et al., 1990], where its regulation can be studied in detail. All of the studies noted above should yield important information about steroid-mediated negative and positive regulation.

B. Regulation by Inhibin and Activin

There is no evidence that either IN or ACT functions at the level of transcription to alter expression of either the FSH-β subunit or the GnRH-Rec gene. One report indicates that ACT increases the stability of FSH-β mRNA in rat pituitary cultures [Carroll et al., 1991]. It may be possible to study IN/ACT actions in primary cell lines, or to develop immortal gonadotrope cell lines that maintain IN/ACT receptors.

XI. EXCITING NEW AREAS OF STUDY INVOLVING "GONADOTROPIN GENES"

Aside from the wealth of information that may be gleaned from understanding gonadotropin and GnRH-Rec gene regulation, there are many new and exciting developments that will rapidly change the face of reproductive studies as we know them. One area involves the cloning and functional expression of ovarian receptors for LH [Segaloff et al., 1990a; Xie et al., 1990] and FSH [Sprengel et al., 1990]. Regulation of these genes is rapidly being investigated [Segaloff et al., 1990b; Wang et al., 1991]. The molecular mechanisms used by LH and FSH receptors to regulate ovarian function are also being studied at a rapid pace.

Cloning of the FSH receptor has now made it possible to rapidly investigate the binding of inhibitors to this receptor. Several potent inhibitors already exist in the form of peptide fragments from the FSH-β subunit [Santa-Coloma et al., 1991]. The three-dimensional structures of these peptides are being determined by nuclear magnetic imaging [Agris et al., 1992], and one future goal of this research is to find orally active molecules that can mimic the effects of the peptide inhibitors. These studies are designed partly to learn more about the mechanistic operations of the FSH receptor. They also have potential for the development of potent inhibitors for the FSH receptor, which could form the basis of a completely new, nonsteroidal approach to contraception. Such an approach should be applicable to males (in whom FSH regulates sperm production) as well as females.

Now that the genes for the LH and FSH subunits have been cloned, they have been stably or transiently expressed in several mammalian cell lines. Mutations have been made in these genes to alter glycosylation [Keene et al., 1989; Matzuk et al., 1989a; Matzuk and Boime, 1989] and subunit combination and secretion [Matzuk et al., 1989b]. As these studies continue they should yield a wealth of knowledge about glycosylation, subunit combination and processing, signal peptide function, and secretion pathways of the gonadotropins.

Finally, immunological techniques have been used for years to analyze LH and FSH in humans with reproductive dysfunction. These rely on interactions of gonadotropin epitopes with antibodies. Recently a defective LH molecule was identified at the gene level [Weiss et al., 1992], although it was perceived as being normal by radioimmunoassay. This is likely to be just the beginning of the use of molecular biology techniques to identify heretofore undetectable defects in genes that are important for reproductive function, such as the genes encoding gonadotropin subunits, gonadotropin receptors, GnRH receptors, GnRH, inhibin, and activin.

ACKNOWLEDGMENTS

I wish to acknowledge the art work of Ms. Norma Pedersen for Figures 1–3. Dr. Basavdutta Ray and Patricia Sullivan are thanked for their aid in proofreading and editing this manuscript. Finally, I thank all of my current and past students for their contributions, which have led to our current concepts of gonadotropin regulation by E, P, IN, and ACT in the ewe.

REFERENCES

Agris PF, Guenther RH, Sierzputowska-Gracz H, Easter L, Smith W, Hardin CC, Santa-Coloma TA, Crabb JW, Reichert Jr LE (1992): Solution structure of a synthetic peptide corresponding to a receptor binding region of FSH (hFSH-β 33–35). J Protein Chem 11:495–507.

Akerblom IE, Slater EP, Beato M, Baxter JD, Mellon PL (1988): Negative regulation by glucocorticoids through interference with a cAMP responsive enhancer. Science 241:350–353.

Alexander DC, Miller WL (1982): Regulation of ovine follicle-stimulating hormone α-chain mRNA by 17β-estradiol in vivo and in vitro. J Biol Chem 257:2282–2286.

Attardi B, Keeping HS, Winters SJ, Kotsuji F, Maurer RA, Troen P (1989): Rapid and profound suppression of messenger ribonucleic acid encoding follicle-stimulating hormone-β by inhibin from primate sertoli cells. Mol Endocrinol 3:280–287.

Attardi B, Miklos J (1990): Rapid stimulatory effect of activin-A on messenger RNA encoding the follicle-stimulating hormone α-subunit in rat pituitary cell cultures. Mol Endocrinol 4:721–726.

Batra SK, Miller WL (1985a): Progesterone decreases the responsiveness of ovine pituitary cultures to luteinizing hormone-releasing hormone. Endocrinology 117:1436–1440.

Batra SK, Miller WL (1985b): Progesterone inhibits basal production of follice-stimulating hormone in ovine pituitary cell culture. Endocrinology 117:2443–2448.

Batra SK, Britt JH, Miller WL (1986): A direct pituitary action of progesterone on basal secretion of follicle-stimulating hormone in ovine cell culture: dependence on ovaries in vivo. Endocrinology 119:1929–1932.

Carroll RS, Corrigan AZ, Gharib SD, Vale W, Chin WW (1989): Inhibin, activin, and follistatin: Regulation of follicle-stimulating hormone messenger ribonucleic acid levels. Mol Endocrinol 3:1969–1976.

Carroll RS, Corrigan AZ, Vale W, Chin WW (1991): Activin stabilizes follicle-stimulating hormone-β messenger ribonucleic acid levels. Endocrinology 129:1721–1726.

Chauchereau A, Savouret J-F, Milgrom E (1992): Control of biosynthesis and post-transcriptional modification of the progesterone receptor. Biol Reprod 46:174–177.

Clarke IJ, Cummins JT (1984): Direct pituitary effects of estrogen and progesterone on gonadotropin secretion in the ovariectomized ewe. Neuroendocrinology 39:267–274.

Clarke IJ, Cummins JT (1987): The significance of small pulses of gonadotrophin-releasing hormone. J Endocrinol 113:413–418.

Clarke IJ, Cummins JT, Crowder ME, Nett TM (1987): Pituitary receptors for gonadotropin-releasing hormone in relation to changes in pituitary and plasma luteinizing hormone in ovariectomized–hypothalamo/pituitary–disconnected EWES. I. Effect of changing frequency of gonadotropin-releasing hormone pulses. Biol Reprod 37:749–754.

Clarke IJ, Cummins JT, Crowder ME, Nett TM (1988): Pituitary receptors for gonadotropin-releasing hormone in relation to changes in pituitary and plasma gonadotropins in ovariectomized hypothalamo/pituitary-disconnected EWES. II. A marked rise in receptor number during the acute feedback effects of estradiol. Biol Reprod 39:349–354.

Clayton RN (1989): Gonadotrophin-releasing hormone: its actions and receptors. J Endocrinol 120:11–19.

Clayton RN, Detta A, Naik SI, Young S, Charlton HM (1985): Gonadotrophin releasing hormone receptor regulation in relationship to gonadotrophin secretion. Steroid Biochem 23:691–702.

Corrigan AZ, Bilezikjian LM, Carroll RS, Bald, LN, Schmelzer CH, Fendly BM, Mason AJ, Chin WW, Schwall RH, Vale W (1991): Evidence for an autocrine role of activin B within rat anterior pituitary cultures. Endocrinology 128:1682–1684.

Cullar MD, Negro-Vilar A (1986): Evidence that pulsatile follicle-stimulating hormone secretion is independent of endogenous luteinizing hormone-releasing hormone. Endocrinology 118:609–612.

Culler MD, Negro-Vilar A (1987): Pulsatile follicle-stimulating hormone secretion is independent of luteinizing hormone-releasing hormone (LHRH): pulsatile replacement of LHRH bioactivity in LHRH-immunoneutralized rats. Endocrinology 120:2011–2021.

Dalkin AC, Haisenleder DJ, Ortolano GA, Ellis TR, Marshall JC (1989): The frequency of gonadotropin-releasing-hormone stimulation differentially regulates gonadotropin subunit messenger ribonucleic acid expression. Endocrinology 125:917–924.

Diamond MI, Miner JN, Yoshinaga SK, Yamamoto KR (1990): Transcription factor interactions: selectors of positive or negative regulation from a single DNA element. Science 249:1266–1272.

Drouin J, Lagacé L, Labrie F (1976): Estradiol-induced increase of the LH responsiveness to LH

releasing hormone (LHRH) in rat anterior pituitary cells in culture. Endocrinology 99:1477–1481.

Esbenshade KL, Vogel MJ, Traywick GB (1986): Clearance rate of luteinizing hormone and follicle stimulating hormone from peripheral circulation in the pig. J Anim Sci 62:1649–1653.

Freeman ME (1988): The ovarian cycle of the rat. In Knobil E, Neill J (eds): The Physiology of Reproduction. New York: Raven Press, pp 1893–1928.

Gharib SD, Wierman ME, Badger TM, Chin WW (1987): Sex steroid hormone regulation of follicle-stimulating hormone subunit messenger ribonucleic acid (mRNA) levels in the rat. J Clin Invest 80:294–299.

Gharib SD, Wierman ME, Shupnik MA, Chin WW (1990): Molecular biology of the pituitary gonadotropins. Endocrinol Rev 11:177–199.

Girmus RL, Wise ME (1991): Direct pituitary effects of estradiol and progesterone on luteinizing hormone release, stores, and subunit messenger ribonucleic acids. Biol Reprod 45:128–134.

Goodman RL (1988): Neuroendocrine control of the ovine estrous cycle. In Knobil E, Neill JD (eds): The Physiology of Reproduction. New York: Raven Press, pp 1929–1969.

Gregg DW, Nett TM (1989): Direct effects of estradiol-17β on the number of gonadotropin-releasing hormone receptors in the ovine pituitary. Biol Reprod 40:288–293.

Gregg DW, Schwall RH, Nett TM (1991): Regulation of gonadotropin secretion and number of gonadotropin-releasing hormone receptors by inhibin, activin-A, and estradiol. Biol Reprod 44:725–732.

Guzman K, Miller CD, Phillips CL, Miller WL (1991): The gene encoding ovine follicle-stimulating hormone β: isolation, characterization, and comparison to a related ovine genomic sequence. DNA Cell Biol 10:593–601.

Haisenleder DJ, Dalkin AC, Ortolano GA, Marshall JC, Shupnik MA (1991): A pulsatile gonadotropin-releasing hormone stimulus is required to increase transcription of the gonadotropin subunit genes: evidence for differential regulation of transcription by pulse frequency in vivo. Endocrinology 128:509–517.

Hall SH, Miller WL (1986): Regulation of ovine pituitary glycoprotein hormone α subunit mRNA by 17β-estradiol in cell culture. Biol Reprod 34:533–542.

Hamernik DL, Crowder ME, Nilson JH, Nett TM (1986): Measurement of messenger ribonucleic acid for luteinizing hormone β-subunit, α-subunit, growth hormone, and prolactin after hypothalamic pituitary disconnection in ovariectomized ewes. Endocrinology 119:2704–2710.

Hamernik DL, Nett TM (1988): Gonadotropin-releasing hormone increases the amount of messenger ribonucleic acid for gonadotropins in ovariec-tomized ewes after hypothalamic–pituitary dis-connection. Endocrinology 122:959–966.

Hirai T, Takikawa H, Kato Y (1990): The gene for the β subunit of porcine FSH: Absence of consensus oestrogen-responsive element and presence of retroposons. J Mol Endocrinol 5:147–158.

Horn F, Bilezikjian LM, Perrin MH, Bosma MM, Windle JJ, Huber KS, Blount AL, Hille B, Vale W, Mellon PL (1991): Intracellular responses to go-nadotropin-releasing hormone in a clonal line of the gonadotropin lineage. Mol Endocrinol 5:347–355.

Hsueh AJW, Erickson GF, Yen SSC (1979): The sen-sitizing effect of estrogens and catechol estrogen on cultured pituitary cells to luteinizing hormone-releasing hormone: its antagonism by progestins. Endocrinology 104:807–813.

Huang ES, Miller WL (1984): Porcine ovarian inhibin preparations sensitize cultured ovine gonadotrophs to luteinizing hormone-releasing hormone. Endocrinology 115:513–519.

Hubert JF, Simard J, Gagné B, Barden N, Labrie F (1988): Effect of luteinizing hormone releasing hormone (LHRH) and [D-Trp^6, Des-Gly-NH$_2$^{10}]LHRH ethylamide on α-subunit and LHβ messenger ribonucleic acid levels in rat anterior pituitary cells in culture. Mol Endocrinol 2:521–527.

Karsch FJ (1987): Central actions of ovarian steroids in the feedback regulation of pulsatile secretion of luteinizing hormone. Annu Rev Physiol 49:365–382.

Karsch FJ, Cummins JT, Thomas GB, Clarke IJ (1987): Steroid feedback inhibition of pulsatile secretion of gonadotropin-releasing hormone in the ewe. Biol Reprod 36:1207–1218.

Kasa-Vubu JZ, Dahl GE, Evans NP, Thrun LA, Moenter SM, Padmanabhan V, Karsch FJ (1992): Progesterone blocks the estradiol-induced gonad-otropin discharge in the ewe by inhibiting the surge of gonadotropin-releasing hormone. Endo-crinology (in press).

Keene JL, Matzuk MM, Boime I (1989): Expression of recombinant human choriogonadotropin in Chinese hamster ovary glycosylation mutants. Mol Endocrinol 3:2011–2017.

Keri RA, Andersen B, Kennedy GC, Hamernik DL, Clay CM, Brace AD, Nett TM, Notides AC, Nilson JH (1991): Estradiol inhibits transcription of the human glycoprotein hormone α-subunit gene de-spite the absence of a high affinity binding site for estrogen receptor. Mol Endocrinol 5:725–733.

Knobil E (1980): The neuroendocrine control of the menstrual cycle. Recent Prog Horm Res 36:53–88.

Knobil E, Plant TM, Wildt L, Belchetz PE, Marshall G (1980): Control of the rhesus monkey menstrual cycle: permissive role of hypothalamic gonadotro-pin-releasing hormone. Science 207:1371–1373.

Kogawa K, Nakamura T, Sugino K, Takio K, Titani K, Sugino H (1991): Activin-binding protein is pres-ent in pituitary. Endocrinology 128:1434–1440.

Labrie F, Drouin J, De Léan A, Lagacé L, Ferland L, Beaulieu M (1977): Mechanism of action of lutein-izing hormone releasing hormone and thyrotropin releasing hormone in the anterior pituitary gland and modulation of their activity by peripheral hor-mones. Adv Exp Med Biol 87:157–179.

Lagacé L, Massicotte J, Labrie F (1980): Acute stim-ulatory effects of progesterone on luteinizing hor-mone and follicle-stimulating hormone release in rat anterior pituitary cells in culture. Endocrinol-ogy 106:684–689.

Lalloz MRA, Detta A, Clayton RN (1988): Gonado-tropin-releasing hormone is required for en-hanced luteinizing hormone subunit gene expres-sion in vivo. Endocrinology 122:1681–1688.

Laws SC, Beggs MJ, Webster JC, Miller WL (1990a): Inhibin increases and progesterone decreases re-ceptors for gonadotropin-releasing hormone in ovine pituitary culture. Endocrinology 127:373–380.

Laws SC, Webster JC, Miller WL (1990b): Estradiol alters the effectiveness of gonadotropin-releasing hormone (GnRH) in ovine pituitary cultures: GnRH receptors versus responsiveness to GnRH. Endocrinology 127:381–386.

Leung K, Kim KK, Maurer RA, Landefeld TD (1988): Divergent changes in the concentrations of gonadotropin α-subunit messenger ribonucleic acid during the estrous cycle of sheep. Mol En-docrinol 2:272–276.

Liu TC, Jackson GL, Gorski J (1976): Effects of syn-thetic gonadotropin-releasing hormone on incor-poration of radioactive glucosamine and amino acids into luteinizing hormone and total protein by rat pituitaries in vitro. Endocrinology 98:151–163.

Loumaye E, Catt KJ (1982): Homologous regulation of gonadotropin-releasing hormone receptors in cultured pituitary cells. Science 215:983–985.

Mason AJ, Pitts SL, Nikolics K, Szonyi E, Wilcox JN, Seeburg PH, Stewart TA (1986a): The hypogonadal mouse: reproductive functions re-stored by gene therapy. Science 234:1372–1378.

Mason AJ, Hayflick JS, Zoeller RT, Young WS III, Phillips HS, Nikolics K, Seeburg PH (1986b): A deletion truncating the gonadotropin-releasing hormone gene is responsible for hypogonadism in the hpg mouse. Science 234:1366–1371.

Massicotte J, Lagacé L, Godbout M, Labrie F (1984): Modulation of rat pituitary gonadotrophin secre-tion by porcine granulosa cell "inhibin," LH re-leasing hormone, and sex steroids in rat anterior pituitary cells in culture. J Endocrinol 100:133–140.

Matzuk MM, Boime I (1989): Mutagenesis and gene transfer define site-specific roles of the gonadotropin oligosaccharides. Biol Reprod 40:48–53.

Matzuk MM, Keene JL, Boime I (1989a): Site specificity of the chorionic gonadotropin N-linked oligosaccharides in signal transduction. J Biol Chem 264:2409–2414.

Matzuk MM, Spangler MM, Camel M, Suganuma N, Boime I (1989b): Mutagenesis and chimeric genes define determinants in the β subunits of human chorionic gonadotropin and lutropin for secretion and assembly. J Cell Biol 109:1429–1438.

McNeilly AS, Swanston IA, Crow W, Tsonis CG, Baird DT (1989): Changes in the plasma concentrations of inhibin throughout the normal sheep oestrous cycle and after the infusion of FSH. J Endocrinol 120:295–305.

Mercer JE (1990): At the cutting edge: pituitary gonadotropin gene regulation. Mol Cell Endocrinol 73:C63–C67.

Mercer JE, Clements JA, Funder JW, Clarke IJ (1987): Rapid and specific lowering of pituitary FSH β mRNA levels by inhibin. Mol Cell Endocrinol 53:251–254.

Mercer JE, Clements JA, Funder JW, Clarke IJ (1988): Luteinizing hormone-β mRNA levels are regulated primarily by gonadotropin-releasing hormone and not by negative estrogen feedback on the pituitary. Neuroendocrinology 47:563–566.

Miller WL, Lin LW, Phillips CL, Wu JC, Laws SC, Beggs MJ, Guzman KG, Milsted A (1988): Regulation of the subunit mRNAs of follicle-stimulating hormone by inhibin, estradiol, and progesterone. In Hodgen GD, Rosenwaks Z, Spieler JM (eds): Nonsteroidal Gonadal Factors: Physiological Roles and Possibilities in Contraceptive Development. Virginia: The Jones Institute Press, pp 110–124.

Miller WL, Knight MM, Grimek HJ, Gorski J (1977): Estrogen regulation of follicle stimulating hormone in cell cultures of sheep pituitaries. Endocrinology 100:1306–1316.

Miller WL, Wu J (1981): Estrogen regulation of follicle-stimulating hormone production in vitro: Species variation. Endocrinology 108:673–679.

Moenter SM, Caraty A, Karsch FJ (1990): The estradiol-induced surge of gonadotropin-releasing hormone in the ewe. Endocrinology 127:1375–1384.

Naor Z, Clayton RN, Catt KJ (1980): Characterization of gonadotropin-releasing hormone receptors in cultured rat pituitary cells. Endocrinology 107:1144–1152.

Naor Z, Fawcett CP, McCann SM (1979): Differential effects of castration and testosterone replacement on basal and LHRH-stimulated cAMP and cGMP accumulation and on gonadotropin release from the pituitary of the male rat. Mol Cell Endocrinol 14:191–198.

Nett TM, Flores JA, Carnevali F, Kile JP (1990): Evidence for a direct negative effect of estradiol at the level of the pituitary gland in sheep. Biol Reprod 43:554–558.

Nilson JH, Nejedlik MT, Virgin JB, Crowder ME, Nett TM (1983): Expression of α subunit and luteinizing hormone β genes in the ovine anterior pituitary. J Biol Chem 258:12087–12090.

Papavasiliou SS, Zmeili S, Herbon L, Duncan-Weldon J, Marshall JC, Landefeld TD (1986): α and luteinizing hormone β messenger ribonucleic acid (RNA) of male and female rats after castration: quantitation using an optimized RNA dot blot hybridization assay. Endocrinology 119:691–698.

Phillips CL, Lin W, Wu JC, Guzman K, Milsted A, Miller WL (1988): 17β-Estradiol and progesterone inhibit transcription of the genes encoding the subunits of ovine follicle-stimulating hormone. Mol Endocrinol 2:641–649.

Rivier C, Vale W (1989): Immunoneutralization of endogenous inhibin modifies hormone secretion and ovulation rate in the rat. Endocrinology 125:152–157.

Rodin DA, Lalloz MRA, Clayton RN (1989): Gonadotropin-releasing hormone regulates follicle-stimulating hormone β-subunit gene expression in the male rat. Endocrinology 125:1282–1289.

Santa-Coloma TA, Crabb JW, Reichert LE Jr. (1991): A synthetic peptide encompassing two discontinuous regions of hFSH-β subunit mimics the receptor binding surface of the hormone. Mol Cell Endocrinol 78:197–204.

Santoro N, Wierman ME, Filicori M, Waldstreicher J, Crowley WF Jr (1986): Intravenous administration of pulsatile gonadotropin-releasing hormone in hypothalamic amenorrhea: effects of dosage. Clin Endocrinol Metab 62: 109–116.

Schule R, Rangarajan P, Kliewer S, Ransone LJ, Bolado J, Yang N, Verma IM, Evans RM (1990): Functional antagonism between oncoprotein c-JUN and the glucocorticoid receptor. Cell 62:1217–1226.

Sealfon SC, Gillo B, Mundamattom S, Mellon PL, Windle JJ, Landau E, Roberts JL (1990a): Gonadotropin-releasing hormone receptor expression in Xenopus oocytes. Mol Endocrinol 4:119–124.

Sealfon SC, Laws SC, Wu JC, Gillo B, Miller WL (1990b): Hormonal regulation of gonadotropin-releasing hormone receptors and messenger RNA activity in ovine pituitary culture. Mol Endocrinol 4:1980–1986.

Segaloff DL, Sprengel R, Nikolics K, Ascoli M

(1990a): Structure of the lutropin/choriogonadotropin receptor. Recent Prog Horm Res 46:261–301.

Segaloff DL, Wang HY, Richards JS (1990b): Hormonal regulation of luteinizing hormone/chorionic gonadotropin receptor mRNA in rat ovarian cells during follicular development and luteinization. Mol Endocrinol 4:1856–1865.

Shupnik MA (1990): Effects of gonadotropin-releasing hormone on rat gonadotropin gene transcription in vitro: requirement for pulsatile administration for luteinizing hormone-β gene stimuation. Mol Endocrinol 4:1444–1450.

Shupnik MA, Rosenzweig BA (1991): Identification of an estrogen-responsive element in the rat LHβ gene. J Biol Chem 266:17084–17091.

Shupnik MA, Weinmann CM, Notides AC, Chin WW (1989): An upstream region of the rat luteinizing hormone b gene binds estrogen receptor and confers estrogen responsiveness. J Biol Chem 264:80–86.

Sprengel R, Braun T, Nikolics K, Segaloff DL, Seeburg PH (1990): The testicular receptor for follicle stimulating hormone: structure and functional expression of cloned cDNA. Mol Endocrinol 4:525–530.

Stanley HF, Lyons V, Obonsawin MC, Bennie J, Carroll S, Roberts JL, Fink G (1988): Regulation of pituitary α-subunit, β-luteinizing hormone and prolactin messenger ribonucleic acid by gonadotropin-releasing hormone and estradiol in hypogonadal mice. Mol Endocrinol 2:1302–1310.

Wang H, Segaloff DL, Ascoli M (1991): Lutropin/-choriogonadotropin down-regulates its receptor by both receptor-mediated endocytosis and a cAMP-dependent reduction in receptor mRNA. J Biol Chem 266:780–785.

Wang QF, Farnworth PG, Findlay JK, Burger HG (1988): Effect of purified 31K bovine inhibin on the specific binding of gonadotropin-releasing hormone to rat anterior pituitary cells in culture. Endocrinology 123:2161–2166.

Wang QF, Farnworth PG, Findlay JK, Burger HG (1989): Inhibitory effect of pure 31-kilodalton bovine inhibin on gonadotropin-releasing hormone (GnRH)-induced up-regulation of GnRH binding sites in cultured rat anterior pituitary cells. Endocrinology 124:363–368.

Ward DN, Bousfield GR, Moore, KH (1991): Gonadotropins. In Cupps PT (ed): Reproduction in Domestic Animals. New York: Academic Press, pp 25–80.

Webster JC, Pedersen NR, Miller WL (1991): Progesterone receptor binds the ovine FSH-beta gene 5 prime-flanking region and can thereby regulate FSH-beta expression. Program and Abstracts of the 73rd Annual Meeting of the Endocrine Society, June 19–22, Washington, DC, No. 927, p 262.

Weiss J, Axelrod L, Whitcome RW, Harris PE, Crowley WF, Jameson JL (1992): Hypogonadism caused by a single amino acid substitution in the β subunit of luteinizing hormone. N Engl J Med 326:179–183.

Wierman ME, Rivier JE, Wang C (1989): Gonadotropin-releasing hormone-dependent regulation of gonadotropin subunit messenger ribonucleic acid levels in the rat. Endocrinology 124:272–278.

Windle JJ, Weiner RI, Mellon PL (1990): Cell lines of the pituitary gonadotrope lineage derived by targeted oncogenesis in transgenic mice. Mol Endocrinol 4:597–603.

Wrathall JHM, Mcleod BJ, Glencross RG, Beard AJ, Knight PG (1990): Inhibin immunoneutralization by antibodies raised against synthetic peptide sequences of inhibin α subunit: effects on gonadotropin concentrations and ovulation rate in sheep. J Endocrinol 124:167–176.

Xie YB, Wang H, Segaloff DL (1990): Extracellular domain of lutropin/choriogonadotropin receptor expressed in transfected cells binds choriogonadotropin with high affinity. J Biol Chem 265: 21411–21414.

Yamamoto KK, Gonzalez GA, Biggs WH III, Montminy MR (1988): Phosphorylation-induced binding and transcriptional efficacy of nuclear factor CREB. Nature 334:494–498.

Yang-Yen HF, Chambard JC, Sun YL, Smeal T, Schmidt TJ, Drouin J, Karin M (1990): Transcriptional interference between c-JUN and the glucocorticoid receptor: mutual inhibition of DNA binding due to direct protein–protein interaction. Cell 62:1205–1215.

Ying SY (1988): Inhibins, activins, and follistatins: gonadal proteins modulating the secretion of follicle-stimulating hormone. Endocrinol Rev 9:267–293.

ABOUT THE AUTHOR

WILLIAM L. MILLER is a professor of Biochemistry at North Carolina State University, where he teaches "Biochemistry of Hormone Action" and a graduate laboratory in Biochemistry/Biotechnology. After receiving his B.S. in Biochemistry from Bucknell University in 1965, he received his Ph.D. in Biochemistry under Dr. James L. Gaylor at Cornell University, where he studied cholesterol biosynthesis in mammalian liver. Dr. Miller worked with Dr. B.P. Doctor on tRNA genes at the Walter Reed Army Institute of Research, Washington, D.C., 1970–1973, and he worked with Dr. Jack Gorski on estrogen action on ovine pituitary gonadotropes at the University of Wisconsin, Madison, 1973–1976. Dr. Miller's current research involves regulation of gonadotropin synthesis by gonadotropin releasing hormone (GnRH), estrogens, progestins, inhibin, and activin. The model system is the ovine pituitary gonadotroph. The research involves isolation and characterization of regulatory elements on the genes encoding the beta subunit of follicle stimulating hormone and the receptor of GnRH. Dr. Miller is currently a NCSU member of the National Center for Infertility Research associated with the University of Michigan.

Genes in Mammalian Reproduction: 271–291

Progesterone and Estrogen

Nancy H. Ing, Sophia Y. Tsai, Ming-Jer Tsai

I. INTRODUCTION

Progesterone and estrogen have profound effects on mammals. The differences between males and females are striking and are primarily the result of the different effects of the male gonadal steroid hormone testosterone and the female's progesterone and estrogen. The differences are manifest in nearly all body tissues, including brain, muscle, and bone, but are most prominent in reproductive tissues. The female mammal has a remarkable ability to adapt her physiology to bear developing concepti and to feed them milk produced by her own body after their birth. Ovarian steroids coordinate the female mammal's production of oocytes, her receptivity to mating, and the uterine receptivity to the conceptus. Often, these estrous cycles are seasonal: Seasonality helps to ensure maximal success of pregnancy, which is important because the energy investment in reproduction by the female is quite large. Successful pregnancies result in great changes in the female's physiology. The uterus grows to be many times larger during pregnancy and, in the human, receives almost 20% of the cardiac output while the blood volume is expanded by 40% [Knobil and Neill, 1988]. After birth, specialized mammary glands may take all of her ingested nutrients, as well as use stored nutrients of the mother's body, in order to produce milk to feed the young. During or after lactation and nurturing of the young, her hormones will direct her to restart estrous cycling and to repeat the reproductive cycle again. Each of these female reproductive stages is directed by a delicate interplay of progesterone and estrogen, as well as polypeptide hormones and neural signals.

The purpose of this chapter is to review the study of progesterone- and estrogen-responsive genes, with insights into the mechanism of steroid hormone action and how these genes are defined. Progesterone- and estrogen-responsive genes in mammals will be reviewed, from initial descriptive studies to sophisticated dissections of cis-elements responsible for hormonal effects. To do this, we focus on the hormonal effects on transcription initiation of genes via their respective receptors. The biochemistry of progesterone and estrogen receptors is also described. Finally, future directions of study are addressed.

II. CURRENT STATUS

A. Study of Progesterone- and Estrogen-Regulated Genes

The female mammal's steroid hormones affect her physiology from brain and behavior to bone structure, as well as control every aspect of reproduction. The challenge facing reproductive endocrinologists is to understand the molecular mechanisms of hormone action. The female gonadal steroids are powerful effectors of tissue growth, differentiation, and remodelling. Secondary effects of the steroids are then multiplied from these primary actions. For example, "estrogen-priming," commonly done as a pretreatment in hormonal action studies, may change the types of cells present. The endometrium of the ovariectomized ewe is a simple flattened epithelium. However, with estrogen and progesterone therapy, the endometrial epithelium increases tremendously in height and area, extending toward the myometrium in long convoluted tubular glands [Ing et al., 1989, and references therein]. The two types of tissues, endometrium from untreated and from steroid-treated animals, must be compared carefully because the cells appear and act differently.

The complexity of mammalian physiology must be considered when defining specific genes activated by progesterone and estrogen. For example, consider steroidogenesis in the ovary. The expression of genes controlling it is regulated by pituitary hormones. There is also evidence of regulation by estrogen. The latter implies autoregulation of steroid synthesis, which makes the system extremely complex since the hormone regulates gene expression of some pituitary hormones and of its own receptor, as well as the enzymes involved in its own synthesis [Hickey et al., 1989]. The effect on the steroidogenic enzymes may be receptor independent by the hormone simply acting as a substrate. The studies are difficult because ovarian tissues are making the steroids that act directly in this regulation. However, in these and other model systems, progress is being made at defining controlling elements, so that primary effects of estrogen and progesterone can be elucidated. These results will allow us to understand some of the steps of the complex reproductive physiology of the female mammal.

There are many publications about progesterone- and estrogen-regulated pro-

teins in the literature. Table I provides an overview of recent studies of proteins whose transcription may be regulated by the female sex steroid hormones. The scope of progesterone and estrogen effects is apparent by scanning the studies in Table I, which are roughly grouped by function. Note that several proteins have no known function but are studied for their progesterone and/or estrogen induction pathway. Table I includes some of the best described progesterone- and estrogen-regulated proteins, as well as some very recently discovered systems. The level of investigation is noted so that the reader has an idea of how specifically the control of the gene by progesterone or estrogen is understood. The early studies describing production of proteins may relate to indirect or nontranscriptional effects of steroid hormones, discussed later. The species and tissue sources are indicated because progesterone- and estrogen-induced genes show great developmental and tissue specificity. In the following sections, the study of progesterone- and estrogen-regulated genes will be described briefly, through all phases of investigation, using the studies listed in Table I as examples.

1. Model systems and reagents. The early investigations of steroid action usually involve qualitative and quantitative studies of protein profiles of tissues and body fluids from animals under different progesterone and estrogen influence. Because animals vary greatly in their individual physiology, differences between reproductive groups, such as pregnant and nonpregnant, may be variable. To control the steroid influences of the subject and maximize differences between groups, in vivo progesterone and estrogen experiments often are preceded by ovariectomizing animals and giving exogenous steroids [e.g., Li et al., 1991]. This lowers basal steroid hormone levels and allows for administration of pharmacological doses of hormone. Such in vivo treatment groups can show clear and

TABLE I. Studies of Gene Regulation by Progesterone and Estrogen*

Steroid receptors				
Progesterone	P−	Human; T47D, MCF7	Rd	Read et al. [1988], Wei et al. [1988], Alexander et al. [1989]
	P−	Rabbit	Df	Misrahi et al. [1988, 1990]
	E+	Human; T47D MCF7	Rd	Read et al. [1988], Wei et al. [1988], Kastner et al. [1990], Vegeto et al. [1990]
	E+	Mouse; mammary (not lactating)	Rd	Shymala et al. [1990]
	E+	Rabbit	Df	Misrahi et al. [1988, 1990]
	E+	Rat; hypothalamus	Re	Romano et al. [1989]
Estrogen	P−	Human; T47D, MCF7	Rd	Alexander et al. [1990]
	E+	Rat; liver, pituitary	Rd	Shupnik et al. [1989]
	E−	Rat; uterus	Rd	Shupnik et al. [1989]
	E−	Rat; hypothalamus	Re	Simerly and Young [1991]
Glucocorticoid	E−	Rat; pituitary	Re	Peiffer and Barden [1987]
Protooncogenes				
c-myc	E+, P+	Mouse; uterus	Pc	Huet-Hudson et al. [1989]
	E+	Rat; uterus	Rd	Persico et al. [1990]
c-jun	E+	Rat; uterus	Rd	Persico et al. [1990], Webb et al. [1990]
c-fos	E+	Rat; uterus	Rd	Persico et al. [1990], Webb et al. [1990]
	E+	Human; T47D, MCF7	Dg	Weisz et al. [1990a, b]
	E+	Rat; brain	Pc	Insel [1990]
Growth factors and their receptors				
EGF	E+	Mouse; endometrium	Rd	Huet-Hudson et al. [1990]
	EP+	Human; breast tumors	R	Dotzlaw et al. [1990]
IGFI	E+	Rat; uterus	Rd	Croze et al. [1990]
	EP+	Pig; uterus	Rd	Simmen et al. [1990]
IGFI-R	E+	Rat; uterus	Rd	Ghahary and Murphy [1989]
IGFI-BP	EP+	Rat; uterus	Rd	Croze et al. [1990]
		Baboon, endometrium	Rd	Fazleabas et al. [1989]
IGFII	E+	Rat; uterus	Rd	Murphy and Ghahary [1990]
TGF-beta2 & 3	E−	Human; T47D, ZR-75-1	R	Arrick et al. [1990]
Peptide hormones				
FSH	E+	Rat; pituitary	Rd	Perheentupa and Huhtaneiei [1990]
Galanin	E+	Rat; pituitary	Pc	Hsu et al. [1990]
GnRH	E+	Rat; hypothalamus	Re	Park et al. [1990]
	E2	Rat; brain	Re	Toranzo et al. [1989]
GnRH-R	P−	Sheep; pituitary	P	Laws et al. [1990]
Insulin	P+	Human; T47D	Rd	Papa et al. [1990]
LH	E+	Rat; pituitary	Rd	Papa et al. [1990]
Müllerian inhibiting substance	E+	Human; ovary	Dg	Guerrier et al. [1990]
Oxytocin	E+	Human	Dg	Richard and Zingg [1990]
Prolactin	E+	Rat	Dg	Waterman et al. [1988]
Proenkephalin	EP+	Rat; ovary, uterus	Rd	Muffly et al. [1988]
	P+	Rat; uterus	Rd	Jin et al. [1988]
	P+	Pig; uterus	Pb	Li et al. [1991]
Proopiomelano-cortin	P+	Rat; brain	Re	Wise et al. [1990]
(Endorphin)	E+	Rhesus; endometrium	Pb	Low et al. [1989]
Neurotensin	E−	Rat; pituitary	Rd	O'Halloran et al. [1990]
Neuropeptide Y	E−	Rat; pituitary	Rd	O'Halloran et al. [1990]
Substance P	E−	Rat; pituitary	Rd	O'Halloran et al. [1990]
Vasoactive intestinal peptide	E+	Rat; pituitary	Rd	O'Halloran et al. [1990]
Proteases				
Kallikrein	E+	Rat; pituitary	Rd	Clements et al. [1990]
PDP (cathepsin L)	P+	Cat; endometrium	Rd	Jaffe et al. [1989]
Proteases				
Procathepsin D	P+	Human; endometrium	Pc	Maudelonde et al. [1990]
	E+	Human; HeLa + ER	Pb	Touitou et al. [1990]
	E+	Human; MCF7	Rd	Capony et al. [1990]

(Continued)

Procollagenase	E+, P+	Guinea pig; cervix	Rd	Rajabi et al. [1991]
Other enzymes and proteins involved in nutrition				
Adenosine phosphoribosyl-transferase	E+	Rat; endometrium	Rd	Cummings [1989]
Alkaline phophatase	E+	Human; Ishikawa	Rd	Albert et al. [1990]
Cholesterol side chain cleavage cytochrome P450scc	E+	Rabbit; corpus luteum	P	Keyes et al. [1990]
Cyclooxygenase	P+	Sheep; endometrium	Re	Eggleston et al. [1990]
Fatty acid synthetase	P+	Human; MCF7, T47D	Rd	Chalbos et al. [1986], Joyeux et al. [1989]
	P+	Human; endometrium	Rd	Escot et al. [1990]
Lactate dehydro-genase	P+, E+	Human; T47D	P	Hissom et al. [1989]
Multidrug resistance P glycoprotein	EP+	Mouse; uterus	Re	Arceci et al. [1990]
Calbindin	E+	Rat; uterus	Rd	L'Horst et al. [1990]
Uteroferrin	P+	Pig; endometrium	Df	Simmen et al. [1989]
Proteins involved in the immune system				
Complement C3	E+	Rat; uterus	Rd	Sundstrom et al. [1990]
Major histocompa-tibility complex (HLA)	P−	Human; T47D	D	Teh and Hui [1989]
	E+	Human; MCF7	D	Teh and Hui [1989]
Structural Proteins				
Osteocalcin, α2 (1) collagen, osteonectin, osteopontin	E+	Rat; bone	Rd	Turner et al. [1990]
Proteins of unknown function				
Uteroglobin	P+	Rabbit; endometrium	P	Mani et al. [1991]
		Rabbit; endometrium	Dg	Cato et al. [1984] Bailly et al. [1986]
		Rabbit; endometrium (chromatin)		Jantzen et al. [1987]
		Rabbit; (transgenics)		DeMayo et al. [1991]
	E+	Rabbit; endometrium	Dg	Slater et al. [1990]
Clara cell 10K or PCB-binding protein	EP+	Rat; uterus	D	Hagen et al. [1990]
UTMP	P+	Sheep; endometrium	Rd	Ing et al. [1989]
UfAP	P+	Pig; endometrium	Rd	Malathy et al. [1990]
Heat shock p90	E+	Mouse; uterus	Rd	Shymala et al. [1989]
μ-Fetoprotein	P−	Rat; fetal liver	Dg	Turcotte et al. [1990]
Pentraxin	E−	Hamster; liver	P	Coe and Ross [1990]
GCDFP-24	E−	Human; ZR-75-1	P	Simard et al. [1990]
pS2	E+	Human; MCF7	Dg	Nunez et al. [1989]
pSyd's	E+	Human; T47D	Rd	Manning et al. [1990]
HMW glycoprotein	E+	Baboon; oviduct	Rd	Donnelly et al. [1991]
LMW protein	P+	Sheep; endometrium	Pc	Kazemi et al. [1990]
???	E+, EP+			
		Baboon; oviduct	P	Fazleabas et al. [1988]

*Gene products, grouped by physiological function, are listed with the effect progesterone (P) and/or estrogen (E) has on their expression (+, induction; −, repression). Also indicated are the species, tissue, and level of investigation: P = protein in a) SDS-polyacrylamide gel electrophoresis, b) Western (immuno-) blots, or c) immunocytochemistry; R = RNA in d) Northern blots or e) in situ hybridization; and D = DNA in f) promoter fragments or g) *cis* element identified.

reproducible responses to progesterone and estrogen.

To dissect away the influences of the steroid-treated animal on the tissue of interest, the tissue may be isolated and cultured in vitro, such as done by Ing et al., [1989]. Although these "organ" cultures treated with steroid contain several cell types, the reactions of the tissues of the rest of the animal do not impinge on the tissue of interest. Paracrine communication between different cell types in tissue culture may be required for some responses. The next reductionist step is to isolate the target cells of the steroid hormone and grow those in culture. Primary cultures of such cells are very time consuming to make and may only be viable for a short time. An exception is a cell culture system developed with rabbit endometrial cells, which are polarized and produce uteroglobin in response to progesterone for several weeks [Mani et al., 1991]. Stable cell lines can be cultured longer, but may lose many of the characteristics of their cellular forebears. However, human breast cancer cell lines MCF-7, T47D, and ZR-75-1 have been invaluable tools for the study of steroid hormone-responsive genes (see Table I).

Specialized reagents are necessary for definitive studies. Proteins may be sampled by directly taking fluid samples or by homogenizing tissues, or they may be purified by schemes tailored to the particular protein. A pure protein, or one resolved from a mixture by SDS-PAGE, may be used to generate antisera [Dunbar and Schwoebel, 1990] and to determine amino acid sequence. Both antisera and DNA probes deduced from amino acid sequence can be used to identify cDNA clones of target genes, the latter by hybridization or polymerase chain reaction priming [Sambrook et al., 1989]. Cloning a protein provides primary sequence data for the protein as well as high-affinity cDNA probes for its mRNA. Additionally, related proteins can be identified by sequence comparison.

Also, there are several systems available to express cloned proteins, sometimes in large amounts for easy purification [Goeddel, 1990]. Genes for target proteins induced by progesterone or estrogen can be cloned with cDNA probes so that the *cis* elements controlling gene expression can be examined. The above tools (pure proteins, antisera, cDNAs, and cloned genes) are sufficient for some of the most sophisticated in vitro studies of steroid hormone gene regulation.

2. Protein induction. Studies of progesterone- and estrogen-treated animals show major changes in protein production of steroid-responsive tissues, often by analyses allowing visualization of many diverse proteins. The use of one- and two-dimensional sodium dodecysulfate-polyacrylamide gel electrophoresis (SDS-PAGE) systems is widespread for this purpose. Proteins separated by molecular size (and charge in two-dimensional gels) migrate to unique positions on slab gels, and protein staining indicates their relative abundance. A very descriptive survey of baboon endometrial oviduct proteins, performed with two-dimensional SDS-PAGE analysis, showed that many estrogen- and progesterone-induced proteins are similar to those of humans [Fazleabas et al., 1988].

With the availability of a high-affinity antiserum, many research avenues open. The unique and selective affinity of an antiserum for a protein can be used to identify and quantitate protein in samples of all stages of purity. Dot and slot blots are some of the quickest techniques by which target proteins are adsorbed on a membrane, washed, and developed by incubation with antiserum followed by a secondary binding reagent [Timmons and Dunbar, 1990]. A more precise, descriptive technique is Western blotting, where proteins separated on SDS-PAGE are blotted to a membrane. After development of the blot, the molecular weight position of the target protein separates it from background signals caused by other proteins. Quantitation may be done by densitometers or β-scanning machines, as they were for the down regulation of the progesterone receptor by progestins in T47D cells as performed by Read et al. [1988]. Antibodies for peptide hormones are used in radioimmunoassays (e.g., for GCDFP-24 [Simard et al., 1990]), while radiolabeled ligands are used in radioreceptor assays (e.g., for progesterone receptor downregulation by retinoic acid [Clarke et al., 1990]). Both techniques require a lot of set up time, but are highly quantitative measures of protein levels. Enzymes can be measured by their activities in specific assays, such as in the estrogen induction of adenosine phosphoribosyltransferase in rat endometrium [Cummings, 1989].

Very precise and descriptive studies of protein induction are also done by antiserum staining in tissue sections. The immunocytochemical approach shows which cells in a tissue produce protein, how many cells produce it, to what extent each cell produces it, and where the protein goes. For example, Huet-Hudson et al. [1989] showed that estrogen increases *c-myc* in mouse uterine epithelium, but progesterone antagonizes the estrogen effect and induces *c-myc* in uterine stroma. Furthermore, they correlated *c-myc* expressing cells with those having enhanced DNA synthesis (thymidine uptake). With immunoelectron microscopy, Kazemi et al. [1990] showed that a low molecular weight, progesterone-modulated protein was associated with crystallin inclusion bodies in sheep endometrium and trophoblast tissues.

3. Transcriptional regulation. At this point, studies turn to molecular biology to address transcriptionally regulated genes. Progesterone- or estrogen-induced mRNAs may be identified with no prior knowledge of protein induction or identities in three ways: differential screening, absorbed probes, and subtracted

libraries. Differential screening is the easiest and most commonly used method, involving hybridization of colonies on replica filters with labeled cDNAs from steroid hormone–treated cells or cDNAs from untreated, control cells. Colonies that differentially hybridize are identified as encoding steroid hormone–regulated genes. This was done for MCF7 cells to identify pS2 [Masiakowski et al., 1982] and fatty acid synthetase or Pg8 [Chalbos et al., 1986], and for T47D cells to find Syd's [Manning et al., 1990].

In the study of a progesterone- or estrogen-induced protein, an increase in its steady-state mRNA levels suggests transcriptional regulation. The mRNA levels can be quantitated on dot or slot blots of total cellular or poly(A)$^+$ RNA from tissue samples probed with radiolabeled cDNA for the gene of interest [DeMayo et al., 1991]. Northern blots of RNA separated on agarose gels are more descriptive because, like Western blots, they describe the molecular size of the RNA that has affinity for the probe. Such blots were quantitated in L'Horset et al.[1990] to describe the estrogen induction of calbindin in the rat uterus. In some cases, many mRNA bands will hybridize with a cDNA probe. This may be due to related gene sequences (sometimes eliminated by high stringency washes of the filter), multiple genes for a protein, or differentially spliced products of a single gene product.

There are other ways to measure levels of a specific mRNA, including analyses by S1 nuclease and RNase A protection, and primer extension [Ausubel et al., 1987]. The first two techniques employ hybridization of single-stranded probes to RNA and digestion with single-strand–specific enzymes. The double-stranded molecules are specifically protected from digestion and, by gel analysis, can quantitate the level of mRNA present in a sample. Primer extension involves annealing a specific primer to an mRNA and synthesizing a cDNA using reverse transcriptase. These methods are more sensitive than Northern blots, although some high background problems may occur with the S1 nuclease technique. In addition to quantitation, all three methods can define the ends of the mRNA of interest. Simmen et al. [1990] showed where the start site of the uteroferrin mRNA was on the gene by primer extension.

As with protein probed with an antiserum in immunocytochemistry, in situ hybridization with antisense RNA probes can identify mRNA in tissues. These studies show precisely which cells transcribe the gene. Such studies were done quantitatively by Escot et al. [1990] for the progesterone induction of fatty acid synthetase in human endometrium during the menstrual cycle and by Park et al. [1990] for the estrogen induction of gonadotrophin-releasing hormone in the rat pituitary.

Drugs that inhibit RNA synthesis may be used to prove transcriptional regulation. All RNA polymerases are inhibited by actinomycin D and cordycepin, among other drugs. The polymerase responsible for mammalian mRNA synthesis, RNA polymerase II, is specifically inhibited by α-amanitin. Muffly et al. [1988] used actinomycin D to prove that the increase in prolactin mRNA after estrogen treatment was due to transcriptional regulation. Criticism of such studies is based on the fact that these inhibitors are poisons with far-reaching effects on cellular metabolism and health.

The best proof of estrogen- and progesterone-induced gene transcription in living cells is by nuclear "run-on" or "run-off" assays. For such assays, nuclei are isolated from the target tissue and then incubated in vitro with radiolabeled ribonucleotide for a short time to extend preinitiated nascent RNA transcripts. The nascent RNA are quantitated by cDNA hybridization and the level of tissue expression of the target

gene is deduced. This was done with uteroglobin early in its study, confirming its transcriptional regulation [Mueller and Beato, 1980].

Cloning genomic DNA for the gene of interest may be the next step in its study. The 5'-flanking region of the first exon usually contains most of the regulatory elements of the gene, although some elements occur within the exon. The promoter, the 100 bp 5' to the transcriptional start site, may contain a TATA box and other elements. However, progesterone and estrogen response elements (PREs and EREs) are usually enhancer elements, existing at positions further away on the gene. Sequencing of the promoter/enhancer can identify regions corresponding to consensus sequences for PREs and EREs (see Fig. 1). However, more definitive proof is required for a functional element.

As proof that the progesterone or estrogen receptor works on the enhancer of a target gene, the enhancer/promoter region can be introduced into a heterologous system, often in a transient transfection. The region of the gene containing the presump-

Fig. 1. *Consensus steroid hormone receptor response elements and important purine contacts. The consensus response elements for progesterone and estrogen receptor action are shown, as determined by comparison with known* cis *elements [Tsai and O'Malley, 1991]. Guanidine bases important for contact with the receptors are indicated with an asterisk. PRE, progesterone response element; ERE, estrogen response element.*

tive controlling elements, or oligonucleo-tides containing the gene elements similar to other known PREs or EREs, is cloned in front of a reporter gene, such as chlor-amphenicol acetyltransferase, luciferase, or β-galactosidase [Ausubel et al., 1987]. Reporter plasmids are transfected into the cells of choice in culture. These cells may contain progesterone and estrogen recep-tors (e.g., MCF7 cells) or be receptor neg-ative (e.g., COS, CV-1, and HeLa cells), in which case progesterone or estrogen re-ceptor expression vectors must be cotransfected with the reporter construct. The appropriate steroid hormone is also required (e.g., estrogen in the study of the oxytocin enhancer by Richard and Zingg [1990]). Cells are usually cultured in char-coal-stripped serum or serum-free condi-tions so that basal levels of reporter gene expression are low. Phenol red, used in most cell culture media, should be avoided with estrogen receptor studies because it contains an estrogenic contaminant [Bindal et al., 1988]. Cells in culture can be induced to take up DNA in transient transfections with calcium phosphate, DEAE dextran, polybrene, or electropora-tion, depending on the cell used and avail-able equipment [Ausubel et al., 1987; Sambrook et al., 1989]. Assays of the re-porter protein may be quantitative for the level of transcription directed by the target gene promoter. Alternatively, RNA tran-scripts may be measured directly, such as by S1 nuclease mapping, used to charac-terize the pS2 gene enhancer [Nunez et al., 1989].

If transfections show that the enhancer is still progesterone or estrogen dependent, the next step is to narrow down the regions responsible for this effect. A full-length en-hancer may be shortened in several 5' and 3' deletion constructs using enzymatic di-gestion (Bal31 or ExoIII coupled with S1 nuclease), deletion with a PCR primer, or by removing parts with convenient restric-tion enzyme sites [Ausubel et al., 1987]. The

last technique was used by Weisz and Rosales [1990] to study the fos enhancer. Transfections are performed as before with these new constructs and the deleted re-gions that eliminate hormone responsive-ness are noted.

The process can be repeated by making linker scanning mutations, which can re-place a small, more defined area of en-hancer DNA with nonsense (linker) DNA [Ausubel et al., 1987]. Linker scanning mu-tants are identical to the native enhancer except in their linker site. Therefore, effects like loss of estrogen or progesterone induc-tion seen with these are usually associated with a change directly in the cis element responsible for induction. Further defini-tion can be obtained by single or multiple point mutations in the enhancer element that are usually oligonucleotide directed. These techniques are difficult but can pro-vide solid proof of the function of individ-ual residues in a cis element.

Another approach is to examine the en-hancer for protein interactions with proges-terone or estrogen receptors in electropho-retic mobility shift (DNA-binding) assays (EMSAs) [see Ausubel et al., 1987]. En-hancer regions or oligonucleotides are end labeled. These are incubated with proges-terone or estrogen receptors, in pure or crude preparations, or with nuclear extract from the tissue of interest [Bagchi et al., 1988]. Reactions are run on native PAGE, and autoradiography reveals retarded bands from receptor–DNA complexes. The specificity of such reactions can be shown by competition of protein–DNA interac-tions with the addition of nonradiolabeled DNA containing the element of interest. After proving that the protein is interacting with a specific element on the DNA, the identity of the protein may be determined by antiserum addition to the EMSA. Monoclonal antiserum AB52 to the human progesterone receptor binds and further retards the receptor–DNA complex [Bagchi et al., 1988], while H222 recognizes the human estrogen receptor [Klein-Hitpass et al., 1989].

DNase1 footprinting of progesterone or estrogen receptors on the target gene enhancers can demonstrate the area bound by the protein. Footprinting requires much protein to bind all sites on the DNA, while limited amounts of DNase1 cleave only portions of the sequence around the *cis* element. When the labeled DNA is analyzed on a sequencing gel, a ladder of fragments is apparent at all positions except those protected by a protein. Such experiments identified several receptor-binding sites in the uteroglobin enhancer [Cato et al., 1984; Bailly et al., 1986]. Addition of a nonlabeled *cis* element DNA can show the specificity of the reaction by competing the footprint away.

Once the *cis* element has been identified, G (and sometimes A) residues in close proximity to the progesterone or estrogen receptors can be demonstrated by methylation interference studies. In these studies, an EMSA probe is treated with dimethyl sulfate prior to use in an EMSA. Complexed and free DNAs are eluted from gels separately and cleaved at Gs (and As) with piperidine. Autoradiography of sequencing gels comparing free and complexed DNAs will show some positions in free DNA missing in complexed DNA because the methylated residues at those positions interfered with protein binding. This method was used by Cato et al. [1984] to more tightly define the regions of the uteroglobin enhancer contracted by receptor.

In vitro transcription is a functional test for both enhancer or *cis* element interaction with progesterone or estrogen receptors, as well as for the resultant effect on transcription. This technology is powerful in being able to dramatically alter transcription conditions and to study its mechanism of action. The target promoter may direct transcripts analyzed by S1 nuclease, RNase protection, or primer extension, described previously. Alternatively, the promoter may direct transcription of a template cut by a restriction enzyme a short distance from the start site. In the presence of high levels of radiolabeled nucleotide, the run-off transcripts created are easily analyzed by sequencing gels and autoradiography. A recently developed assay involves the target promoter directing discrete transcripts of a G-free cassette [Sawadogo and Roeder, 1985]. GTP is omitted from reactions and 3'-O-methyl-GTP is included so that transcripts terminate a finite distance from initiation. T1 RNase, which cleaves after G, is included to reduce background. These in vitro transcription assays use nuclear extracts of cells to provide highly active transcriptional machinery. If absent, addition of progesterone or estrogen receptors is required for transcription from the target promoter [Klein-Hitpass et al., 1990]. Some sources of the receptors are still hormone-responsive in vitro [Bagchi et al., 1991]. Transcription effects of the progesterone or estrogen receptors can be inhibited by the addition of oligonucleotides containing the specific *cis* element recognized by the receptor.

B. Receptors as Transducers of Progesterone and Estrogen Action

1. Mechanism of progesterone and estrogen action. Since the initial discovery of steroid hormone effects in animals, dissecting the molecular mechanisms has been a high-priority goal. From viewing gross anatomical changes to being able to characterize changes in tissue proteins, the field of steroid biology jumped ahead with the discovery of specific, high-affinity receptors for progesterone and estrogen. The receptor proteins were purified, characterized, and intensively studied to discern their mode of action [for reviews, see Tsai and O'Malley, 1991; Beato, 1989; Evans, 1988]. There is a long running debate as to where the receptors exist in the cell: in the cytoplasm or the nucleus. It is known at least that without hormone the receptors exist in large complexes with heat shock proteins. And in the presence of hormone, which because of its nonpolar nature freely permeates cell membranes, the receptors become "activated," existing as smaller forms, presumably by shedding their heat shock proteins (see Fig. 2). The receptors then dimerize. In the nucleus, the ligand-activated progesterone and estrogen receptors bind to their respective steroid response elements of target genes. When bound, they stimulate the formation of stable protein complexes involved in transcription initiation [Klein-Hitpass et al., 1990]. These complexes include RNA polymerase II and the general transcription factors TFIIB, TFIID, TFIIE, and TFIIF, which then function by transcribing the gene.

The above mechanism of steroid hormone action, discussed later in detail, does not account for all effects of progesterone and estrogen. Some steroid hormone effects on transcription are negative, that is, they repress transcription of specific genes (see Table I). Transcriptional repression of some genes by glucocorticoids has been shown to involve the interaction of

glucocorticoid receptor with members of the *fos-jun* oncogene/transcription factor family [Diamond et al., 1990; Lucibello et al., 1990; Schule et al., 1990; Yang-Yen et al., 1990]. Progesterone repression of α-fetoprotein may act through a similar mechanism [Turcotte et al., 1990].

The steroid hormones have also been shown to have posttranscriptional effects. Estrogen receptors in MCF-7 cells suppress estrogen receptor mRNA posttranscriptionally [Saceda et al., 1989]. The stability of mRNAs for fatty acid synthetase and calbindin are increased by progesterone and estrogen, respectively, along with the transcription rates of the genes [Joyeux et al., 1989, L'Horset, 1990]. Steroid hormones can alter rates of protein synthesis. Both progesterone and estrogen increase myelin basic protein translation and decrease cyclic nucleotide 3'-phosphodiesterase synthesis in vitro, while estrogen decreases estrogen receptor translation [Verdi and Campagnoni, 1990]. The effect on the myelin basic protein was found to require a steroid modulatory element in the 5'-untranslated region, pre-

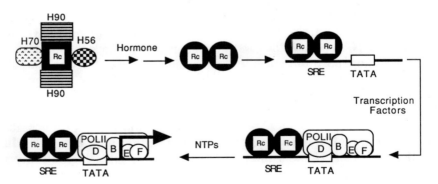

Fig. 2. *The mechanism of progesterone receptor action. The intracellular steroid hormone receptor (Rc) exists in a complex with heat shock proteins (H) 90, 70, and 56. Hormone transforms the receptor to an active state where it dimerizes and binds to specific* cis *elements of genes, called* steroid response elements (SREs). *DNA-bound receptors stabilize the transcription initiation complex, composed of RNA polymerase II (POLII) and general transcription factors ((TFII-)B, D, E, and F). In the presence of nucleotide triphosphates (NTPs),* transcription initiates (large arrow). *Thus the progesterone receptor transduces the hormone signal into an increased rate of gene expression.*

sumed to interact with receptor. Protein secretion and turnover may be affected by steroid hormones. For example, secretion of cathepsin D by MCF7 cells is enhanced by estrogen [Capony et al., 1990]. Estrogen from 11-day-old pig concepti are believed to trigger exocytosis of material in endometrial secretory vesicles into the uterine lumen [Roberts and Bazer, 1988, and references therein]. These secretion and turnover effects may be related to steroid hormone regulation of glycosylation. Estrogen has been shown to stimulate N-linked glycoprotein assembly in uteri of ovariectomized mice, while progesterone inhibits the effect [Dutt et al., 1986]. Progesterone has been shown to increase turnover of estrogen receptors in hamsters uterus [Takeda and Leavitt, 1986]. Steroid hormones can have receptor-independent effects also. Progesterone binds a multidrug resistance gene product on cell membranes and alters permeability [Yang et al., 1990]. The above nontranscriptional effects are important and growing areas of research.

2. Functional analysis of progesterone and estrogen receptors. Progesterone receptors are produced in A (94 kD for the human) and B (120 kD, human) forms, while the estrogen receptor is synthesized as a single polypeptide form (66 kD, human) [Evans, 1988]. The progesterone and estrogen receptors were among the first proteins cloned. The primary amino acid structures of the estrogen and progesterone receptors revealed that they were related to each other, as well as to an ever-growing superfamily of steroid/thyroid hormone receptors relatives. Included in the superfamily are receptors for androgens, glucocorticoids, mineralocorticoids, thyroid hormone, retinoic acid, vitamin D, and several "orphan receptors," such as the chicken ovalbumin upstream promoter-transcription factor (COUP-TF) which has no known ligand (Fig 3). The superfamily members are highly homologous in three

domains, one of which is a DNA-binding domain that functions in sequence specific DNA binding. The other two conserved domains are contained within the large, C-terminal ligand-binding domain. Other functions of estrogen and progesterone receptors include activation of transcription, nuclear localization, and dimerization. The function of the progesterone and estrogen receptors largely exist in domains that can be transferred to other proteins in "domain swap" experiments (see Fig. 4). However, some functions are interdependent in vivo. For example, the transactivation function requires that ligand- and DNA-binding domains and nuclear localization and dimerization activities be functional. The functional domains will be described here [for detailed discussion of the works done to define them, see Tsai and O'Malley, 1991; Carson-Jurica et al., 1990; Beato, 1989].

The conserved DNA-binding domain of the steroid hormone receptor superfamily contains two zinc finger structures. Many superfamily members bind similar DNA cis elements: progesterone, glucorticoid, androgen, and mineralocorticoid receptors bind (PREs and GREs), while estrogen, thyroid hormone, retinoic acid, and vitamin D receptors and COUP-TFs bind similar elements. By comparing the identified cis elements of glucocorticoid-activated genes, a consensus was determined (Fig. 2). The progesterone receptor was found to bind these sites also with high affinity. The starred residues in Figure 2 are important contact points as determined by methylation interference. The element exhibits partial dyad symmetry, and the progesterone receptors bind the DNA as dimers. The 5' half site is more variable than the 3' and may have lower affinity. The ERE consensus has perfect dyad symmetry. The GGTCA half-site is present in response elements of some other receptors mentioned above. However, the spacing between the half-sites is crucial for estrogen-

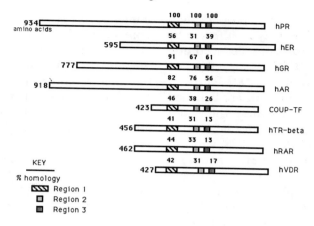

Fig. 3. *Conservation of domains among members of the steroid hormone receptor superfamily. Amino acid sequences of selected members of the steroid hormone receptor superfamily are shown as cartoons with the amino-terminus at left and carboxy-terminus at right. The number of total residues is indicated at left. The percent homologies in conserved regions 1, 2, and 3, relative to the human progesterone receptor (hPR) are indicated above each sequence [Carson-Jurica et al., 1990]. Other members of the steroid hormone receptor superfamily shown are the human estrogen receptor (hER), human glucocorticoid receptor (hGR), human androgen receptor (hAR), chicken ovalbumin upstream promoter-transcription factor (COUP-TF), human thyroid hormone receptor-β (hTR-β), human retinoic acid receptor (hRAR), and human vitamin D receptor (hVDR).*

receptor binding, which occurs also as dimers. Very sophisticated studies using point mutations in the DNA-binding domains have shown that mutation of three amino

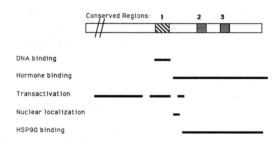

Fig. 4. *Functional domains of the progesterone receptor. The progesterone receptor is diagrammed as in Figure 3. Regions of the receptor responsible for the various functions listed are mapped below the amino acid sequences [Carson-Jurica, et al., 1990]. HSP90, heat shock protein 90.*

acids of the estrogen receptor to those of the progesterone receptor can alter the specificity of receptor binding to DNA from an ERE to a PRE and vice versa [reviewed by Berg, 1989].

Ligand binding domains are not well conserved between receptors of the steroid hormone receptor superfamily and are defined loosely over a large C-terminal region. Both estrogen and progesterone receptors bind more than one ligand, and considerable research was and is devoted to the study of natural and synthetic ligands. Some ligands are quite powerful agonists and/or antagonists and are used to alter mammalian physiology for reproductive and cancer control.

Transactivation domains of the estrogen and progesterone receptors are a little better described than ligand-binding domains. These domains are defined by deletion analysis: deletion mutants must bind DNA and ligand but not transactivate in cotransfection experiments. Two progesterone receptor activation domains exist close to and on either side of the DNA-binding domain. The estrogen receptor contains two activation domains, transactivating function-1 and -2 (TAF-1 and TAF-2), one upstream of the DNA binding domain and one in the ligand-binding domain [Tora et al., 1989]. N-terminal domains also contribute to the transactivation function of the receptors. The transactivation domains are thought to interact with other proteins of the transcriptional machinery. With purified components, in vitro transcription has shown that progesterone receptors increase transcription by increasing stable preinitation complex formation. The progesterone and estrogen receptors also carry nuclear localization signals in the "hinge" region between DNA- and ligand-binding domains.

The progesterone and estrogen receptors are known to form dimers in solution and to bind DNA as dimers. The estrogen receptor has a dimerization domain in within the ligand-binding domain and is believed to be composed of a seven base pair repeat of hydrophobic residues [Fawell et al., 1990]. Regions around the DNA-binding domain also have dimerization function. Heat shock protein 90 binds to progesterone and estrogen receptors in regions of their ligand-binding domains.

3. Biology of the progesterone and estrogen receptors. After the discussion of the complex nature of the estrogen and progesterone receptors and their interactions with themselves, other proteins, steroid ligands, and DNA, their physiology only seems more complex when the regulation of their production is discussed. Since they are the transducers for the steroid hormones, the biology of the receptors must be described in order to understand their effects in a cell, tissue, or animal. The progesterone and estrogen receptor genes are turned on and off just as are the genes they regulate. In fact, not only do both estrogen and progesterone receptors autoregulate their own production, but also they regulate the production of each other (see Table I and references therein). These effects are transcriptional and possibly translational, but are not equivalent between receptors or tissues. The estrogen receptor can up- or downregulate its receptor, depending on the tissue; in rat liver and pituitary, sheep endometrium, and mouse decidual tissue, it increases levels of its own receptor, while in rat uterus it paradoxically decreases it. Progesterone acts to downregulate estrogen receptors in most systems studied, as it does its own receptor. However, estrogen upregulates progesterone receptor levels. These effects occur through EREs and PREs on the progesterone receptor enhancer, which are similar to consensus sequences and have been confirmed by DNase1 footprinting [Misrahi et al., 1990]. This physiological interplay explains why several progesterone-responsive genes require priming doses of estrogen on the tissue in order to see strong induction by progesterone. Receptor synthesis is regulated by other factors as well, including other members of the steroid hormone receptor superfamily: Androgens downregulate estrogen receptors in ZR-75-1 cells [Poulin et al., 1989], as does retinoic acid for progesterone receptors in T47D cells [Clarke et al., 1990], while thyroxine increases levels of estrogen receptors in rat liver [Freyschuss and Eriksson, 1988] and pituitary [Fujimoto et al., 1988]. For the future, we need to understand the complex biology of the steroid hormone receptors thoroughly, including what factors control their production and turnover and how their production is controlled through developmental and reproductive stages.

III. FUTURE DIRECTIONS

We must understand the regulation of receptor activity. Exciting evidence has shown that regulation of the progesterone receptors are dependent on its state of phosphorylation. The progesterone and estrogen receptors are phosphoproteins whose activities are controlled by phosphorylation at specific sites [see Migliaccio et al., 1989, and references below]. In addition, molecular weight changes in SDS-PAGE believed to be the result of phosphorylation are seen only when receptor is allowed to bind its specific response element DNA (M.K. Bagchi, unpublished data). Evidence with the chicken progesterone receptor, highly homologous to the human's, has recently shown that phosphorylation is important, because 8-Br-cAMP and okadaic acid can mimic the progesterone effect in cotransfection experiments [Denner et al., 1990]. Another startling discovery is that dopamine, a neurotransmitter and precursor of norepinephrine and epinephrine, can also activate the progesterone receptor to enhance target gene transcription [Power et al., 1991]. Thus the hormonal effects of progesterone have been traced through the progesterone receptor, a transcription factor, which may share a pathway for stimulation with a neurotransmitter. In the future, we may understand how the body's physiological systems intercommunicate to regulate a healthy female mammal smoothly through complex life changes.

In the case of progesterone receptors, we must consider both A and B translation products. In addition, no other member of the steroid hormone receptor superfamily is synthesized similarly. No distinct function for the sequence carried only by the larger A form has been identified, although some functional differences between A and B forms of the progesterone receptor have been suggested by Kastner et al. [1990]. The A and B forms of the progesterone receptor may have different phosphorylation, transactivation, and regulation in the female.

A question basic to the study of transcription factors is how the activation domains work. The hormone-regulatable progesterone and estrogen receptors are ideal models, because the polypeptides are inert until transformation by ligand. Ligand-induced changes in conformation have been proposed to release heat shock proteins and allow nuclear localization and DNA binding. Recent evidence using progesterone receptors free of heat shock proteins has shown that ligand is still required for in vitro DNA binding and transcription [Bagchi et al., 1991]. Generally, activation domains of transcription factors are thought to provide surfaces for protein–protein interactions with other parts of the transcriptional machinery. It has been proposed that the two activation domains of the estrogen receptor (TAF-1 and TAF-2) transactivate by different mechanisms [Tora et al., 1989]. A likely explanation is that the two regions contact different proteins involved in transcription. The transcription field is focusing on these protein–protein interactions in order to understand transactivation. Many systems have shown that proteins other than previously described DNA-binding transcription factors are required for transactivation, such as so-called adaptor proteins [Berger et al., 1990; Kelleher et al., 1990; Pugh and Tjian, 1990]. The steroid hormone receptor–related COUP-TF requires a non-DNA-binding transcription factor, S300-II, to transactivate the ovalbumin gene in vitro [Tsai et al., 1987]. Progesterone and estrogen receptors compete for a limiting transcription factor required for transactivation in vivo and in vitro to suggest that they too require an adaptor protein [Meyer et al., 1989]. And how do multiple DNA-bound transcription factors have a synergistic effect on transcription? In natural enhancers, PREs and EREs exist in arrays of other *cis* elements. The pS2 enhancer, for example, has responsive elements for estrogen, epidermal growth factor, and phorbol esters [Nunez et al., 1989]. There is good evidence that the receptors can synergize with other transcription factors to enhance transcription, but the molecular mechanisms remain to be defined [Schule et al., 1988; Strahle et al., 1988; Allan et al., 1991].

Another obvious but often ignored fact is that a cell with activated estrogen or progesterone receptors only increases transcription of a subset of target genes. This silencing of genes with *cis* elements activated in other tissues or at other developmental times may be due to negative regulatory transcription factors or chromosomal structures that do not allow the receptors access to the DNA. Chromatin regulation of uteroglobin gene expression was investigated in rabbit endometrium by Jantzen et al. [1987]. Our laboratory is investigating progesterone and estrogen receptor effects on minichromosomes containing PRE and ERE-controlled genes in yeast. Hormone addition has been shown by DNase1 analyses to induce footprints and hypersensitive sites specifically in the response element regions of the promoters [Pham et al., 1991]. This signifies chromatin conformational change, presumably due to binding of the receptors. Therefore, activated steroid hormone receptors may destabilize histones and other chromatin proteins to make genes accessible to proteins involved in transcription [Beato, 1989].

Finally, the most well-studied genes are being reintroduced into mammalian genomes to study their effects. Mice made transgenic for a rabbit uteroglobin gene construct showed appropriate tissue-specific and hormone-responsive regulation [DeMayo et al., 1991]. Thus we have come full circle: from analysis of steroid-induced proteins from natural genes to analysis of transgenes introduced from one species to another. The success of the uteroglobin transgenics proves that the studies of pro-

gesterone and estrogen regulated proteins are physiologically significant and useful for the manipulation of mammalian physiology.

As more steroid-responsive genes are defined, the complexity between events leading to physiological changes is more and more obvious. The studies of progesterone- and estrogen-induced genes in Table I show that the female gonadal steroids induce growth factors, oncogenes, and transcription factors for growth regulation, neuroendocrine signals and metabolic proteins for functional alteration, as well as proteases, protease inhibitors, and structural proteins for remodeling in responsive tissues. It is perhaps not surprising that mammalian reproductive tissues are common sites of cancer, since they have the highest growth potential at some stages in the physiology of the normal animal. When we understand the molecular mechanisms involved in these processes, we will then be able to control the effects of progesterone and estrogen on responsive tissues in health and disease.

BIBLIOGRAPHY

There are many interesting aspects to the subjects mentioned in this chapter that were only dealt with briefly. For references on related topics that complement this work, please see

Clarke CL, Sutherland RL (1990): Progestin regulation of cellular proliferation. Endocr Rev 11:266–301.
Sato GH, Steven JL (1990): Molecular endocrinology and steroid hormone action. Prog Clin Biol Res 322.
Milligan SR (1978–1990): Oxford Reviews of Reproductive Biology. New York: Oxford University Press.

For techniques, see
Ausbel FM, Brent R, Kingston RE, Moore DD, Seidman JG, Smith JA, Struhl K (1987): Current Protocols in Molecular Biology. New York, John Wiley & Sons.

Sambrook J, Fritsch EF, Maniatis T (1989): Molecular Cloning: A Laboratory Manual, 2nd ed. Cold Spring Harbor, N.Y.: Cold Spring Harbor Laboratory Press.

Abelson JN, Simon MI (eds.): Methods in Enzymology. New York: Academic Press.

Including

Berger SL, Kimmel AR (1987): Guide to molecular cloning techniques. Methods Enzymol 152.

Goeddel DV (1990): Gene expression technology. Methods Enzymol 185.

REFERENCES

Albert JL, Sundstrom SA, Lyttle CR (1990): Estrogen regulation of placental alkaline phosphatase gene expression in a human endometrial adenocarcinoma cell line. Cancer Res 50:3306–3310.

Alexander IE, Clarke CL, Shine J, Sutherland RL (1989): Progestin inhibition of progesterone receptor gene expression in human breast cancer cells. Mol Endocrinol 3:1377–1386.

Alexander IA, Shine J, Sutherland RL (1990): Progestin regulation of estrogen receptor messenger RNA in human breast cancer cells. Mol Endocrinol 4:821–828.

Allan GF, Ing NH, Tsai SY, Srinivasan G, Weigel NW, Thompson EB, Tsai M-J, O'Malley BW (1991): Synergism between steroid response and promoter elements during cell-free transcription. J Biol Chem 266:5905–5910.

Arceci RJ, Baas F, Raponi R, Horwitz SB, Housman D, Croop JM (1990): Multidrug resistance gene expression is controlled by steroid hormones in the secretory epithelium of the uterus. Mol Reprod Dev 25:101–109.

Arrick BA, Korc M, Derynck R (1990): Differential regulation of expression of three transforming growth factor beta species in human breast cancer cell lines by estradiol. Cancer Res 50:299–303.

Ausubel FM, Brent R, Kingston RE, Moore DD, Seidman JG, Smith JA, Struhl K (1987): Current Protocols in Molecular Biology. New York: John Wiley & Sons.

Bagchi MK, Elliston JF, Tsai SY, Edwards DP, Tsai M-J, O'Malley BW (1988): Steroid hormone-dependent interaction of human progesterone receptor with its target enhancer element. Mol Endocrinol 2:1221–1229.

Bagchi MK, Tsai SY, Tsai M-J, O'Malley BW (1991): Progesterone enhances target gene transcription by receptor free of heat shock proteins hsp90, hsp56 and hsp70. Mol Cell Biol 11:4998–5004.

Bailly A, LePage C, Rauch M, Milgrom E (1986): Sequence-specific DNA binding of the progesterone receptor to the uteroglobin gene: effects of hormone, antihormone and receptor phosphorylation. EMBO J 5:3235–3241.

Beato M (1989): Gene regulation by steroid hormones. Cell 56:335–344.

Berg JM (1989): DNA binding specificity of steroid receptors. Cell 57:1065–1068.

Berger SL, Cress WD, Cress A, Treizenberg SJ, Guarente L (1990): Selective inhibition of activated but not basal transcription by the acidic activation domain of VP16: evidence for transcription adaptors. Cell 61:1199–1208.

Berger SL, Kimmel AR (1987): Guide to molecular cloning techniques. Methods Enzymol 152.

Bindal RD, Carlson KE, Katzenellenbogen BS, Katzenellenbogen JA (1988): Lipophilic impurities, not phenolsulfophthalein, account for the estrogenic activity in commercial preparations of phenol red. J Steroid Biochem 31:287–293.

Capony F, Cavailles V, Rochefort H (1990): Estradiol increases the secretion by MCF7 cells of several lysosomal pro-enzymes. Biochem Biophys Res Commun 171:972–978.

Carson-Jurica MA, Schrader WT, O'Malley BW (1990): Steroid receptor family: structure and functions. Endocr Rev 11:201–220.

Cato ACB, Geisse S, Wenz M, Westphal HM, Beato M (1984): The nucleotide sequences recognized by the glucocorticoid receptor in the rabbit uteroglobin gene region are located far upstream from the initiation of transcription. EMBO J 3:2771–2778.

Chalbos D, Westley B, May F, Alibert C, Rochefort H (1986): Cloning of cDNA sequences of a progestin-regulated mRNA from MCF7 human breast cancer cells. Nucleic Acids Res 14:965–982.

Clarke CL, Roman SD, Graham J, Koga M, Sutherland RL (1990): Progesterone receptor regulation by retinoic acid in the human breast cancer cell line T47D. J Biol Chem 265:12694–12700.

Clarke CL, Sutherland RL (1990): Progestin regulation of cellular proliferation. Endocr Rev 11:266–301.

Clements JA, Matheson BA, Funder JW (1990): Tissue-specific developmental expression of the kallikrein gene family in the rat. J Biol Chem 265:1077–1081.

Coe JE, Ross MJ (1990): Armenian hamster female protein: A pentraxin under complex regulation. Am J Physiol 259:R341–349.

Croze F, Kennedy TG, Schroder IC, Friesen HG, Murphy LJ (1990): Expression of insulin-like growth factor-1 and insulin-like growth factor–binding protein-1 in the rat uterus during decidualization. Endocrinology 127:1995–2000.

Cummings AM (1989): The role of steroid hormones

and decidual induction in the regulation of adenosine diphosphoribosyl transferase activity in rat endometrium. Endocrinology 124:1408–1416.

DeMayo FJ, Damak S, Hansen TN, Bullock DW (1991): Expression and regulation of the rabbit uteroglobin gene in transgenic mice. Mol Endocrinol 5:311–318.

Denner LA, Weigel NL, Maxwell BL, Schrader WT, O'Malley BW (1990): Regulation of progesterone receptor-mediated transcription by phosphorylation. Science 250:1740–1743.

Diamond MI, Miner JN, Yoshinaga SK, Yamamoto KR (1990): Transcription factor interactions: Selectors of positive or negative regulation from a single DNA element. Science 249:1266–1272.

Donnelly KM, Fazleabas AT, Verhage HG, Mavrogianis PA, Jaffe RC (1991): Cloning of a recombinant complementary DNA to a baboon (*Papio anubis*) estradiol-dependent oviduct-specific glycoprotein. Mol Endocrinol 5:356–364.

Dotzlaw H, Miller T, Karvelas J, Murphy LC (1990): Epidermal growth factor gene expression in human breast cancer biopsy samples: relationship to estrogen and progesterone receptor gene expression. Cancer Res 50:4204–4208.

Dunbar BS, Schwoebel ED (1990): Preparation of polyclonal antibodies. Methods Enzymol 182:663–670.

Dutt A, Tang J-P, Welply JK, Carson DD (1986): Regulation of N-linked glycoprotein assembly in uteri by steroid hormones. Endocrinology 118:661–673.

Eggleston DL, Wilken C, Van Kirk EA, Slaughter RG, Ji TH, Murdoch WJ (1990): Progesterone induces expression of endometrial messenger RNA encoding for cyclooxygenase (sheep). Prostaglandins 39:675–683.

Escot C, Joyeux C, Mathieu M, Maudelonde T, Pages A, Rochefort H, Chalbos D (1990): Regulation of fatty acid synthetase ribonucleic acid in the human endometrium during the menstrual cycle. J Clin Endocrinol Metab 70:1319–1324.

Evans RM (1988): The steroid and thyroid hormone receptor superfamily. Science 240:889–894.

Fawell SE, Lees JA, White R, Parker MG (1990): Characterization and colocalization of steroid binding and dimerization activities in the mouse estrogen receptor. Cell 60:953–962.

Fazleabas AT, Jaffe RC, Verhage HG, Waites G, Bell SC (1989): An insulin-like growth factor-binding protein in the baboon (*Papio anubis*) endometrium: synthesis, immunocytochemical localization, and hormonal regulation. Endocrinology 124:2321–2329.

Fazleabas AT, Miller JB, Verhage HG (1988): Synthesis and release of estrogen- and progesterone-dependent proteins by the baboon (*Papio anubis*) uterine endometrium. Biol Reprod 39:729–736.

Freyschuss B, Eriksson H (1988): Evidence for a direct effect of thyroid hormones on the hepatic synthesis of estrogen receptors in the rat. J Steroid Biochem 31:247–249.

Fujimoto N, Roy B, Watanabe H, Ito A (1988): Increase of estrogen receptor level by thyroxine in estrogen dependent pituitary tumor (MTT/F84) in rats. Biochem Biophys Res Commun 152:44–48.

Gharary A, Murphy LJ (1989): Uterine insulin-like growth factor-I receptors: regulation by estrogen and variation throughout the estrous cycle. Endocrinology 125:597–604.

Goeddel DV (1990): Gene expression technology. Methods Enzymol 185.

Guerrier D, Boussin L, Mader S, Josso N, Kahn A, Picard J-Y (1990): Expression of the gene for anti-müllerian hormone. J Reprod Fertil 88:695–706.

Hagen G, Wolf M, Katyal SL, Singh G, Beato M, Suske G (1990): Tissue-specific expression, hormonal regulation and 5'-flanking gene region of the rat Clara cell 10 kDa protein: Comparison to rabbit uteroglobin. Nucleic Acids Res 18:2939–2946.

Hickey GJ, Oonk RB, Hall PF, Richards JS (1989): Aromatase cytochrome P450 and cholesterol side-chain cleavage cytochrome P450 in corpora lutea of pregnant rats: Diverse regulation by peptide and steroid hormones. Endocrinology 125:1673–1682.

Hissom JR, Bowden RT, Moore MR (1989): Effects of progestins, estrogens, and antihormones on growth and lactate dehydrogenase in the human breast cancer cell line T47D. Endocrinology 125:418–423

Hsu DW, El-Azouzi M, Black PM, Chin WW, Hedley-Whyte ET, Kaplan LM (1990): Estrogen increases galanin immunoreactivity in hyperplastic prolactin-secreting cell in Fisher 344 rats. Endocrinology 126:3159–3167.

Huet-Hudson YM, Andrews GK, Dey SK (1989): Cell type-specific localization of *c-myc* protein in the mouse uterus: modulation by steroid hormones and analysis of the periimplantation period. Endocrinology 125:1683–1690.

Huet-Hudson YM, Chakraborty C, De SK, Suzuki Y, Andrews GK, Dey SK (1990): Estrogen regulates the synthesis of epidermal growth factor in mouse uterine epithelial cells. Mol Endocrinol 4:510–523.

Ing NH, Francis H, McDonnell JJ, Amann JF, Roberts RM (1989): Progesterone-induction of the uterine milk proteins; major secretory proteins of sheep endometrium. Biol Reprod 41:643–654.

Insel TR (1990): Regional induction of *c-fos*-like protein in rat brain after estradiol administration. Endocrinology 126:1849–1853.

Jaffe RC, Donnelly KM, Mavrogianis PA, Verhage HG (1989): Molecular cloning and characterization of a progesterone-dependent cat endometrial secretory protein complementary deoxyribonucleic acid. Mol Endocrinol 3:1807–1814.

Jantzen K, Fritton HP, Igo-Kemenes T, Espel E, Janich S, Cato ACB, Mugele K, Beato M (1987): Partial overlapping of binding sequences for steroid hormone receptors and DNaseI hypersensitive sites in the rabbit uteroglobin gene region. Nucleic Acids Res 15:4535–4552.

Jin DF, Muffly KE, Okulicz WC, Kilpatrick DL (1988): Estrous cycle- and pregnancy-related differences in expression of the proenkephalin and proopiomelanocortin genes in the ovary and uterus. Endocrinology 122:1466–1471.

Joyeux C, Rochefort H, Chalbos D (1989): Progestin increases gene transcription and messenger ribonucleic acid stability of fatty acid synthetase in breast cancer cells. Mol Endocrinol 3:681–686.

Kastner P, Krust A, Turcotte B, Stropp U, Tora L, Gronemeyer H, Chambon P (1990): Two distinct estrogen-regulated promoters generate transcripts encoding the two functionally different human progesterone receptor forms A and B. EMBO J 9:1603–1614.

Kazemi M, Amann JF, Keisler DH, Ing NH, Roberts RM, Morgan G, Wooding FBP (1990): A progesterone-modulated, low-molecular-weight protein from the uterus of the sheep is associated with crystalline inclusion bodies in uterine epithelium and embryonic trophectoderm. Biol Reprod 43:80–96.

Kelleher RJ, Flanagan PM, Kornberg RD (1990): A novel mediator between activator proteins and the RNA polymerase II transcription apparatus. Cell 61:1209–1215.

Keyes PL, Kostyo JL, Hales DB, Chou SY, Constantino CX, Payne AH (1990): The biosynthesis of cholesterol side-chain cleavage cytochrome P-450 in the rabbit corpus luteum depends upon estrogen. Endocrinology 127:1186–1193.

Klein-Hitpass L, Tsai SY, Greene GL, Clark JH, Tsai M-J O'Malley BW (1989): Specific binding of estrogen receptor to the estrogen response element. Mol Cell Biol 9:43–49.

Klein-Hitpass L, Tsai SY, Weigel NL, Allan GF, Riley D, Rodriguez R, Schrader WT, Tsai M-J, O'Malley BW (1990): The progesterone receptor stimulates cell-free transcription by enhancing the formation of a stable preinitiation complex. Cell 60:247–257.

Knobil E, Neill JD (1988): The Physiology of Reproduction, Vol 2. New York: Raven Press. pp. 2145–2148.

Laws SC, Beggs MJ, Webster JC, Miller WL (1990): Inhibin increases and progesterone decreases receptors for gonadotropin-releasing hormone in ovine pituitary culture. Endocrinology 127:373.

L'Horset F, Perret C, Brehier A, Thomasset M (1990): 17B-Estradiol stimulates the calbindin-D9k (CaBP9k) gene expression at the transcriptional and posttranscriptional levels in the rat uterus. Endocrinology 127:2891–2897.

Li W-I, Sung L-C, Bazer FW (1991): Immunoreactive methionine-enkephalin secretion by procine uterus. Endocrinology 128:21–26.

Low KG, Nielsen CP, West NB, Douglass J, Brenner RM, Maslar IA, Melner MH (1989): Proenkephalin gene expression in the primate uterus: regulation by estradiol in the endometrium. Mol Endocrinol 3:852–857.

Lucibello FC, Slater EP, Jooss KU, Beato M, Muller R (1990): Mutual transrepression of fos and the glucocorticoid receptor: involvement of a functional domain in fos which is absent in fos B. EMBO J 9:2827–2834.

Malathy P-V, Imakawa K, Simmen RCM, Roberts RM (1990): Molecular cloning of the uteroferrin-associated protein, a major progesterone-induced serpin secreted by the procine uterus, and the expression of its mRNA during pregnancy. Mol Endocrinol 4:428–440.

Mani SK, Power RF, Decker GL, Glasser SR (1991): Hormonal responsiveness by immature rabbit uterine epithelial cells polarized in vitro. Endocrinology 128:1563–1573.

Manning DL, Archibald LH, Ow KT (1990): Cloning of estrogen-responsive messenger RNAs in the T-47D human breast cancer cell line. Cancer Res 50:4098–4104.

Masiakowski P, Breathnach T, Bloch J, Gannon F, Krust A, Chambon P (1982): Cloning of cDNA sequences of hormone-regulated genes from the MCF-7 human breast cancer cell line. Nucleic Acids Res 10:7895–7903.

Maudelonde T, Martinez P, Brouillet J-P, Laffargue F, Pages A. Rochefort H (1990): Cathepsin-D in human endometrium: induction by progesterone and potential value as a tumor marker. J Clin Endocrinol Metab 70:115–121.

Meyer M-E, Gronemeyer H, Turcotte B, Bocquel M-T, Tassel D, Chambon P (1989): Steroid hormone receptors compete for factors that mediate their enhancer function. Cell 57:433–442.

Migliaccio A, Di Domenico M, Green S, De Falco A, Kajtaniak EL, Blasi F, Chambon P, Auricchio F (1989): Phosphorylation on tyrosine of in vitro synthesized human estrogen receptor activates its hormone binding. Mol Endocrinol 3:1061–1069.

Milligan SR (1978–1990): Oxford Reviews of Reproductive Biology. New York: Oxford University Press.

Misrahi M, Loosfelt H, Atger M, Meriel C, Zerah V,

Dessen P, Milgrom E (1988): Organisation of the entire rabbit progesterone receptor mRNA and of the promoter and 5' flanking region of the gene. Nucleic Acids Res 16:5459–5472.

Misrahi M, Loosfelt H, Atger M, Guiochon-Mantel A, Applanat M, Bailly A, Vu Hai-Luu Thi MT, Lescop P, Lorenzo F, Bouchard P, Milgrom E (1990): Structural and functional studies of mammalian progesterone receptors. Horm Res 33:95–98.

Mueller H, Beato M (1980): RNA synthesis in rabbit endometrial nuclei: hormonal regulation of transcription of the uteroglobin gene. Eur J Biochem 112:235–241.

Muffly KE, Jin DF, Okulicz WC, Kilpatrick DL (1988): Gonadal steroids regulate proenkephalin gene expression in a tissue-specific manner within the female reproductive system. Mol Endocrinol 2:979–985.

Murphy LJ, Ghahary A (1990): Uterine insulin-like growth factor-1: regulation of expression and its role in estrogen-induced uterine proliferation. Endocr Rev 11:443–453.

Nunez A-M, Berry M, Imler J-L, Chambon P (1989): The 5' flanking region of the pS2 gene contains a complex enhancer region responsive to oestrogens, epidermal growth factor, a tumour promoter (TPA), the *c-Ha-ras* oncoprotein and the *c-jun* protein. EMBO J 8:823–829.

O'Halloran DJ, Jones PM, Ghatei MA, Domin H, Bloom SR (1990): The regulation of neuropeptide expression in rat anterior pituitary following chronic manipulation of estrogen status: A comparison between substance P, neuropeptide Y, neurotensin, and vasoactive intestinal peptide. Endocrinology 127:1463–1469.

Papa V, Reese CC, Brunetti A, Vigneri R, Siiteri PK, Goldfine ID (1990): Progestins increase insulin receptor content and insulin stimulation of growth in human breast carcinoma cells. Cancer Res 50:7858–7862.

Park O-K, Gugneja S, Mayo KE (1990): Gonadotropin-releasing hormone gene expression during the rat estrous cycle: effects of pentobarbital and ovarian steroids. Endocrinology 127:365–372.

Peiffer A, Barden N (1987): Estrogen-induced decrease of glucocorticoid receptor messenger ribonucleic acid concentration in rat anterior pituitary gland. Mol Endocrinol 1:435–440.

Perheentupa A, Huhtaniemi I (1990): Gonadotropin gene expression and secretion in gonadotropin-releasing hormone antagonist-treated male rats: effect of sex steroid replacement. Endocrinology 126:3204–3209.

Persico E, Scalona M, Cicatiello, Sica V, Bresciani F, Weisz A (1990): Activation of "immediate-early" genes by estrogen is not sufficient to achieve stimulation of DNA synthesis in rat uterus. Biochem Biophys Res Commun 171:287–292.

Pham TA, Elliston JF, Nawaz Z, McDonnell DP, Tsai M-J, O'Malley BW (1991): Antiestrogen can establish nonproductive receptor complexes and alter chromatin structure at target enhancers. Proc Natl Acad Sci USA 88:3125–3129.

Poulin R, Simard J, Labrie C, Petitclerc L, Dumont M, Lagace L, Labrie F (1989): Down-regulation of estrogen receptors by androgens in the ZR-75-1 human breast cancer cell line. Endocrinology 125:392–399.

Power RF, Lydon RF, Conneely OM, O'Malley BW (1991): Dopamine activation of an orphan of the steroid receptor superfamily. Science 252:1546–1548.

Pugh BF, Tjian R (1990): Mechanism of transcriptional activation by Sp1: evidence for coactivators. Cell 61:1187–1197.

Rajabi M, Solomon S, Poole AR (1991): Hormonal regulation of interstitial collagenase in the uterine cervix of the pregnant guinea pig. Endocrinology 128:863–871.

Read LD, Snider CE, Miller JS, Greene GL, Katzenellenbogen BS (1988): Ligand-modulated regulation of progesterone receptor messenger ribonucleic acid and protein in human breast cancer cell lines. Mol Endocrinol 2:263–271.

Richard S, Zingg HH (1990): The human oxytocin gene promoter is regulated by estrogens. J Biol Chem 265:6098–6103.

Roberts RM, Bazer FW (1988): The functions of uterine secretions. J Reprod Fertil 82:875–892.

Romano GJ, Krust A, Pfaff DW (1989): Expression and estrogen regulation of progesterone receptor mRNA in neurons of the mediobasal hypothalamus: An in situ hybridization study. Mol Endocrinol 3:1295–1300.

Saceda M, Lippman ME, Lindsey RK, Puente M, Martin MB (1989): Role of an estrogen receptor-dependent mechanism in the regulation of estrogen receptor mRNA in MCF-7 cells. Mol Endocrinol 3:1782–1787.

Sambrook J, Fritsch EF, Maniatis T (1989): Molecular Cloning: A Laboratory Manual, 2nd ed. Cold Spring Harbor NY: Cold Spring Harbor Laboratory.

Sato GH, Steven JL (1990): Molecular endocrinology and steroid hormone action. Prog Clin Biol Res 322.

Sawadogo M, Roeder RG (1985): Factors involved in specific transcription by human RNA polymerase II: Analysis by a rapid and quantitative in vitro assay. Proc Natl Acad Sci USA 82:4394–4398.

Schule R, Muller M, Kaltschmidt C, Renkawitz R (1988): Many transcription factors interact synergistically with steroid receptors. Science 242:1418–1420.

Schule R, Rangaran P, Kliewer S, Ransone LJ, Boldo J, Yang N, Verma I, Evans RM (1990): Functional antagonism between oncoprotein c-jun and the glucocorticoid receptor. Cell 62:1217–1226.

Shupnik MA, Gordon MS, Chin WW (1989): Tissue-specific regulation of rat estrogen receptor mRNAs. Mol Endocrinol 3:660–665.

Shymala S, Gauthier Y, Moore SK, Catelli MG, Ullrich SJ (1989): Estrogenic regulation of murine uterine 90-kilodalton heat shock protein gene expression. Mol Cell Biol 9:3567–3570.

Shymala G, Schneider W, Schott D (1990): Developmental regulation of murine mammary progesterone receptor gene expression. Endocrinology 126:2882–2889.

Simard J, Dauvios S, Haagensen DE, Levesques C, Merand Y, Labrie F (1990): Regulation of progesterone-binding breast cyst protein GCDFP-24 secretion by estrogens and androgens in human breast cancer cells: A new marker of steroid action in breast cancer. Endocrinology 126:3223–3231.

Simerly RB, Young BJ (1991): Regulation of estrogen receptor messenger ribonucleic acid in rat hypothalamus by sex steroid hormones. Mol Endocrinol 5:424–432.

Simmen RCM, Simmen FA, Hofig A, Farmer SJ, Bazer FW (1990): Hormonal regulation of insulin-like growth factor gene expression in pig uterus. Endocrinology 127:2166–2174.

Simmen RCM, Srinivas V, Roberts RM (1989): cDNA sequence, gene organization, and progesterone induction of mRNA for uteroferrin, a porcine uterine iron transport protein. DNA 8:543–554.

Slater EP, Redeuihl G, Theis K, Suske G, Beato M (1990): The uteroglobin promoter contains a non-canonical estrogen responsive element. Mol Endocrinol 4:604–610.

Strahle U, Schmid W, Schutz G (1988): Synergistic action of the glucocorticoid receptor with transcription factors. EMBO J 7:3389–3395.

Sundstrom SA, Komm BS, Xu Q, Boundy V, Lyttle CR (1990): The stimulation of uterine complement component C3 gene expression by anti-estrogens. Endocrinology 126:1449–1456.

Takeda A, Leavitt WW (1986): Progestin-induced down regulation of nuclear estrogen receptor in uterine decidual cells: Analysis of receptor synthesis and turnover by the density-shift method. Biochem Biophys Res Commun 135:98–104.

Teh M, Hui KM (1989): Modulation of MHC gene expression in human breast carcinoma cells by hormones. J Immunogenet 16:397–405.

Timmons TM, Dunbar BS (1990): Protein blotting and immunodetection. Methods Enzymol 182:679–688.

Tora L, White J, Brou C, Tasset D, Webster N, Scheer E, Chambon P (1989): The human estrogen receptor has two independent nonacidic transcriptional activation functions. Cell 59:477–487.

Toranzo D, Dupont E, Simard J, Labrie C, Couet J, Labrie F, Pelletier G (1989): Regulation of pro-gonadotropin-releasing hormone gene expression by sex steroids in the brain of male and female rats. Mol Endocrinol 3:1748–1756.

Touitou I, Mathieu M, Rochefort H (1990): Stable transfection of the estrogen receptor cDNA into HeLa cells induces estrogen responsiveness of endogenous cathepsin D gene but not of cell growth. Biochem Biophys Res Commun 169:109–115.

Tsai M-J, O'Malley BW (1991): Mechanisms of regulation of gene transcription by steroid receptors. In Folkes JG, Cohen P (eds): Hormonal Regulation of Transcription. Amsterdam: Elsevier-Biomedical Press, pp. 101–116.

Tsai SY, Sagami I, Wang H, Tsai M-J, O'Malley BW (1987): Interactions between a DNA-binding transcription factor (COUP) and a non-DNA binding factor (S3000-II). Cell 50:701–709.

Turcotte B, Meyer M-E, Bocquel M-T, Belanger L, Chambon P (1990): Repression of the alpha-fetoprotein gene promoter by progesterone and chimeric receptors in the presence of hormones and antihormones. Mol Cell Biol 10:5002–5006.

Turner RT, Colvard DS, Spelsberg TC (1990): Estrogen inhibition of periosteal bone formation in rat long bones: down-regulation of gene expression for bone matrix proteins. Endocrinology 127:1346–1351.

Vegeto E, Cocciolo MG, Raspagliesi F, Piffanelli A, Fontanelle R, Maggi A (1990): Regulation of progesterone receptor gene expression. Cancer Res 50:5291–5295.

Verdi JM, Campagnoni AT (1990): Translational regulation by steroids: identifications of a steroid modulatory element in the 5′-untranslated region of the myelin basic protein messenger RNA. J Biol Chem 265:20314–20320.

Waterman ML, Alder S, Nelson C, Greene GL, Evans RM, Rosenfield MG (1988): A single domain of the estrogen receptor confers deoxyribonucleic acid binding and transcriptional activation of the rat prolactin gene. Mol Endocrinol 2:14–21.

Webb DK, Moulton BC, Khan SA (1990): Estrogen induced expression of the c-Jun proto-onocogene in the immature and mature rat uterus. Biochem Biophys Res Commun 168:721–726.

Wei LL, Krett NL, Francis MD, Gordon DF, Wood WM, O'Malley BW, Horwitz KB (1988): Multiple human progesterone receptor messenger ribonucleic acids and their autoregulation by progestin agonists and antagonists in breast cancer cells. Mol Endocrinol 2:62–72.

Weisz A, Cicatiello L, Persico E, Scalona M, Bresciani F (1990a): Estrogen stimulates transcription of *c-Jun* protooncogene. Mol Endocrinol 4:1041–1050.

Weisz A, Rosales R (1990b): Identification of an estrogen response element upstream for the human *c-Fos* gene that binds the estrogen receptor and the AP-1 transcription factor. Nucleic Acids Res 18:5097–5106.

Wise PM, Scarborough K, Weiland NG, Larson GH (1990): Diurnal pattern of proopiomelanocortin gene expression in the arcuate nucleus of proestr-

ous, ovariectomized, and steroid-treated rats: A possible role in cyclic luteinizing hormone secretion. Mol Endocrinol 4:886–892.

Yang C-PH, Cohen D, Greenberger LM, Hsu SI-H, Horwitz SB (1990): Differential transport properties of two *mdr* gene products are distinguished by progesterone. J Biol Chem 265:10282–10288.

Yang-Yen H-S, Chambard J-C, Sun Y-L, Smeal T, Schmidt TJ, Drouin J, Karin M (1990): Transcriptional interference between *c-jun* and the glucocorticoid receptor: mutual inhibition of DNA binding due to direct protein–protein interaction. Cell 62:1205–1215.

ABOUT THE AUTHORS

NANCY H. ING is a postdoctoral fellow at Baylor College of Medicine in the Department of Cell Biology. After receiving a B.S. degree in Zoology from the University of Florida, she remained there to begin the first joint D.V.M/Ph. D. degree program. Her thesis work in the Department of Biochemistry and Molecular Biology involved the characterization and cloning of the major progesterone-regulated proteins of the ovine uterus, the uterine milk proteins. She followed her major professor, R. Michael Roberts, to the University of Missouri at Columbia, and completed her thesis there. Dr. Ing is currently studying progesterone-regulated gene expression in the laboratory of Bert W. O'Malley. She recently submitted two manuscripts describing work on the identification of a general transcription factor that is the target through which steroid hormone receptors activate transcription and the inhibition of progesterone receptor function by triplex-forming oligonucleotides. After being elected to Phi Beta Kappa in 1979, she has received the Rita McTigue O'Connell Award, the Graduate Fellowship for Women Entering Non-Traditional Careers, and the American Medical Association-ERF Award.

SOPHIA Y. TSAI is an Associate Professor of Cell Biology at Baylor College of Medicine, where she teaches structure and function of the eukaryotic nucleus. After receiving her B.S. and M.S. degrees in Chemistry from the University of Wisconsin, Madison, in 1965 and 1966 respectively, she obtained her Ph.D. in 1970 under Dr. Richard S. Criddle at the University of California, Davis. Her thesis involved studies of the biogenesis of yeast mitochondria. Dr. Tsai pursued postdoctoral research at Cornell University in the laboratory of Gottfried Schatz, characterizing mitochondrial ribosomal RNA in yeast petite mutants. She is currently studying the regulation of expression of the insulin receptor gene and the mechanism by which the steroid hormone receptor is controlling the expression of its responsive genes. Her research papers have appeared in such journals as *Cell, Nature, Molecular and Cellular Biology,* and the *Journal of Biological Chemistry.*

MING-JER TSAI is Professor of Cell Biology at Baylor College of Medicine, where he teaches molecular genetics of eukaryotes. After receiving his B.S. from National Taiwan University in 1967, he received his Ph.D. in 1971 under Dr. Richard S. Criddle at the University of California, Davis, where he worked on the RNA polymerases in yeast mitochondria. Dr. Tsai pursued postdoctoral research at the University of Texas M.D. Anderson Cancer Center, studying transcriptional regulation during tumorgenesis under Dr. Grady F. Saunders. Dr. Tsai's current research involves regulation of gene expression at the transcriptional level, especially the insulin gene regulation and mechanism of steroid hormone regulation of target gene expression. His research papers have appeared in such journals as *Cell, Nature, Science, Molecular and Cellular Biology,* and the *Journal of Biological Chemistry.*

Index